Before the Deluge

To Ann, Holly and Daniel

John Briggs

Before the Deluge

Photographs of Cardiff's Docklands communities in the Seventies

Foreword by Neil Sinclair

seren

Seren is the book imprint of Poetry Wales Press Ltd
Nolton Street, Bridgend, Wales
www.seren-books.com

ISBN 1-85411-315-1
A CIP record for this title is available from the British Library

The publisher works with the financial assistance of the Arts Council of Wales

Front cover: Family group, Garesfield Street
Back cover: Ship Lane

Printed in Perpetua by Gomer Press, Llandysul

Contents

Foreword

The images of American photographer John Briggs came to my attention in May 2000 when *Before the Deluge*, his extensive photographic essay on Butetown, was first exhibited at the gallery of Butetown History and Arts Centre in lower Bute Street in Cardiff's docklands. Having been chair of the BHAC for four years and editor of its quarterly magazine *The Voice of the Tiger* I had had a long association with the organisation's aim in promoting a faithful portrait of the dockland community and its colourful past. I was most pleased with its acceptance of the photographer's work. Intriguingly, like ambassadors to each other's native country, Briggs had travelled from his home in Minnesota to train as a teacher and photograph Cardiff during the mid-70s while I, on the other hand, sojourned in Southern California studying for a Masters Degree at the University of California Los Angeles. When I returned to Wales at the beginning of the 90s, I found my much-beloved community of Tiger Bay totally transformed from the one I grew up in during the 40s and 50s. But fortunately for both of us, Butetown History and Arts Centre had established its foundations as an institution devoted to Butetown's past and was playing host to his exhibition when we eventually met in its gallery.

Despite promises of improvement by the powers that were, the notorious past of my home community seemed to survive the grotesque metamorphosis from run-down but characterful Victorian architectural townscape laid out and built by the Second Marquis of Bute to modern lacklustre council estate. Having left Tiger Bay *intacto* during the 60s I was fortunate not to see our cherished streets succumb to the wrecking ball. That destructive period of demolition which all too hastily began, and then continued through the 70s and 80s, inspired John Briggs to capture on film the remnants of the familiar structures which my memory still has the power to reconstruct each night as I sleep in Loudoun Square. Unfortunately for me, much of what I would have asked him to record had already disappeared. The Upper Docks of Butetown, known to those in the know as Tiger Bay, had been almost completely obliterated by the time of his arrival. He did manage to capture on film the Nur Ul Mosque, Britain's first, in Peel Street during its last days. But the rest of the street, so important to my family's early twentieth century Caribbean and Welsh origins was no longer in existence. He was just in time to photograph the first Cardiff Board School in Eleanor Street before its demise, but gone was the imposing structure of South Church Street Board School which also bordered Peel Street where my grandparents and other relatives once resided. No magnificent Loudoun Square or mysterious Hodges Row, the boundary between the Tiger Bay and Docks communities where people once lived on houseboats and where I played as a child. Sadly, all were gone.

Nonetheless, what John Briggs was able to capture in this photographic collage has the undeniable power to warm the hearts of those who remember the docklands communities and to be of interest to those who were not there but who cherish the historic development of our towns and cities. Still surviving and awaiting his perceptive eye were the remnants of the Lower Docks. Streets such as South William and Louisa were yet to be demolished as was George Street, once known as Little Madrid because of the Spanish and Basques who had their homes and businesses there. Poignantly, one of the photographs presented here shows the George Street Public Library in a dilapidated state which makes a stark contrast with my memory of its more vibrant days. Yet on the same street, closer to the James Street corner, the Cardiff Castle Hotel still maintained a dignified appearance in his photograph prior to its demolition.

Broadening its artistic stroke, this collection of images harbours not only reminiscences of the streets of Tiger Bay and the residential Docks but also instances of the life, maritime activity and architecture of the original Bute Docks. This will clearly be an inspiration to those who worked there and to seamen all over the world who chanced to dock or even jump ship in the old port of Cardiff. Furthermore, although not on display at the original exhibition of *Before the Deluge*, other communities which verge on the edge of dockland such as Adamsdown, the Irish community of Newtown and the lower Splottlands are also included. In the latter area, the image of the old East Moors steelworks towers over its streets and dear old 'Portmama' (Portmanmoor) Road where Shirley Bassey once lived reminds me of friends from my school days. Inevitably, for those with memories of a more vibrant time, there is a sadness in the moribund aspect of his depictions of the lifeless streets and derelict dockland edifices.

As a resident of Newport since 1976 John Briggs has witnessed the regeneration of south Cardiff undertaken by the now-defunct Cardiff Bay Development Corporation. In the text accompanying his photographs, he resists mention of the new Cardiff Bay out of respect for the docklands communities' past and in the firm belief, which I also state in *Endangered Tiger*, a soon to be published manuscript, that 'Cardiff Bay' as a name does not replace 'Tiger Bay' as perhaps some would hope. In itself, the old Bay was a mere one square mile in geographic area, whereas Cardiff Bay, extending from the River Ely in the west to the River Rumney in the east, is one sixth of the entire city. Under its far-flung umbrella we still find in existence the communities of Splott, the remnants of Newtown and Adamsdown, even parts of Tremorfa and Pengam, not to mention Portway Marina, the newly emerging community of Penarth Docks. But it should not be expected that the oldest community among them all should relinquish its identity. I would like to pay tribute to John Briggs' work in helping to perpetuate this fact and in furthering the awareness, through his own visual language, of all the docklands communities' vital contribution to Cardiff's essential history and unique character. Thanks to his efforts, to those of Butetown History and Arts Centre, to mine and to those of many others, their names, and particularly that of Tiger Bay will still exist after that of Cardiff Bay may have vanished into the mists of some future antiquity.

Neil Sinclair, M.A., Tiger Bay, 2001

Introduction

It was on January 12th, 1974 that I arrived in south Wales and, as it happened, began to make it my permanent home. Why Wales and, for a time, why Cardiff? For that I've got to go back another 30 years to 1944, the year my father, an officer in the American army, was based in this country for several months during the build-up to the D-Day invasion. It was not in Cardiff but rather at the American army camp in Malpas, Newport, that he was stationed and where he first experienced the British way of life, albeit in wartime. As with so many other American officers and GIs, that experience left a lasting impression on him and was one from which I too was to benefit years later. But unlike many American servicemen who met and married their war-brides in south Wales, my father found and wed his wife in Liège, Belgium, where his regiment was based in the final months of World War II. In 1946 he returned to America with his French-speaking Belgian wife. They made their home in St. Paul, Minnesota, my birthplace in 1947 and the town where I grew up speaking both French and English. Being brought up bilingually was to have a vital influence on my future, as well as my interest in photography which began in my boyhood days.

Photography was a favourite hobby by the time I was 12 and I set up a darkroom with a friend, wanting to develop and print pictures 'as good as they do at the drugstore'. The interest continued to grow during my teens as I went through high school in St. Paul. And then one day in my seventeenth year I was very surprised to be asked by my father if I would like to spend a year in Britain – quite an unusual idea for the time but one which no doubt stemmed from his lasting and positive memories of this country twenty years before. Apprehensive, but not wanting to miss out on such an opportunity, I agreed and soon found myself winging towards the UK and, like my father before me, towards south Wales. The final destination in my case was not, of course, an army camp as his had been, but rather one of Wales' most prestigious educational institutions – Atlantic College, 20 miles to the west of Cardiff. There I spent two years doing A-levels from 1964-66 and, just as importantly, broadened my experience of life, another culture and my photographic horizons.

My involvement in photography became increasingly intense as I met other boys at Atlantic College with similar interests but technically and aesthetically ahead of me. Thanks to them I first became acquainted with the work of the British and French photographers of the 60s and figures such as Henri Cartier-Bresson and Tony Ray-Jones became my inspiration. But just as significant was my occasional contact with Cardiff, the urban magnet for students at the weekend and for me the unfamiliar but alluring city that provided the prospect of photographing its people and its streets with my first SLR. In fact two of my images in this collection are from those days, but from the Hayes area in central Cardiff rather than Butetown or other communities near the Docks. All I knew about such places in 1964-65 was through rumours about the dangers of Bute Street ('coppers go in twos') and stories brought back to college by one or two boys who had been to visit 'girls' and clubs in this 'notorious' avenue. Just the kind of stories that made you want to rush down there with your camera to take pictures, but I was some way from translating that sort of thought into action in those days.

Had I acquainted myself more extensively with Cardiff in the mid-1960s I might have seen Tiger Bay as described by Neil Sinclair in his Foreword and as he has described for me so enchantingly during our conversations, vital in helping me identify the people and places from his own and other nearby communities, my photographic subjects in the mid-1970s. I might also have seen the terraced streets of Newtown before they were obliterated and perhaps too the whole of Lower Splott, on the brink of disappearing by the time I finally reached it in 1974-75. As it was, Butetown, Tiger Bay, Newtown, Adamsdown and Splott remained unknown quantities for the time being. In fact Cardiff might have become only a distant memory because, returning to America in 1966 at the end of my two years in Wales, I committed myself to university studies at the University of Minnesota in Minneapolis.

My chosen field of study was French although it was not long before I found myself keen to take up the photographic opportunities that came my way. The flexibility of courses in American universities is such that one can do certain modules (in my case Photojournalism) that don't necessarily relate to one's 'major' subject. So I did just that and also landed a part-time job (as students at American universities usually do) as staff photographer for the *Minnesota Daily*. This was campus newspaper for this huge educational institution of 50,000 and its urban student communities. The mid and late-60s were a time of great social change and political activism in America instigated by the younger generation. Anti-big-corporation protests, sit-ins, draft-card and flag-burning, peace rallies and demonstrations against the war in Vietnam happened frequently on the university campus and in the streets and parks of Minneapolis as in other American cities. It was an exciting time to be a fledgling news photographer and one which provided much incentive for visually documenting my experience of the era. But by the mid-70s much of this social and political excitement had come to an end. I had graduated with a B.A. in French, gained some professional experience and formal training in photojournalism, and the Vietnam war was thankfully at an end, but university graduates were plentiful, America was in recession, jobs were few and so were my career prospects. So once again I set my sights on Wales where, as I discovered, there were career opportunities for Modern Language teachers.

Coming to train in Cardiff in 1974 it was then that I also began to systematically explore the city, camera in hand – an exploration that quickly became a source of fascination and one that has never really ended. Just as in Minnesota during the 1960s the times they were a changin' (to quote a famous fellow-Minnesotan), so too were they changing in Cardiff during the 1970s, although in a very different way. Having no inkling of this when I first arrived but wanting to acquaint myself with the city that was now my home, I started with the old Bute Docks and lower Butetown. I was truly astonished by what I saw and had never seen anything quite like it before: the remains of a once-great port and its dependent community, a maritime wasteland and ghost-town waiting to be wiped off the face of the earth yet still with enough stark and striking evidence even in the midst of mind-blowing dereliction to create an indelible impression of a life that had once been. I can't remember how many times I covered the length of Bute Street to the Pier Head, stopping to see what remained of Tiger Bay before carrying on to the streets of the Lower Docks, or how many times I wandered along the Pier Head, Roath Basin and East Dock wharves, or how many times I walked along the old Junction Canal path and into Herbert Street past the last traces of Newtown before making towards Splott. But everywhere I went I found scenes that were as emotionally and visually mesmerising as they were desolate. There was also enough human life in the streets, most often children, as well as in the houses, shops and pubs to provide me with the inspiration and the desire to commit to film the diversity and character of those people who seemed to be the last witnesses of all that had gone before them. It was obvious who and what I photographed would not be there for much longer and I wanted to make very sure that it did not go unrecorded, not least for my own memory. Especially from 1974-1976 but occasionally afterwards this was my photographic domain and I must say I revelled in it, returning to photograph in certain favourite places after I finished my teacher-training, left Cardiff and took up my profession as a French teacher in Newport in 1976.

Over time these images of Cardiff and their settings remained important to me even if I did not think in terms of exhibiting or trying to publish them until quite recently. In fact it was not until I left teaching in 1999 and began my involvement with the Butetown History and Arts Centre that this really became possible. At first I took a few images along to that organisation thinking that they might be a useful contribution to the archive collection. When, soon after, it was suggested to me that I apply for an exhibition in the BHAC Gallery in Bute Street I jumped at the chance. It was with delight soon followed by a good deal of anxiety that I reacted to the news that my images had been accepted for exhibition in the summer of 2000. Now 25 years on from when the photographs had been taken I had to re-familiarise myself with hundreds and hundreds of frames, choose then print a selection of images most of which had remained only as negatives in my files: while I was teaching I had never had sufficient time to print more than a couple of dozen.

As I began to review what I had taken during my photographic meanderings, the number of images for the exhibition grew and grew. In the end *Before the Deluge* consisted of more than 160 photographs, the result of my determination to put together a varied, coherent and detailed set of pictures. In addition, the exhibition was divided into three sections, which have been retained for this book. As for the order of photographs in each section, they are intended to follow the paths I took through the different communities around Cardiff Docks. The sequences should make particular sense to those who are familiar with or who have lived in these parts of Cardiff. For those who are less familiar or unfamiliar, the collection should acquaint them with the location and characteristics of Cardiff's docklands communities as they were 25 years ago before great changes began to take place. It must also be kept in mind that the people and places towards which I was drawn were the result of personal preference and not, as it were, in answer to the demands of a photographic assignment for a newspaper or magazine. My approach was to photograph what attracted my personal attention in a disappearing world: buildings, everyday 'slice of life' scenes and people encountered at random. Inevitably there are certain omissions as well as a few curious inclusions. Those who know my photographic territory will perhaps wonder why there are not more photographs of significant locations such as the Hamadryad Seamen's Hospital, the 'Big Windsor' or the Coal Exchange in Mount Stuart Square. Buildings still in everyday use such as these I tended to leave unphotographed. And being large and diverse, the area that I covered was difficult to photograph with absolute consistency. I do think, however, that my images of these Cardiff communities are numerous and varied enough to present a picture true to the people, places and times in which they were taken.

When *Before the Deluge* as an exhibition took place in the summer of 2000, first in the Bute Street gallery of BHAC and then in abbreviated form in the Norwegian Church Arts Centre, I was very much heartened by the kind of interest it aroused amongst local people and visitors alike. In addition to their aesthetic appeal the photographs evoked intellectual and emotional responses, much more than just nostalgia, from past and present inhabitants who were encountering images of their surroundings from former times. They were pleased to have an extensive and coherent reminder, a sort of visual biography of their past human and physical environment from 25 years ago, now so thoroughly transformed by redevelopment. With them I share the conviction that the appreciation of positive elements from the past needs to be safeguarded, reaffirmed and perpetuated. The immediacy and dramatic intensity of black-and-white photography is, I think, best suited to this. I would like to think that the presentation of these images in book form will carry on the work of promoting the contribution of these communities to modern Cardiff's history and essential character. They are of fundamental value and too important not to be given greater recognition. Many Cardiffians are of the opinion, and quite justifiably so, that the significance of these communities to Cardiff's heritage has not been given the profile it deserves in the present scheme of things. This book aims to help achieve that goal.

For visitors coming to see the exhibition of *Before the Deluge* in the summer of 2000 but not acquainted with these parts of Victorian Cardiff, the reaction to the visual evidence presented was often one of surprise sometimes bordering on bewilderment that so much of substance had been there and that so much had been done away with. Some wondered, for instance, at how much Cardiff dockland's former human makeup and architectural identity had been altered when looking at my photographs of the Lower Docks and then comparing them to the new Mermaid Quay complex at the bottom of Bute Street. The question was often asked: Why such an unsympathetic contrast to the remains of elegant Victorian architecture nearby? With new commercial activity catering blatantly to visitors or well-heeled residents of the 'brand new' Cardiff Bay, what has happened to the ethnically diverse but poorer docklands communities? Have they simply been forgotten?

If these images can help create lasting impressions of the nature and importance of these historic docklands communities as they were 'before the deluge' of redevelopment at the end of the Millennium, they will, I think, have accomplished a valid task in the much-changed city that is Cardiff at the beginning of the twenty-first century.

Part 1

Bute Street
From the Hayes to the Pier Head

In former times when Cardiff Docks, bursting at the seams with ships, received seafaring men in waves and propelled them towards the delights of Bute Street, it might have been more appropriate to turn the wording of the title the other way around. A relentless flow of human and commercial traffic starting at the dock gates worked its way towards the city for a mile via Bute Street, one of the world's most famous, or infamous, thoroughfares. But gone are those days and now more so than ever, traffic flows from the city towards the docks and the new Cardiff Bay.

It did in my case back in the mid-1970s, when I set off on photographic sorties bound for the docklands. The way to the Pier Head was usually by the well-trodden route – starting from the Hayes, at one time at the northern extremity of Bute Street. Here the open square embraced a multitude of folk, then as it does now, going about their daily business, some stopping to replenish themselves at the Hayes Island Snackbar, or Bwyty Hayes as it is known in these days of Welsh revival. There was the occasional Johnny/Shoni from Brittany selling his strings of onions and a fruit and vegetable barrow nearby. No St. David's Hall yet or Working Street pedestrianisation. Down Hayes Bridge Road (the top of Bute Street until 1937) sometimes making a detour into Bridge Street and the ones off it – Mary Ann Street, Millicent Street, Frederick Street, etc. to have a look at all those funny little shops – Fry's Dart's, Huxley's Medical Supplies, the Antiquarian Bookshop, or down Mill Lane past the fruit and veg stalls towards Custom House Street. Whatever the diversion, the eventual destination was always the same: Bute Street. Then follow it under the railway bridge and, still as in former times when Tiger Bay really was Tiger Bay, you were in a different world.

Even if little remained of the original by the mid-1970s there was enough left to tell you that you couldn't be anywhere else. At Crichton Street there was the Custom House, the pub where 'the girls' were and The Quebec where Vic 'Narker' Parker from Tiger Bay still played jazz guitar. Crossing Wharf Road and carrying on took me past the once-teeming Rainbow Club for the Bay's multiethnic children, Mahfood's cafe, typical of so many that had gone before, and then to Bute Bridge over the abandoned remains of the West Junction Canal. St. Mary the Virgin Church and the Salvation Army hostel came next. Across Bute Street was the bus stop, still there today, with as racially-mixed a group of locals waiting in the queue as might be found anywhere in the world. Walking past West Indians, Somalis and others I could still see the remains of faded old hand-painted signs on cement facades along the railway wall. I couldn't help but notice the street names to which the Butes had given a thoroughly feminine touch in honour of their women – Maria, Christina, Angelina, Sophia, Hannah – and the Muslim mosque, Britain's first, standing amongst the 1960s houses that had replaced Tiger Bay's terraces ten years earlier. Towering above them the high-rises of Loudoun Square asserted themselves as the most visible of Tiger Bay's landmarks. Well over halfway to the Pier Head now I crossed the time-honoured but invisible dividing line at Hodges Row signifying passage from Tiger Bay into the Lower Docks area. To the right West Bute Street angled towards Mount Stuart Square but straight on Bute Street made its way inexorably towards the Pier Head, the harbour and the sea. Both right and left were fine rows of commercial buildings, thankfully today spared the ravages of redevelopment: structures such as Dock Chambers, former offices for Cardiff's millionaire shipping entrepreneurs and the National Provincial Bank, once the second largest in the land outside London. Still in business or not they commanded attention. On to the junction with James Street I waited to cross in front of the Cory Building, the headquarters of Cardiff's greatest-ever ship-owning dynasty, but also the site of 'penniless point' for sailors and dockers failing to find work as the fortunes of the shipping industry rose and fell like the tide. Across James Street turn left it's the docks (with ships), turn right it's the Docks (with streets) but the Pier Head, the harbour and the sea can't be far; the names tell you. There's Silver's seamen's oufitters, the National Union of Seamen headquarters, the Packet Hotel, Pier Head Chambers and the Pier Hotel. After 1975 there was also the Industrial and Maritime Museum. But most of all there was the prospect of the sea. You couldn't come to the end of Bute Street without stopping for it. And standing and staring, just as W.H. Davies, Newport's tramp poet advised us to do. And in my case I stood and stared from the Pier Head many a time through the camera's eye.

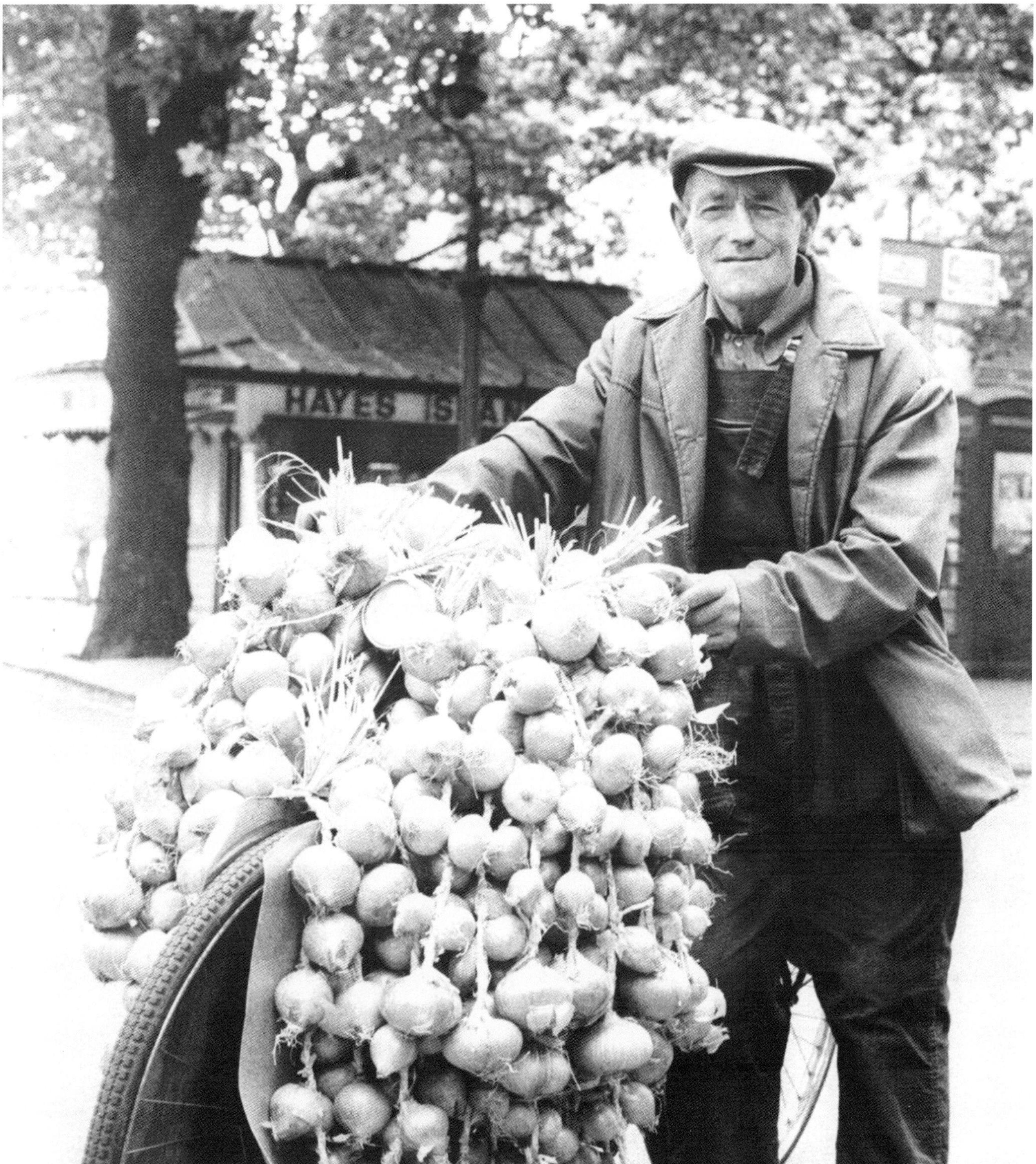
HAYES

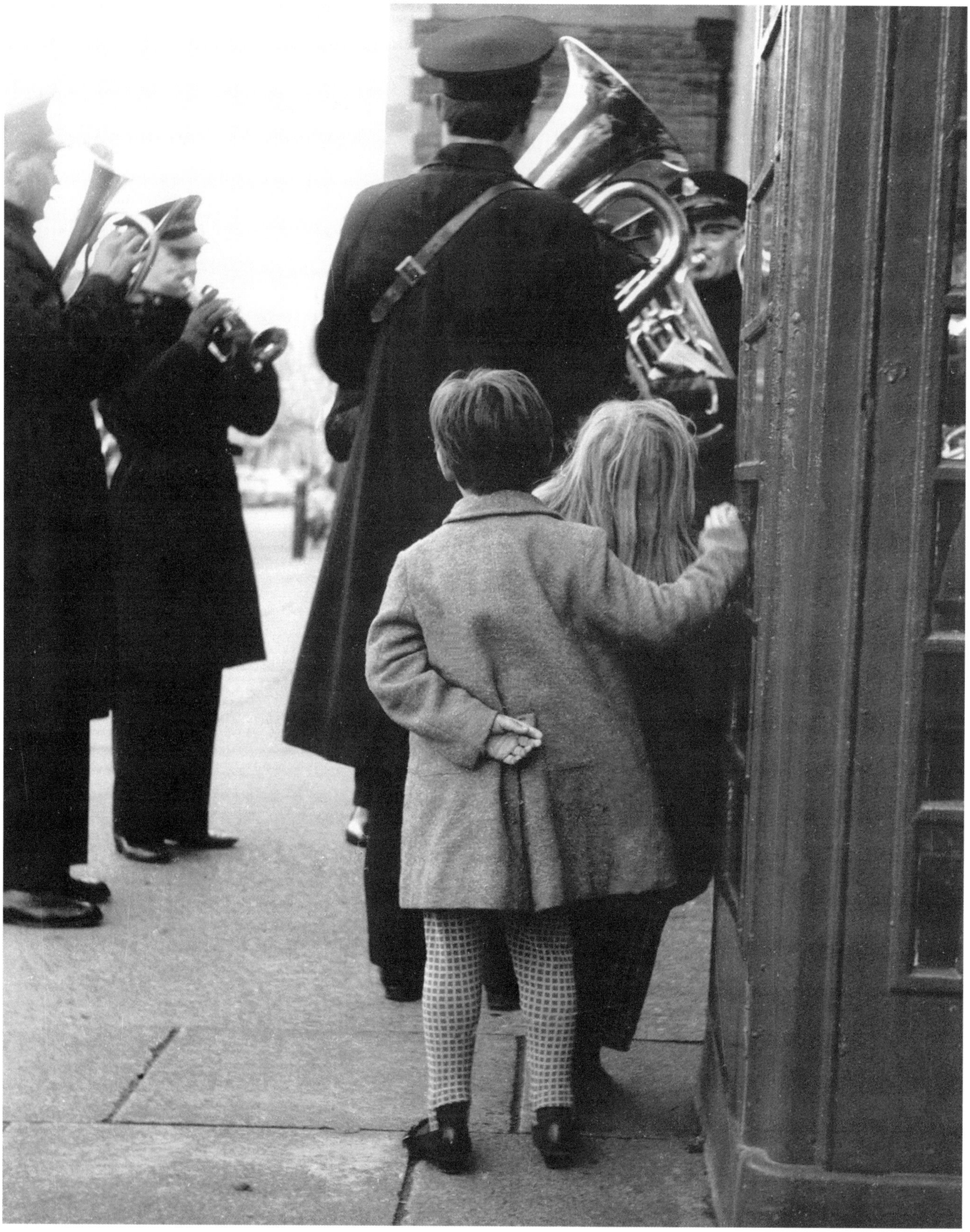

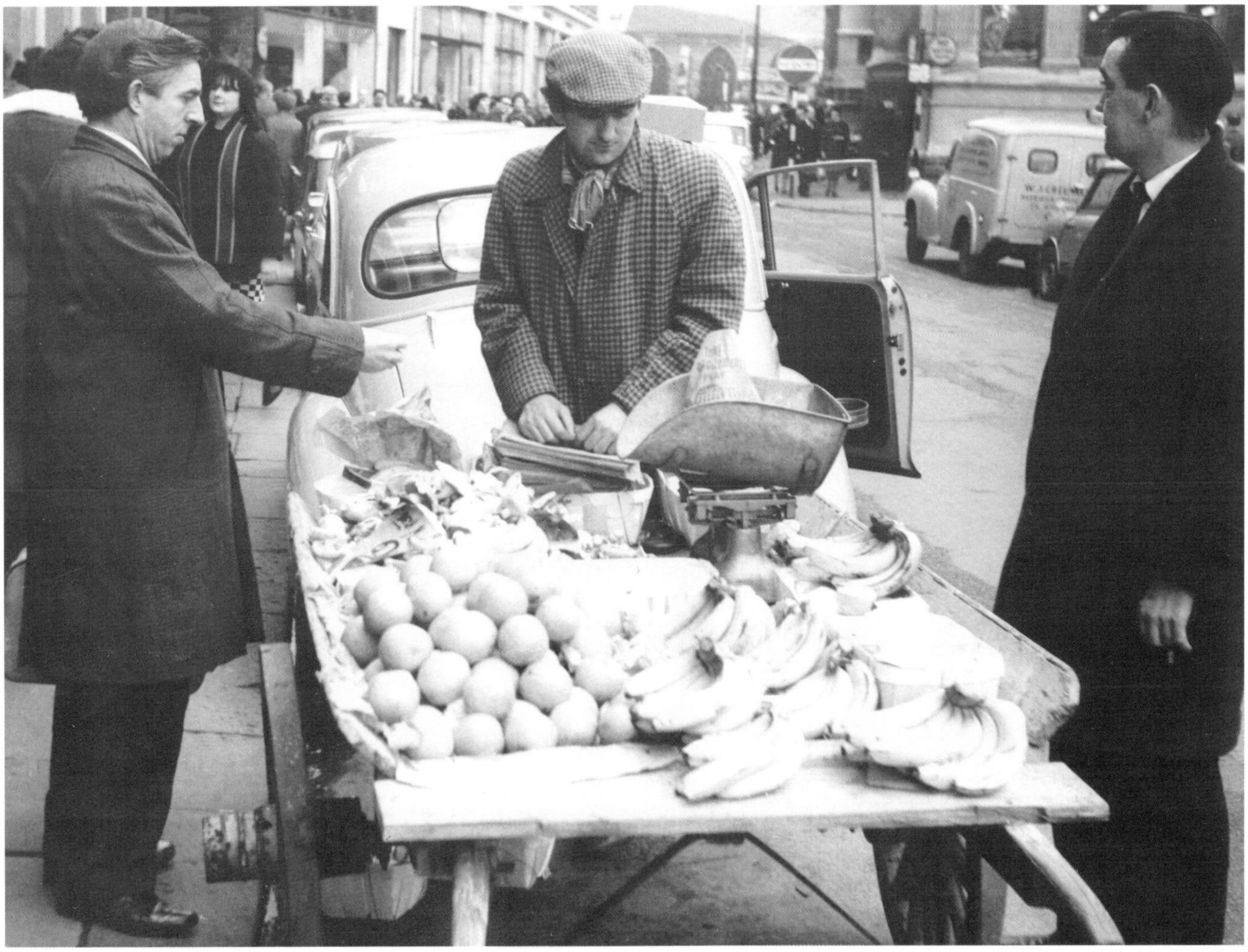

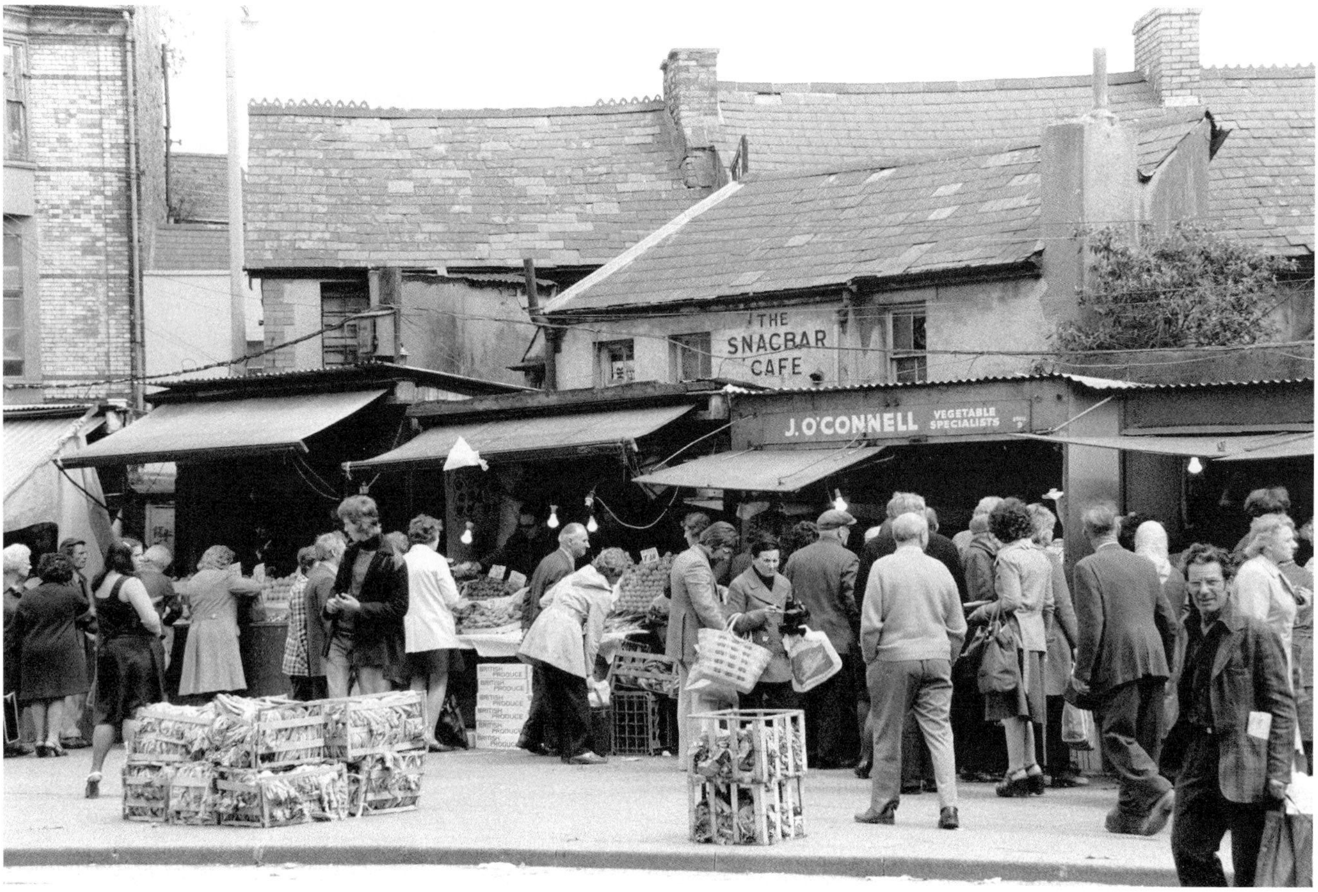
THE
SNACBAR
CAFE
J. O'CONNELL
VEGETABLE
SPECIALISTS

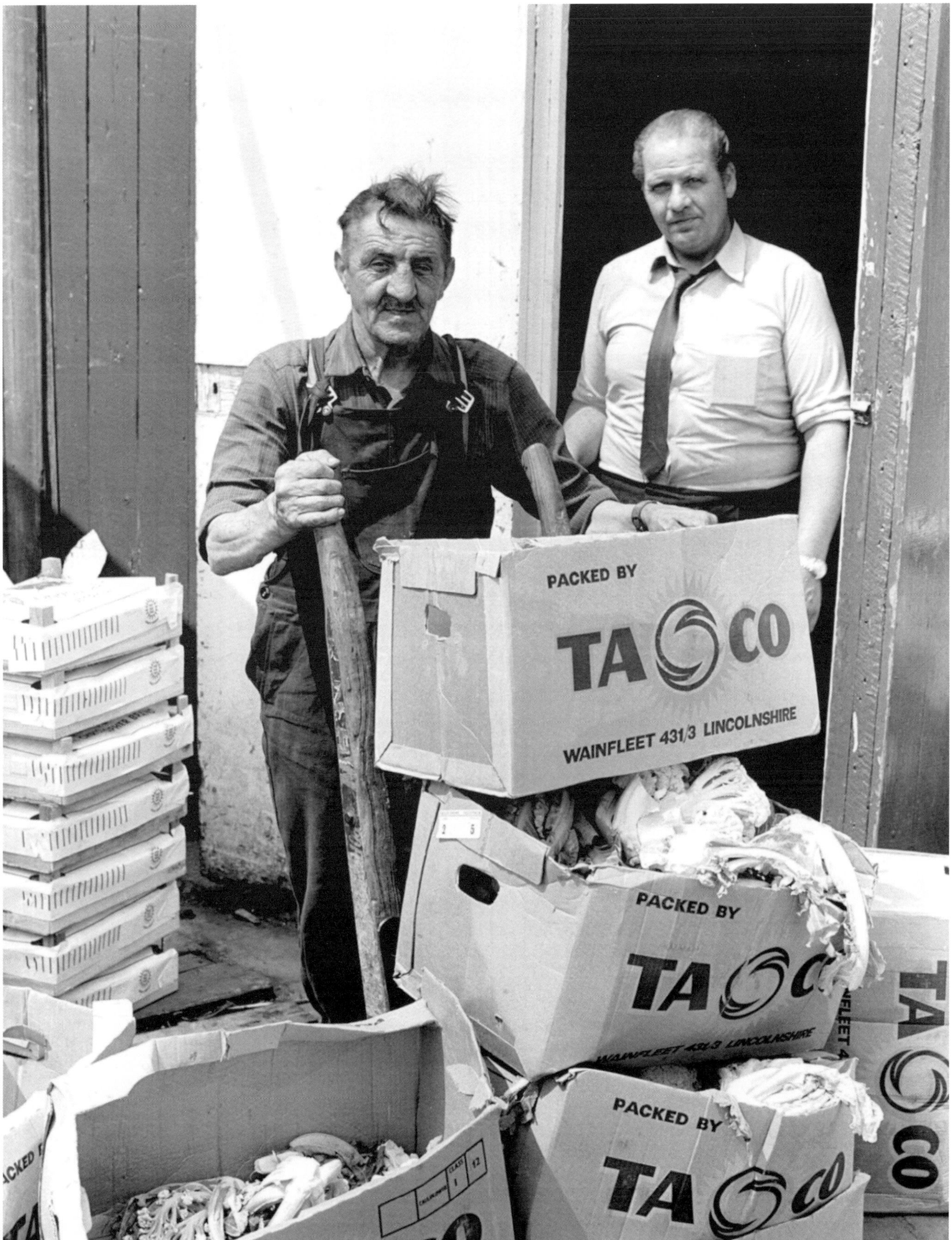
PACKED BY
TA CO
WAINFLEET 431/3 LINCOLNSHIRE
PACKED BY
TA CO
PACKED BY
TA CO
PACKED BY
TA CO
PACKED BY
TA CO

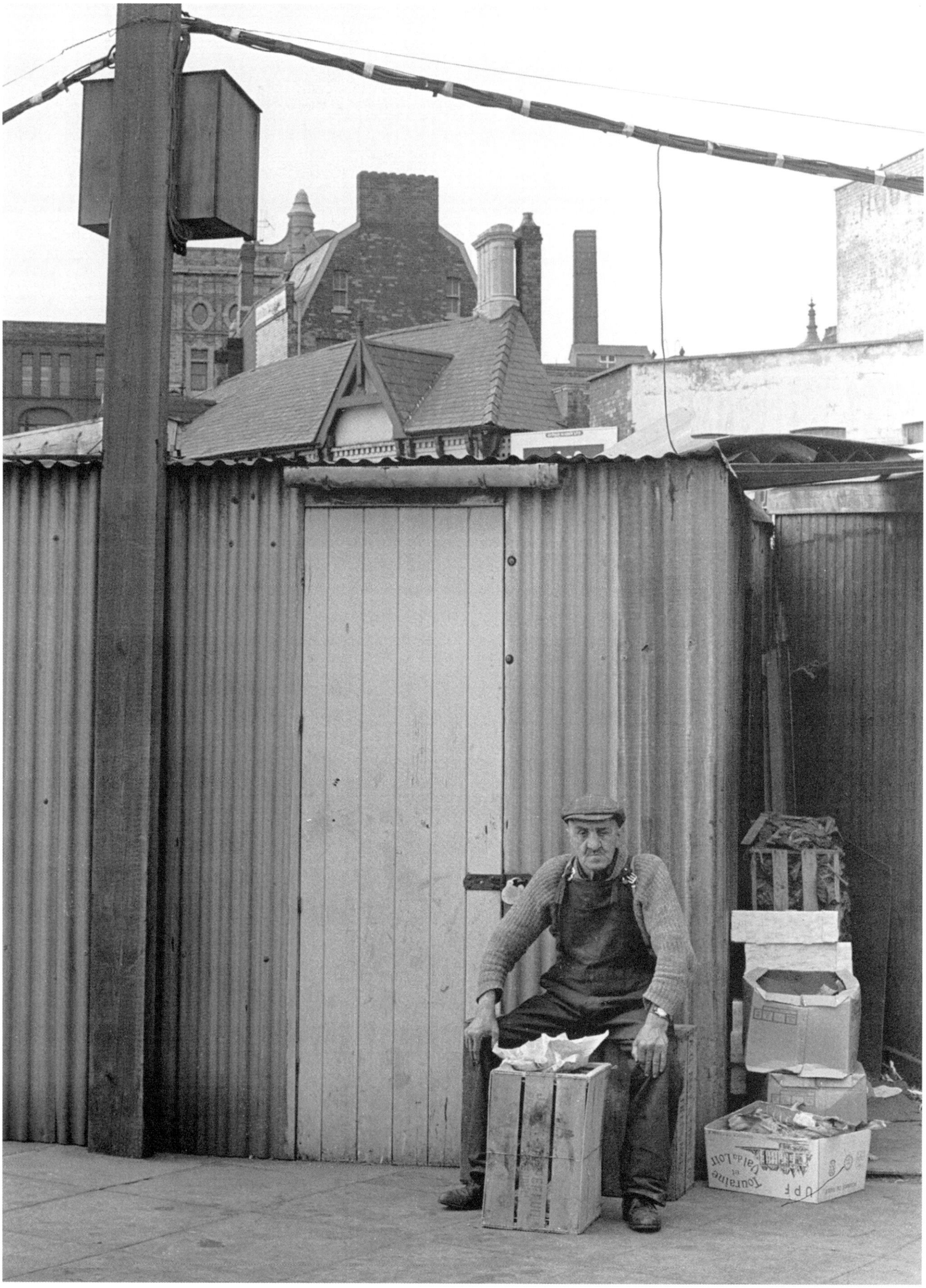

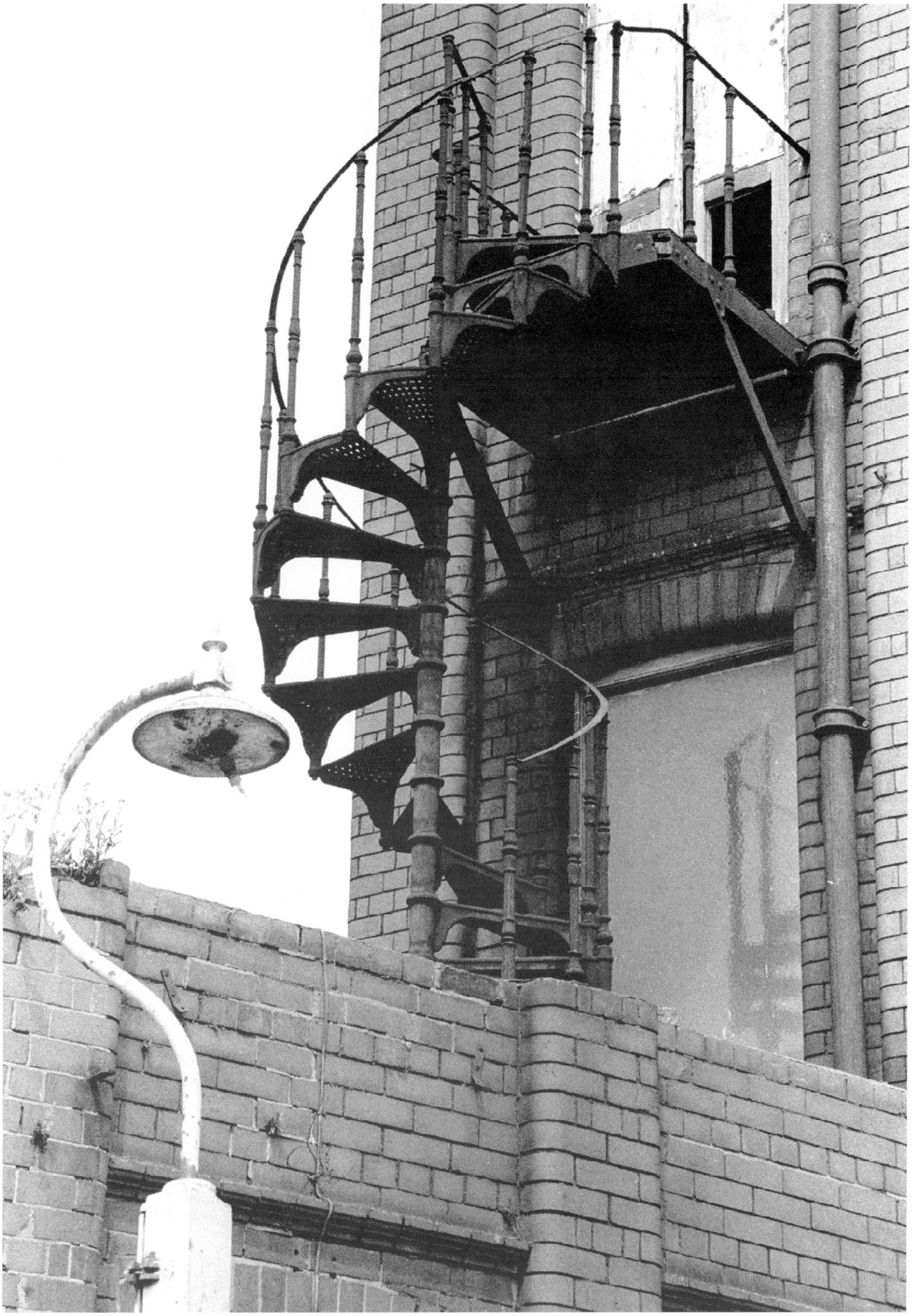

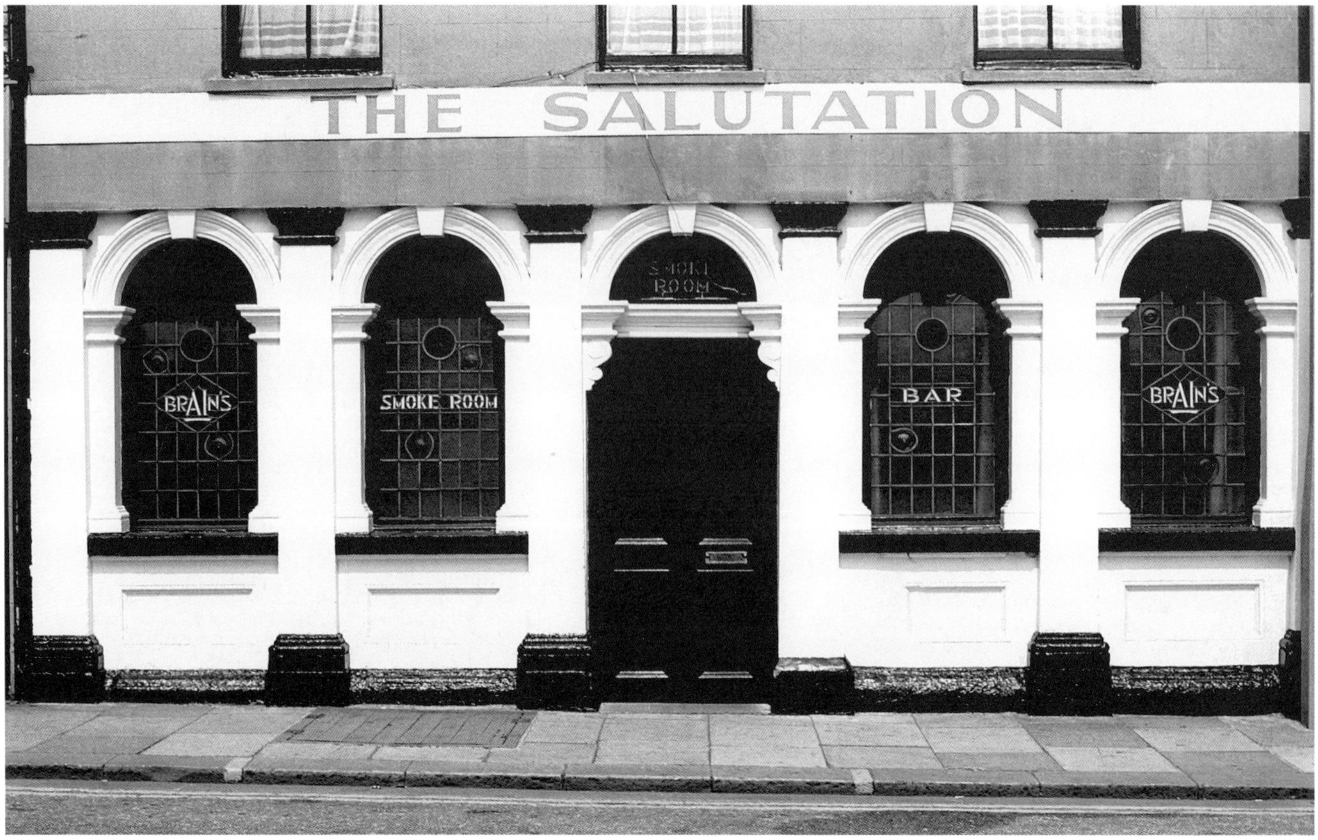
THE SALUTATION
SMOKE ROOM
BRAINS
SMOKE ROOM
BAR
BRAINS

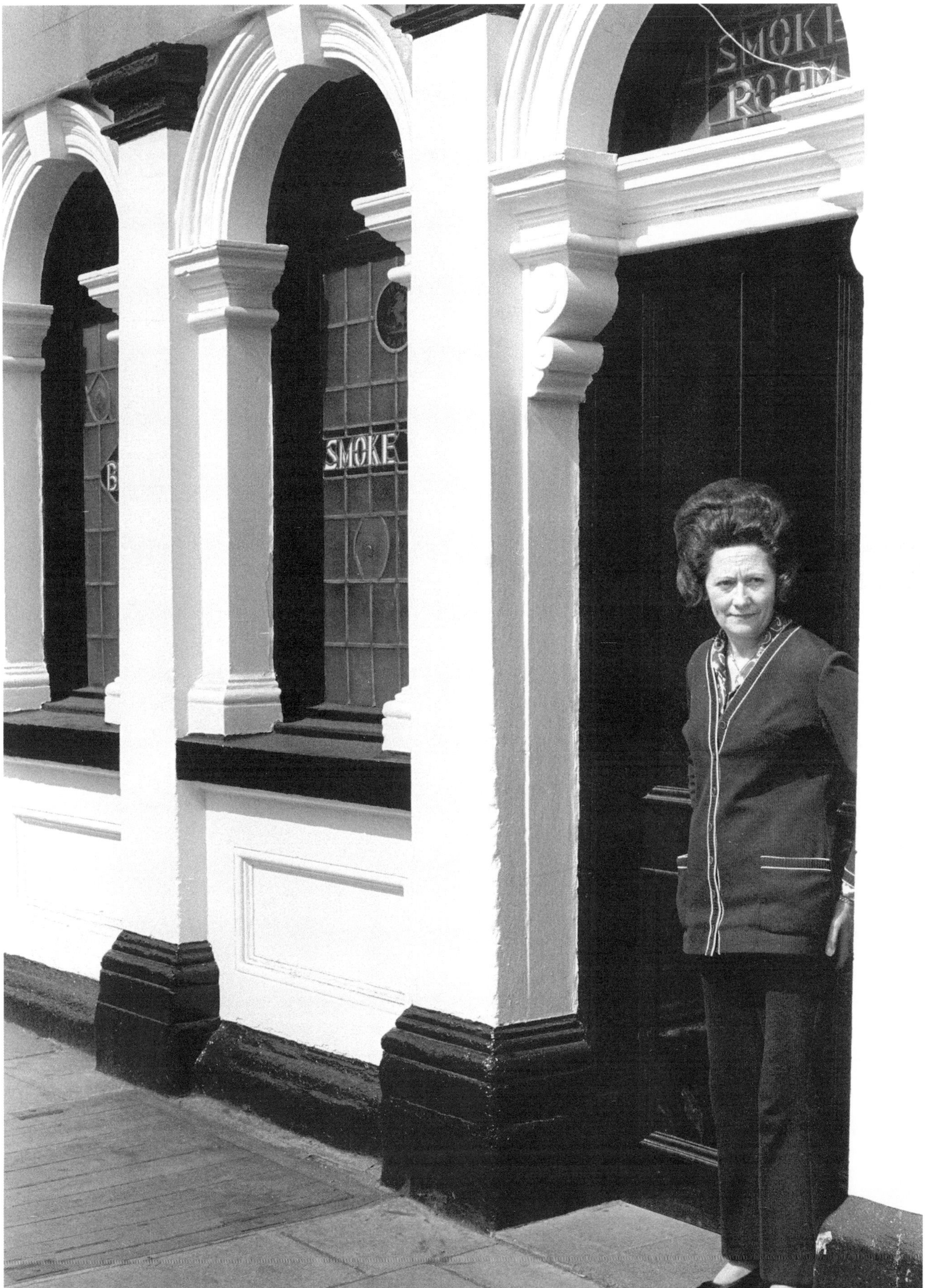
SMOKE
ROOM
SMOKE

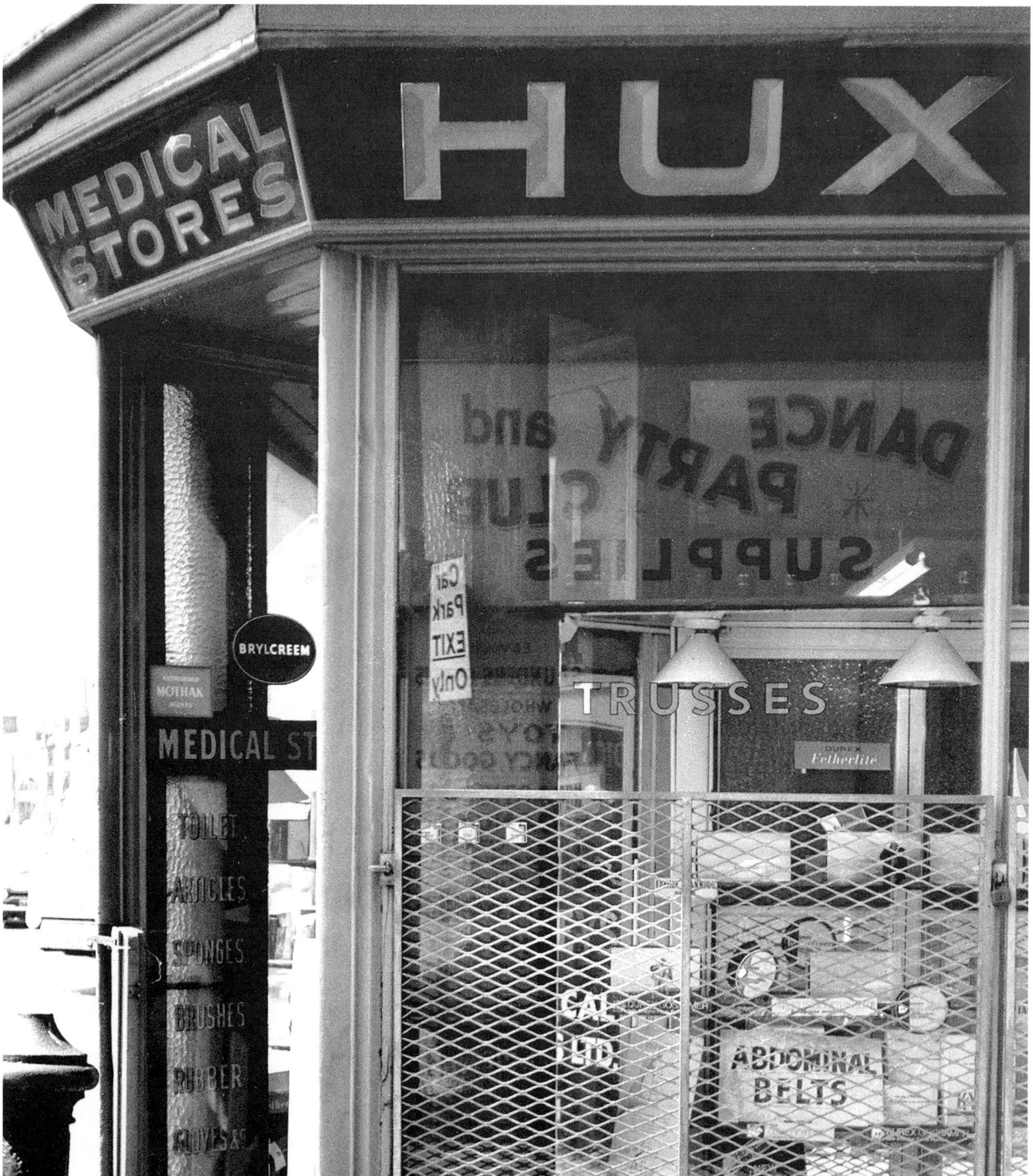
MEDICAL STORES
HUX
Car Park EXIT Only
BRYLCREEM
MOTHAK
MEDICAL ST
TOILET
ARTICLES
SPONGES
BRUSHES
RUBBER
GLOVES &c
TRUSSES
DUREX
Fetherlite
ABDOMINAL BELTS
RUPTURE TRUSSES

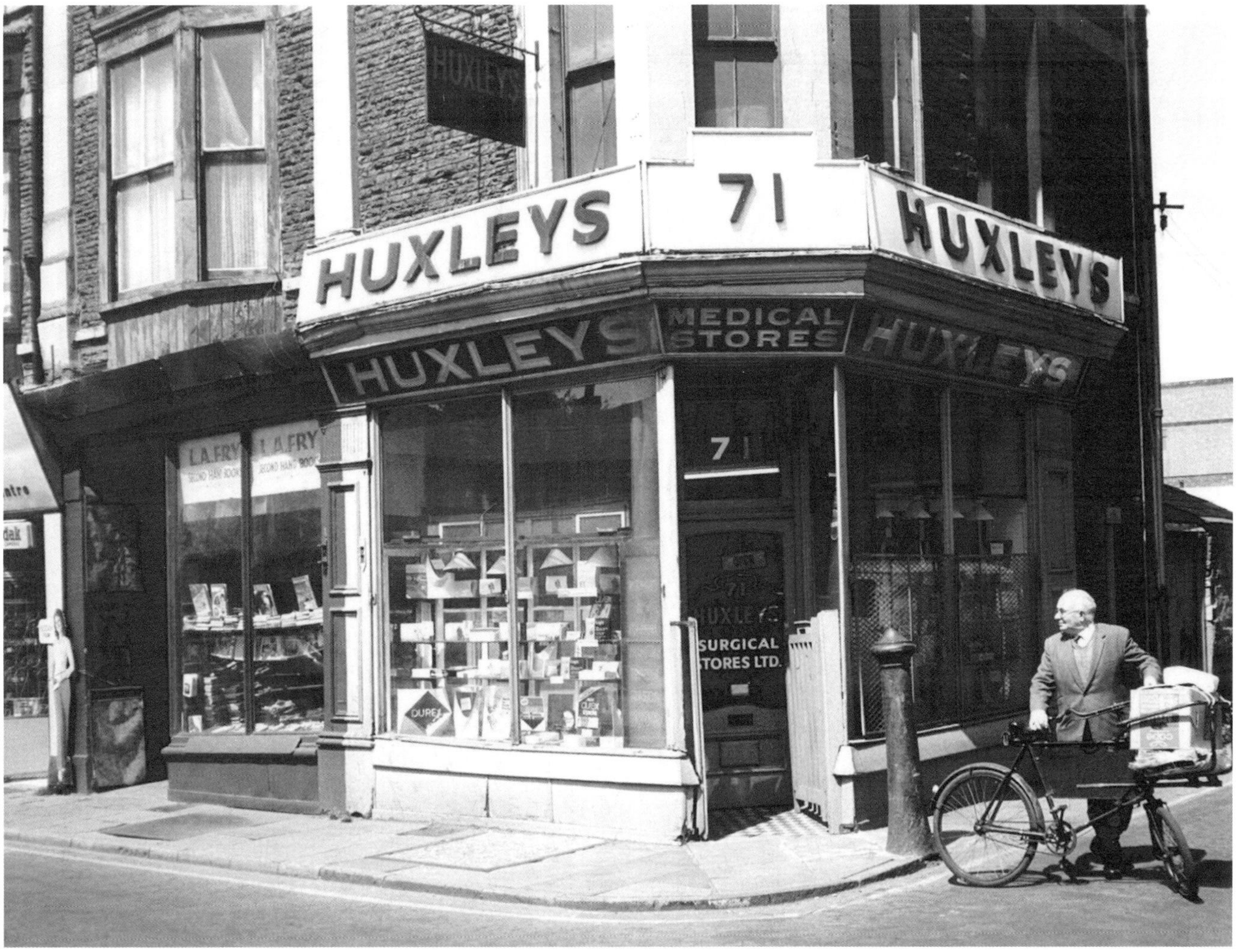
HUXLEYS
71
HUXLEYS
HUXLEYS
MEDICAL
STORES
HUXLEYS
L.A.FRY
L.A.FRY
71
SURGICAL
STORES LTD.

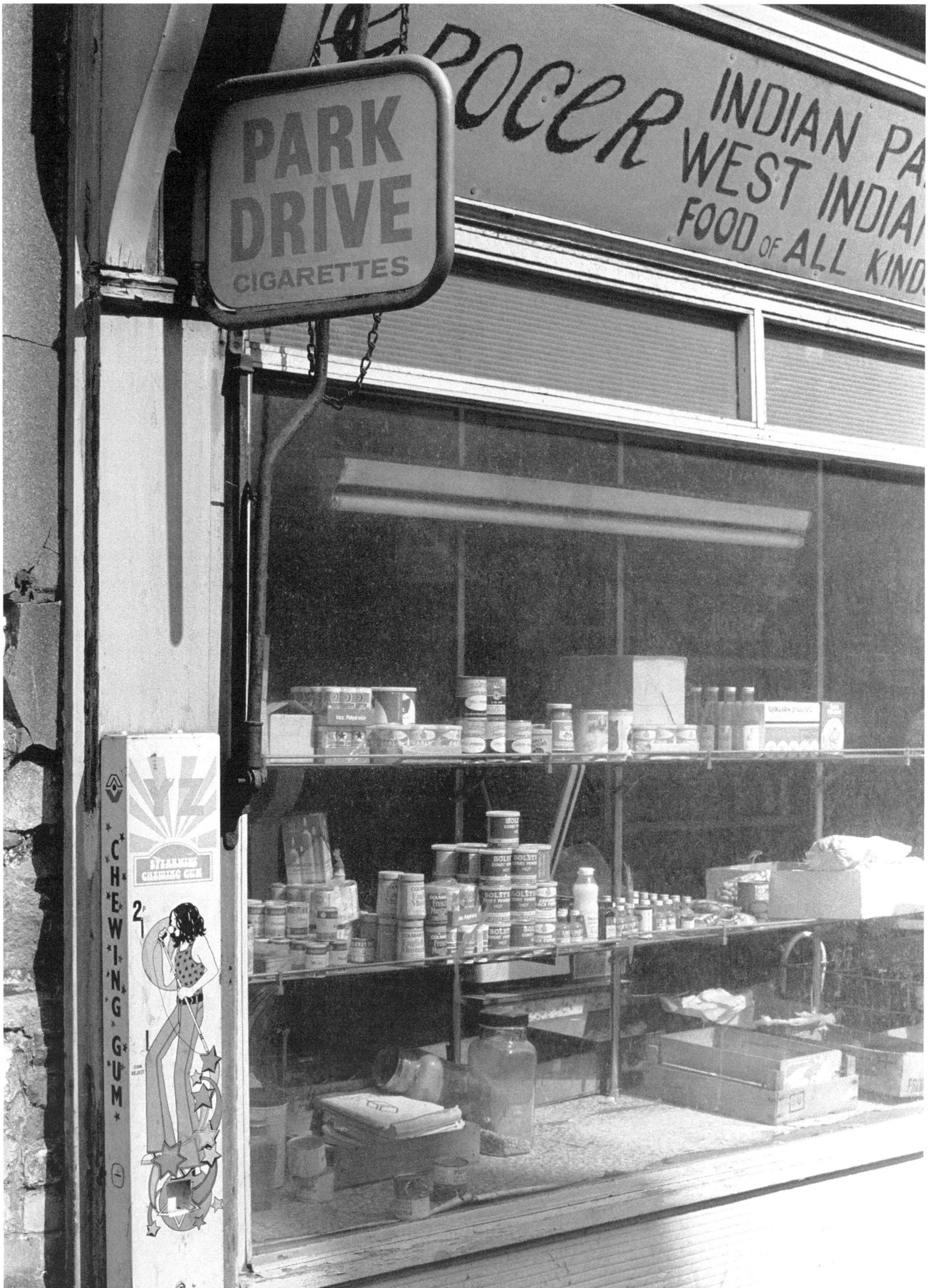
PARK
DRIVE
CIGARETTES
INDIAN
WEST INDIAN
FOOD OF ALL
CHEWING GUM

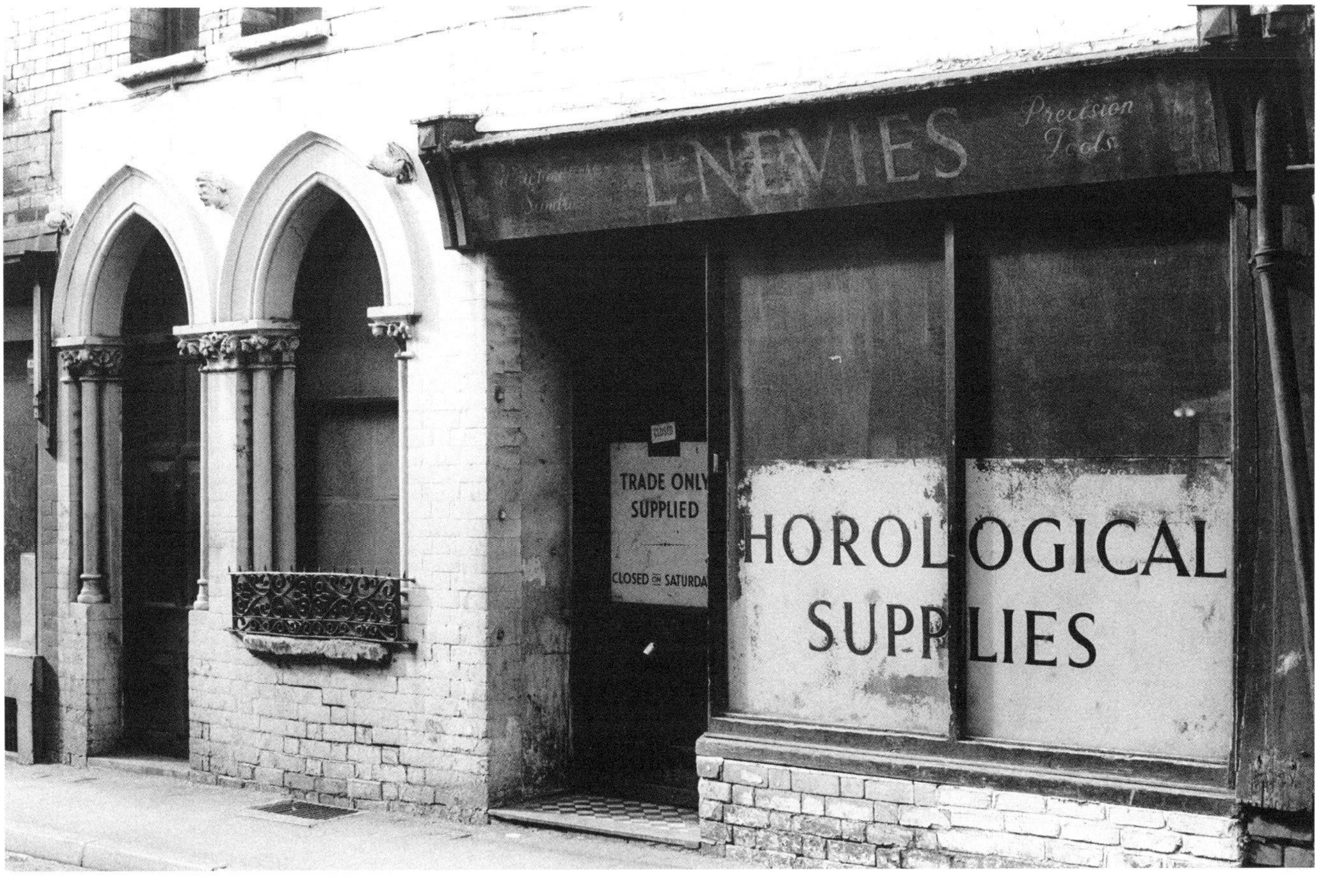
L.NEVIES
Precision Tools
TRADE ONLY SUPPLIED
CLOSED ON SATURDA
HOROLOGICAL SUPPLIES

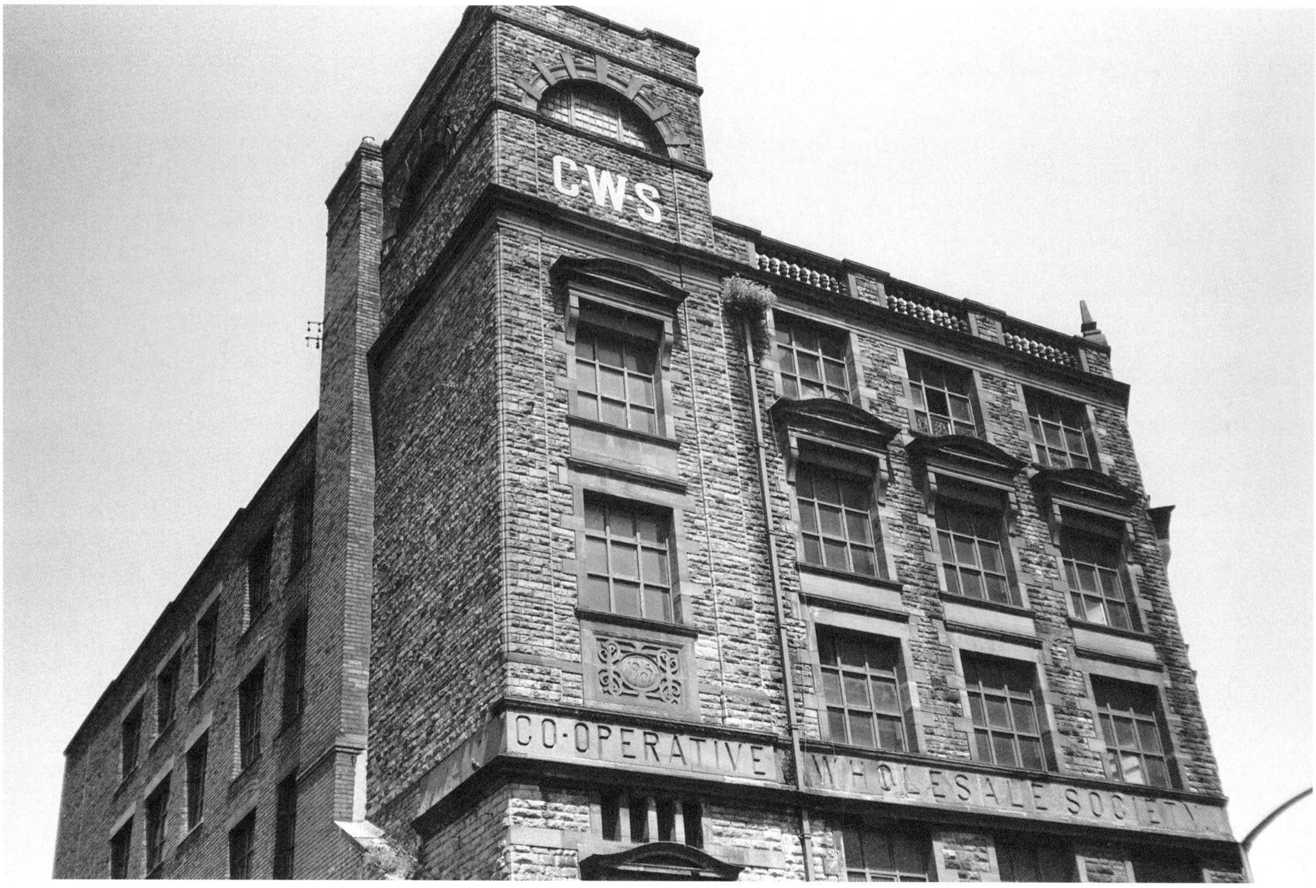
C·W·S
CO-OPERATIVE WHOLESALE SOCIETY

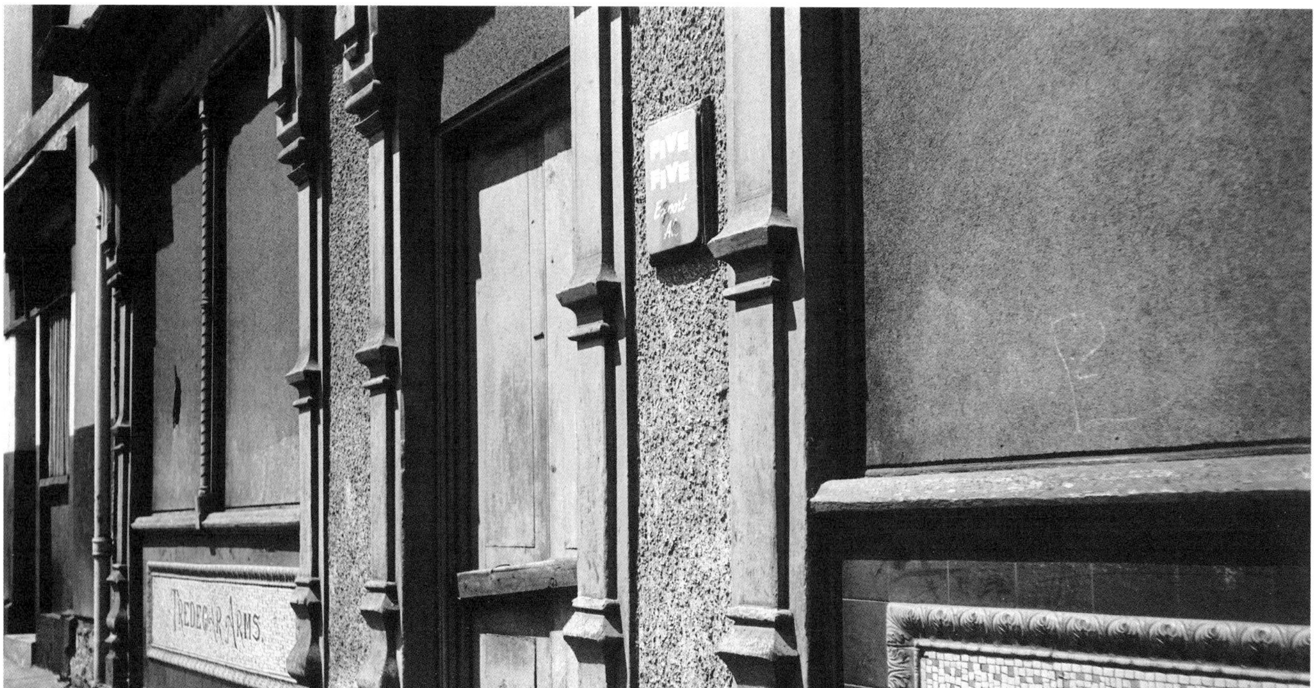
FIVE
FIVE
TREDEGAR ARMS.

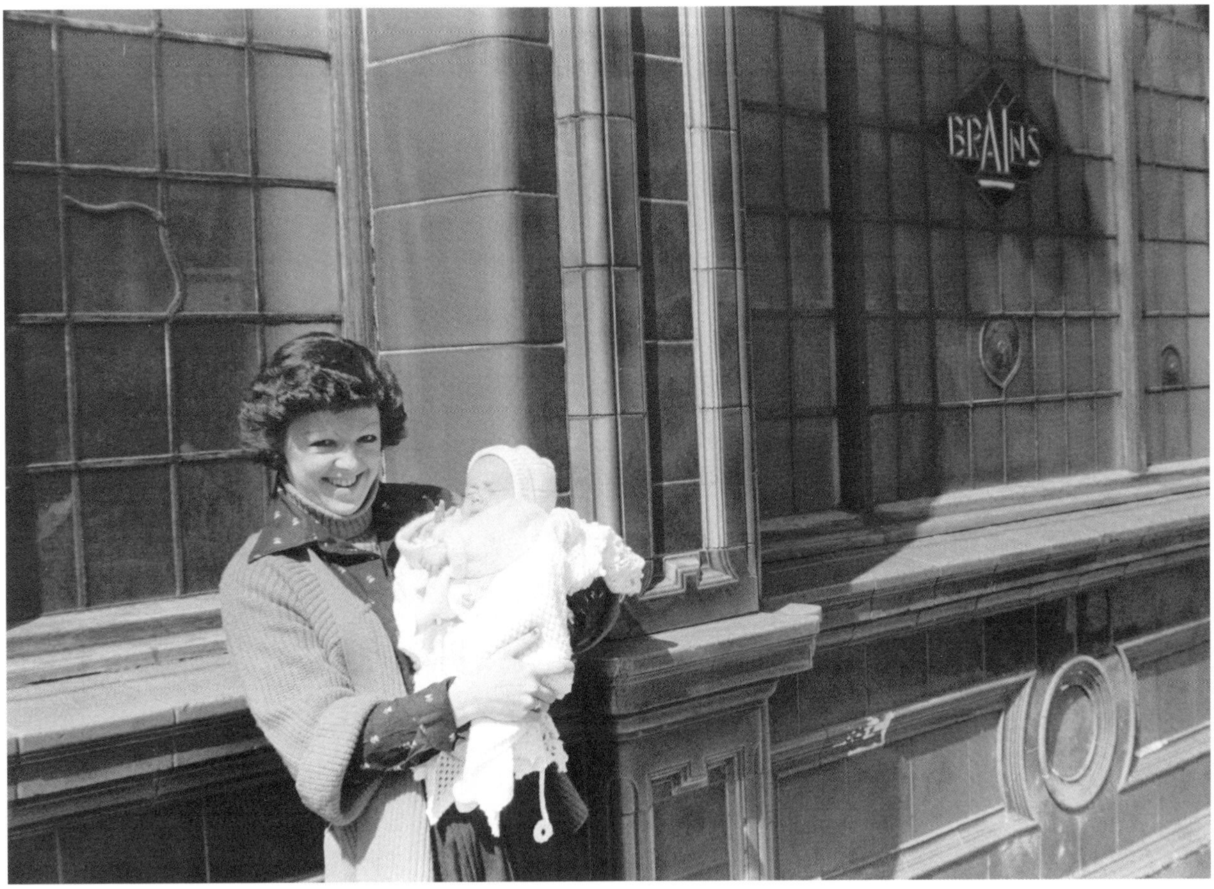
BRAINS

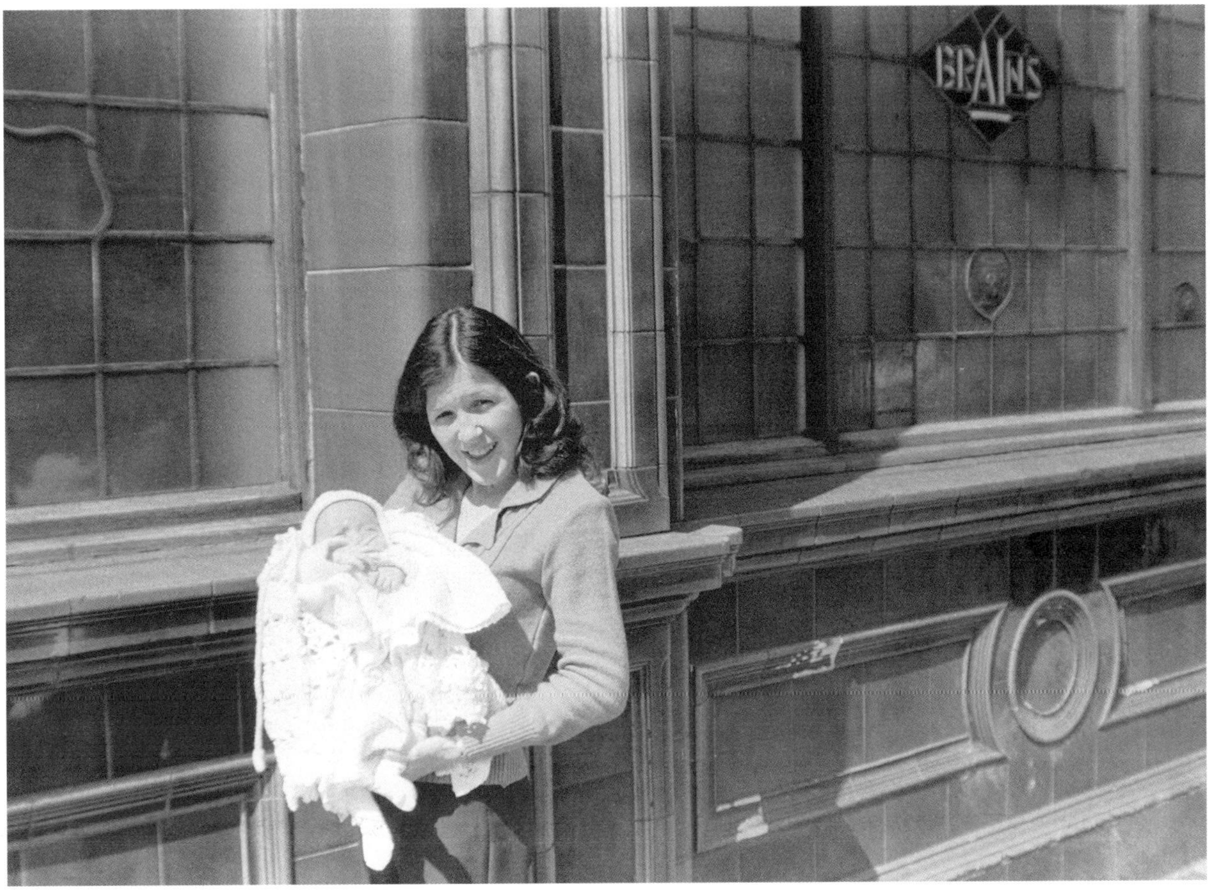
BRAINS

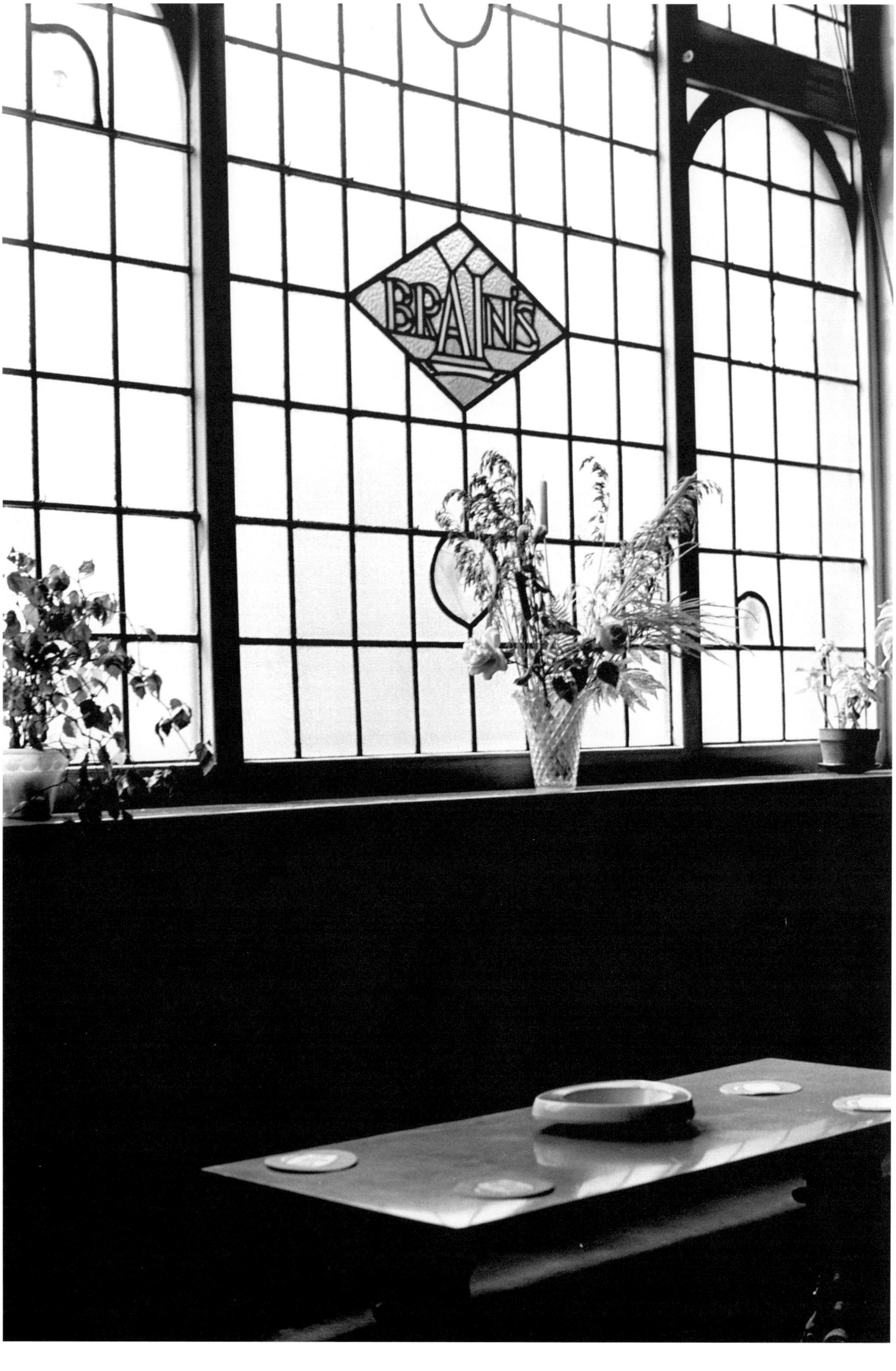
BRAIN'S

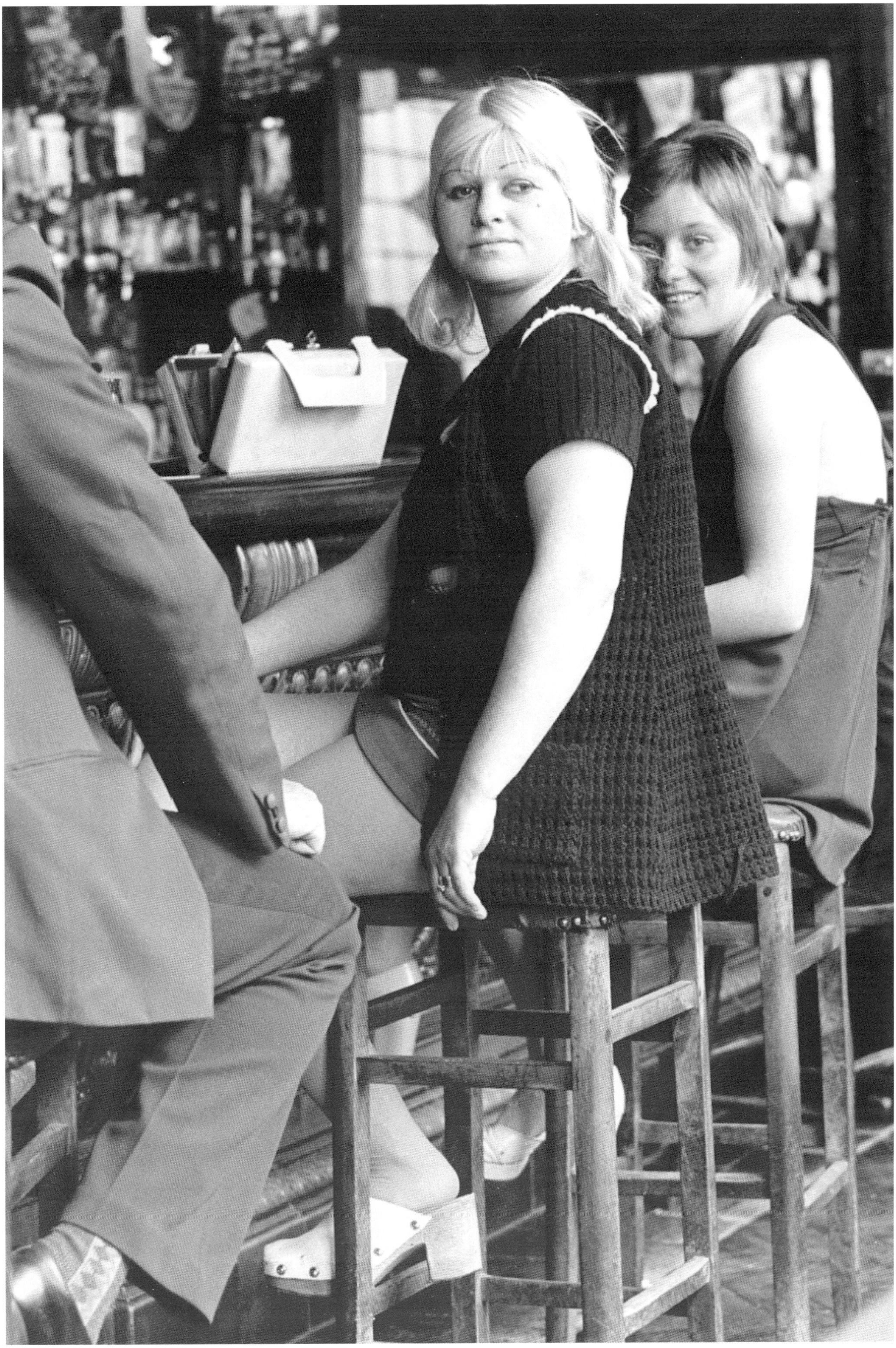

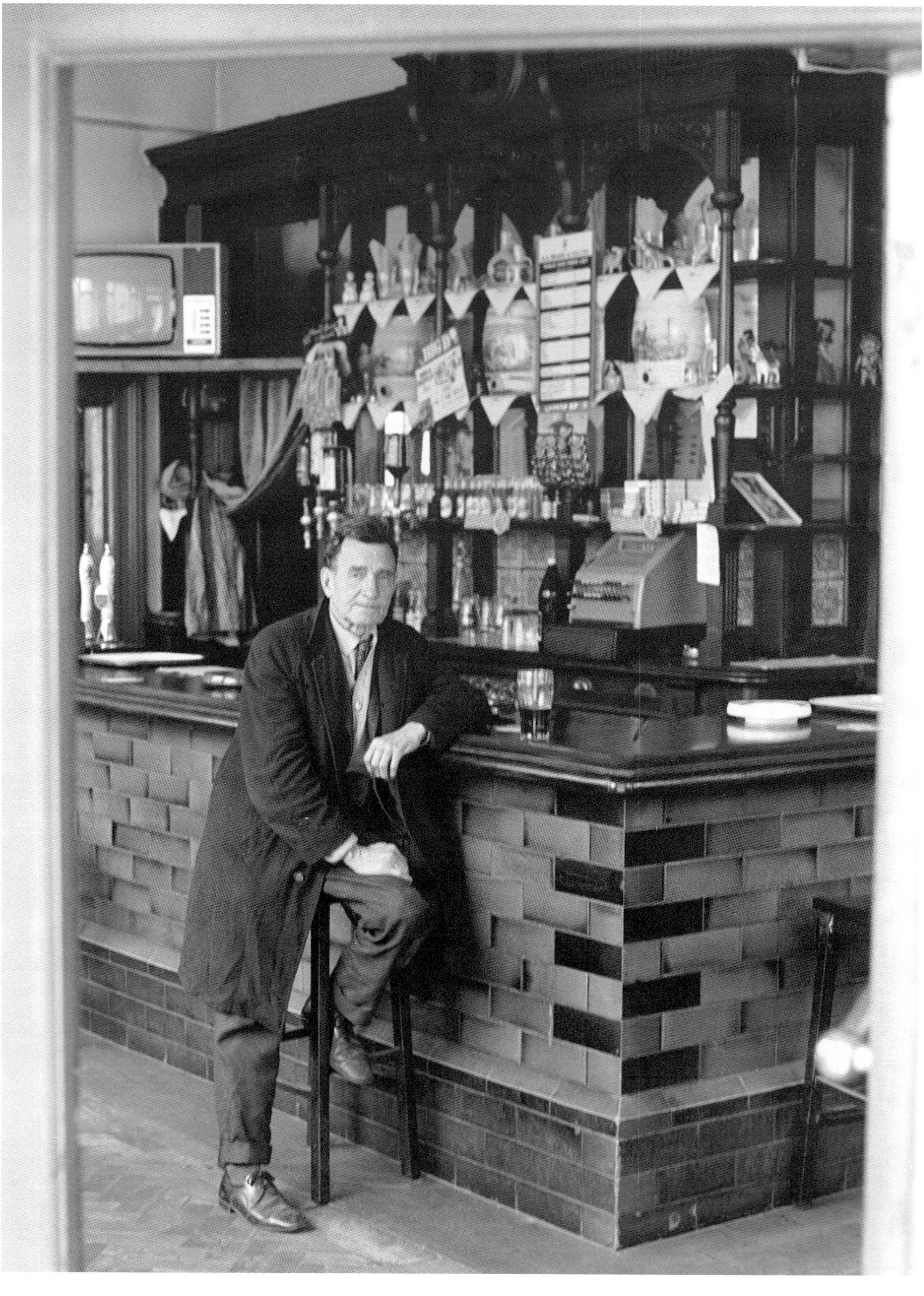

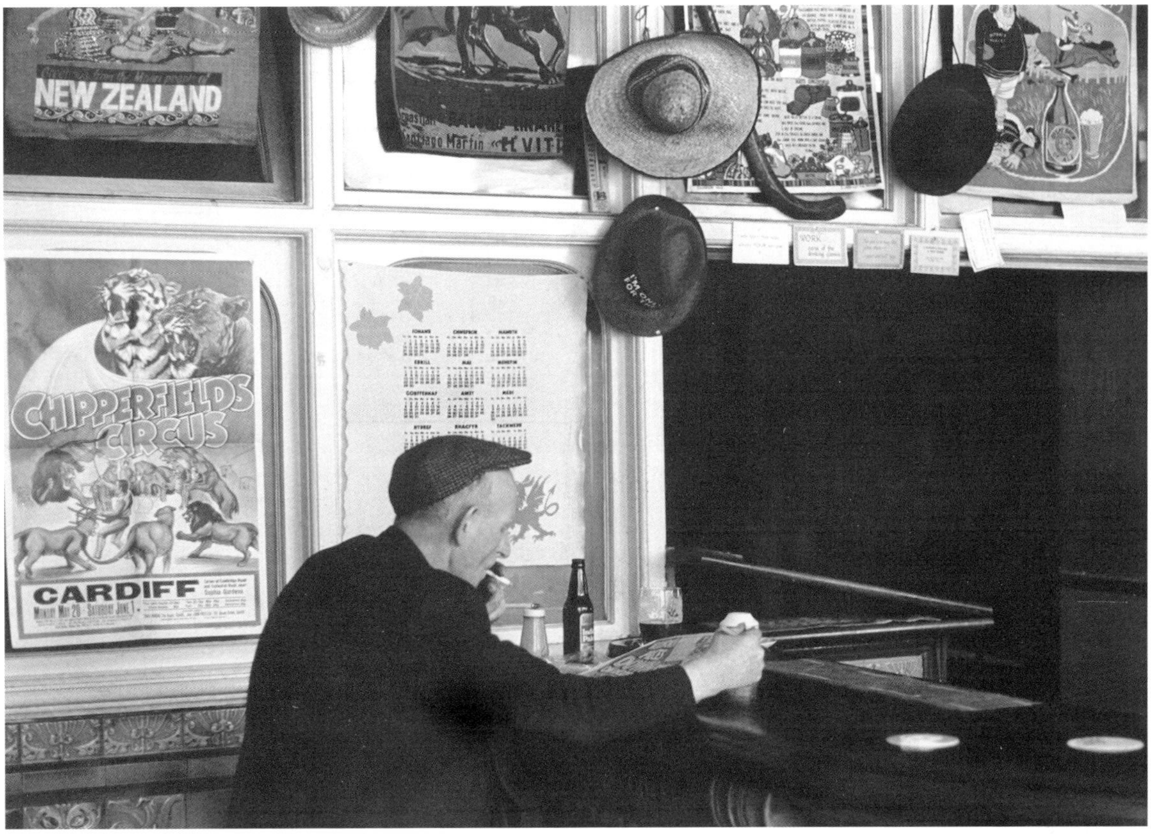

NEW ZEALAND
CHIPPERFIELDS
CIRCUS
CARDIFF

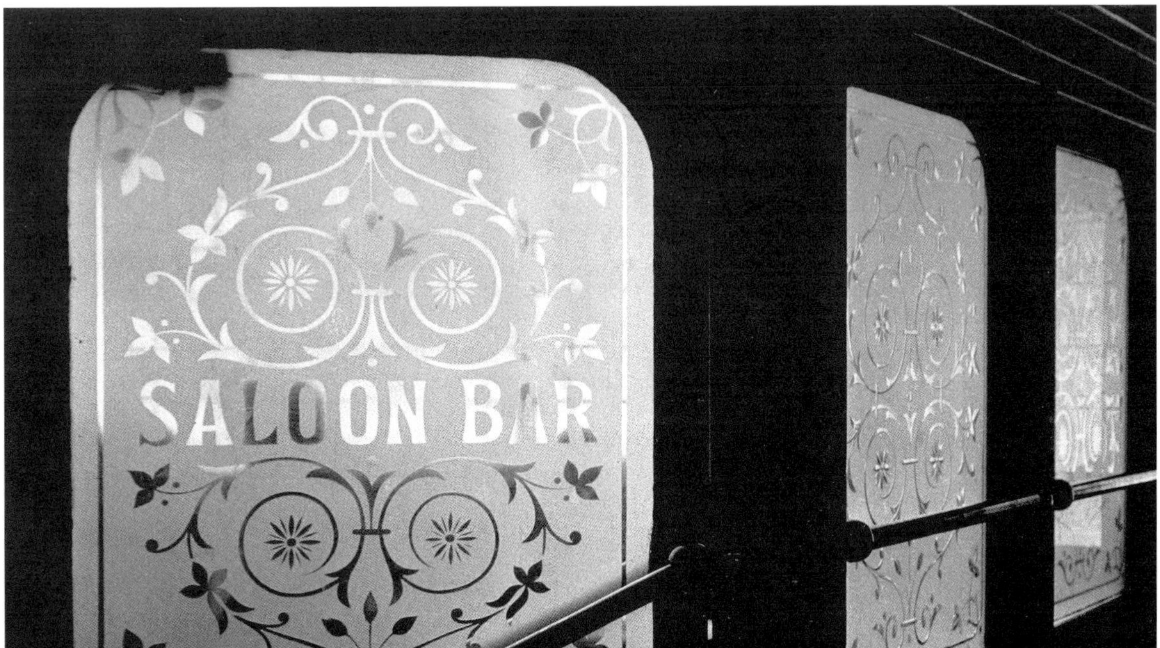

SALOON BAR

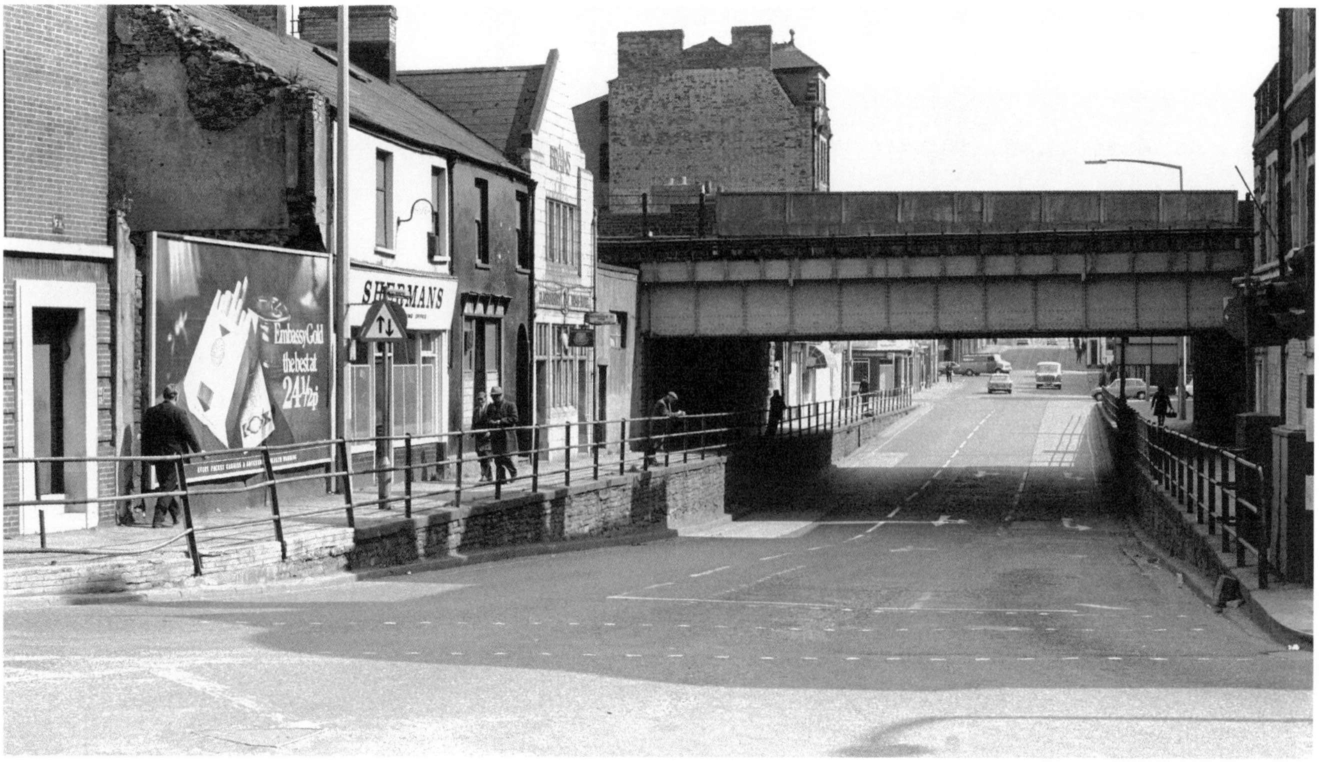
SHERMANS
Embassy Gold
the best at
24½p

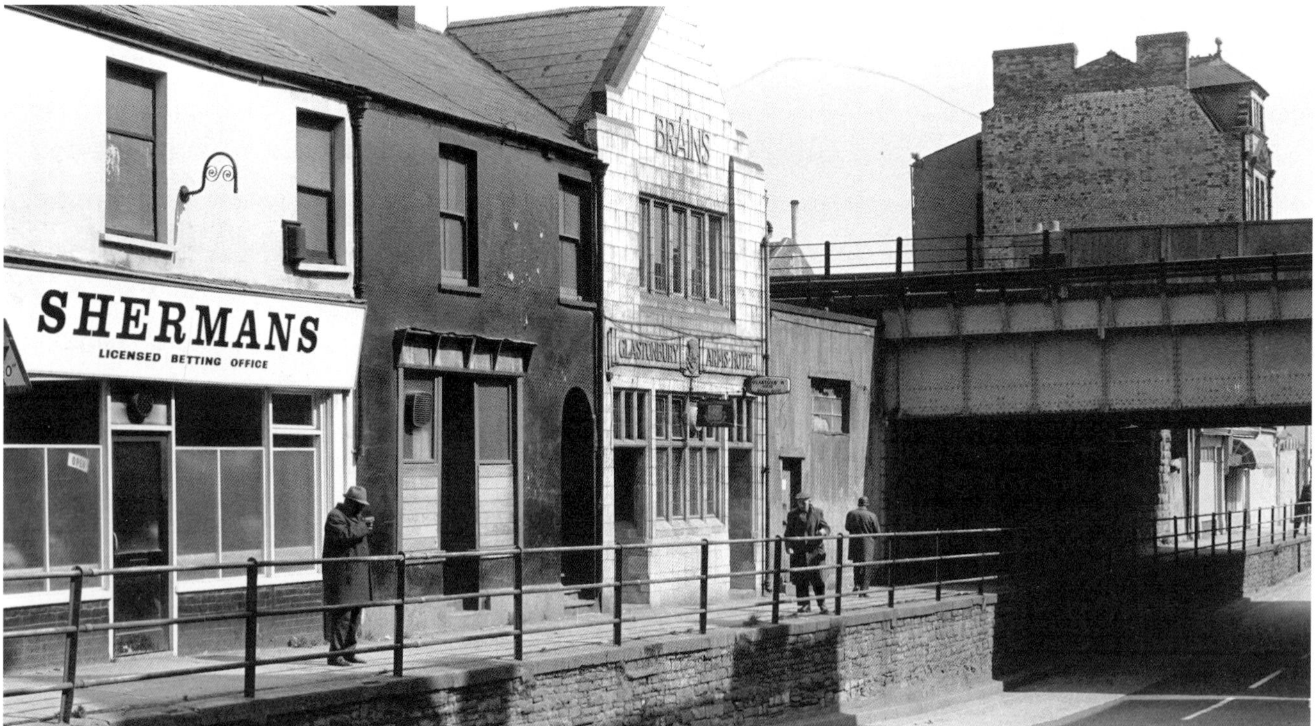
SHERMANS
LICENSED BETTING OFFICE
BRAINS
GLASTONBURY ARMS HOTEL

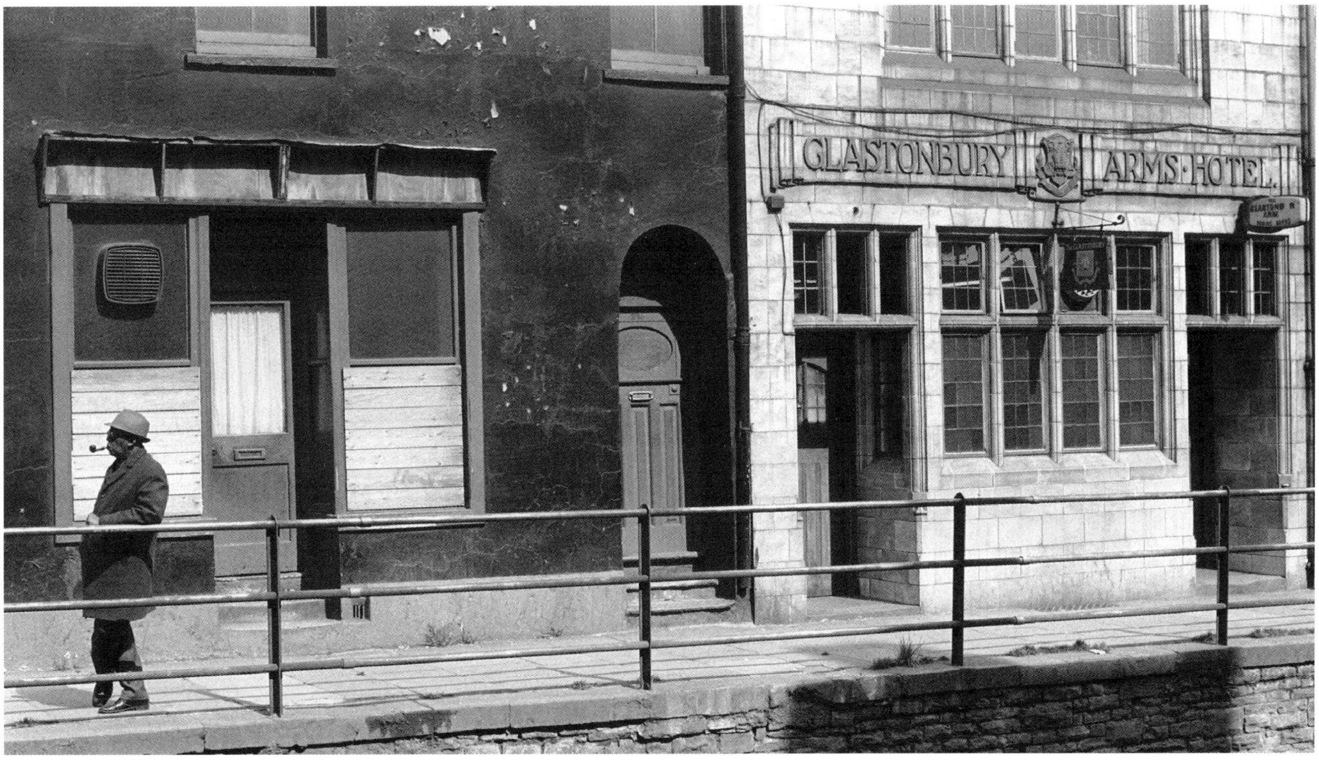
GLASTONBURY
ARMS·HOTEL

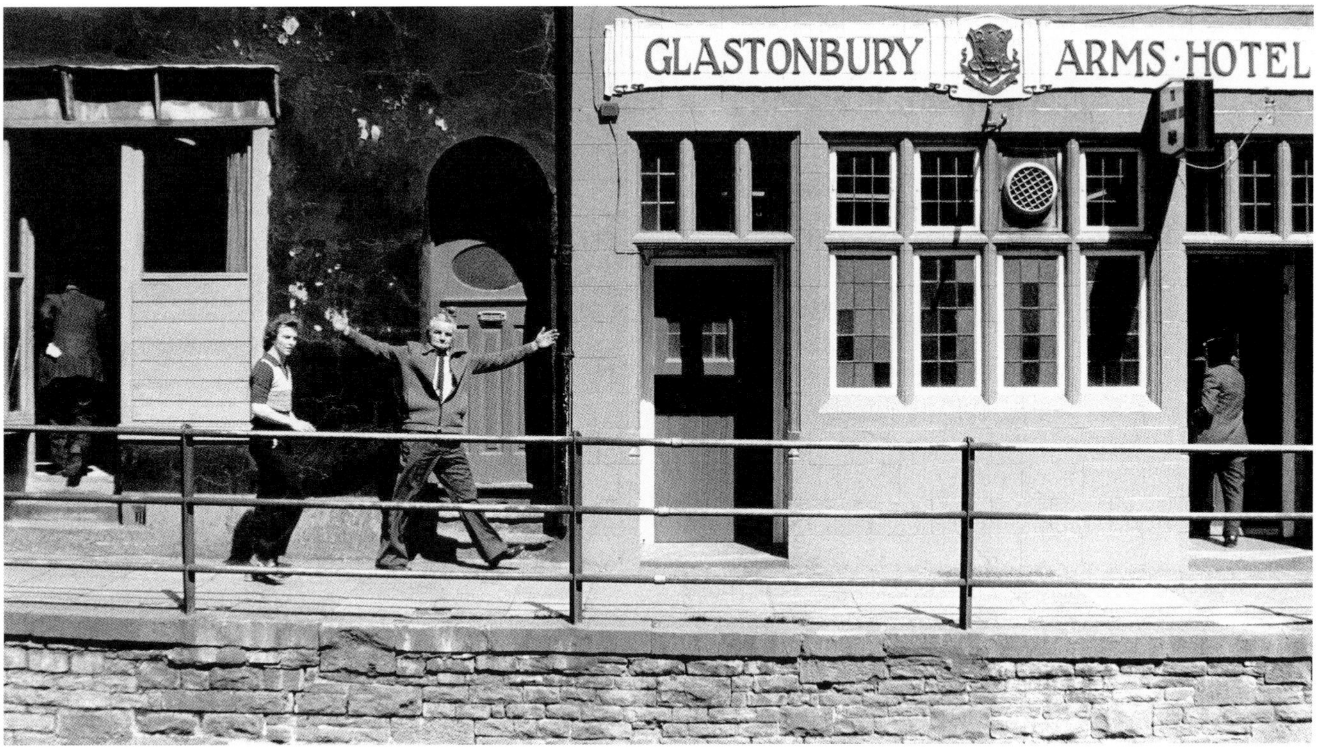
GLASTONBURY
ARMS·HOTEL

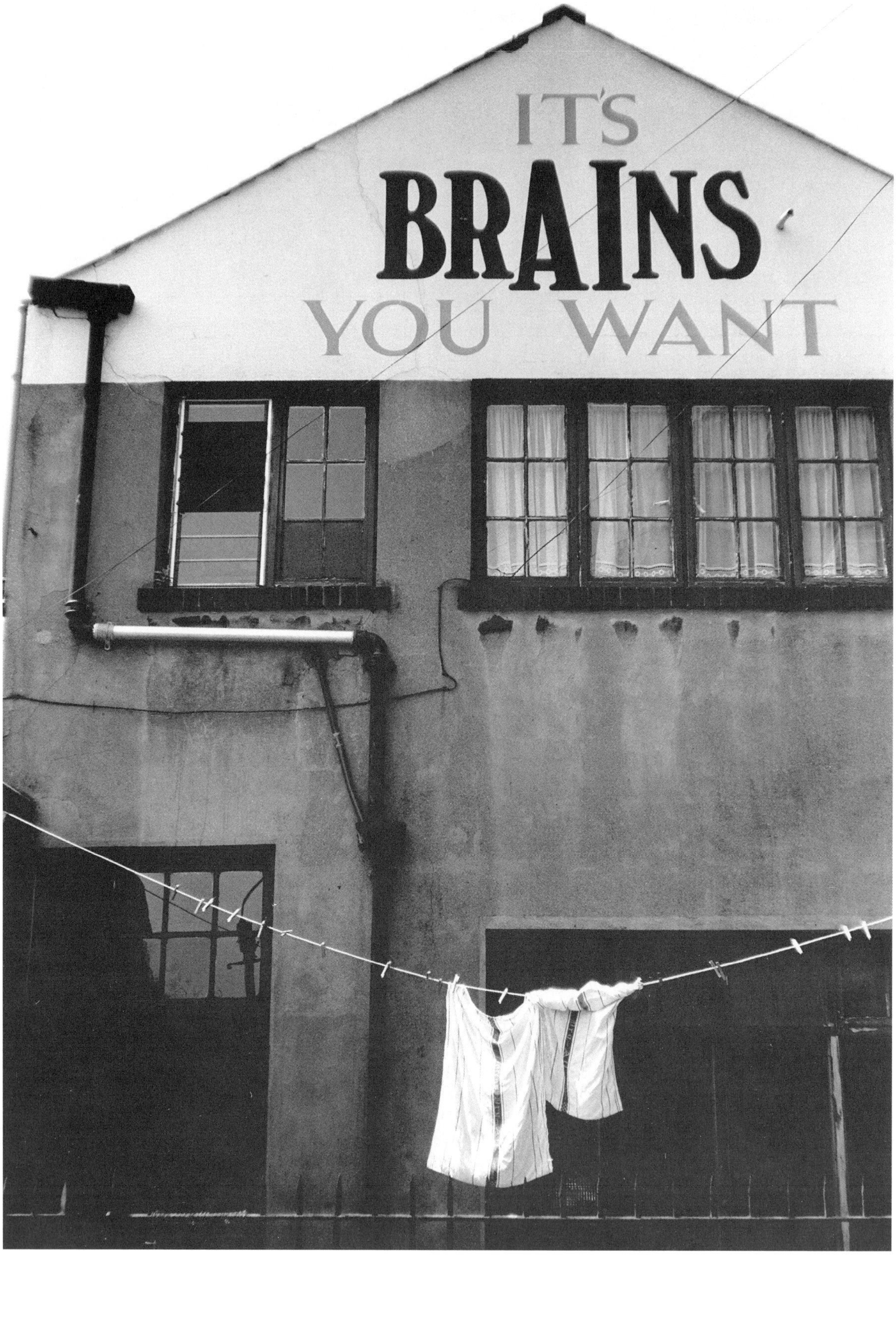
IT'S
BRAINS
YOU WANT

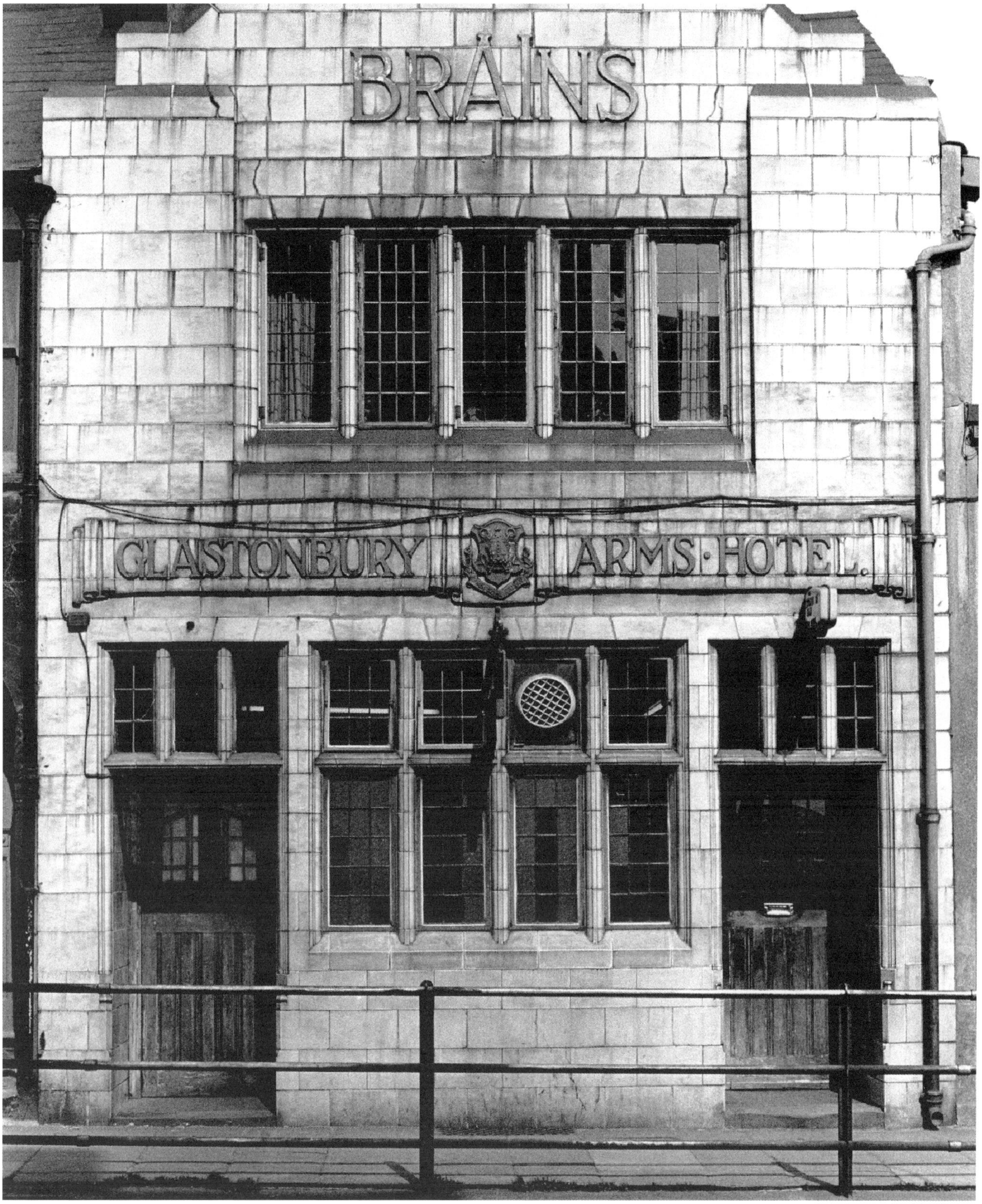
BRAINS
GLASTONBURY ARMS·HOTEL.

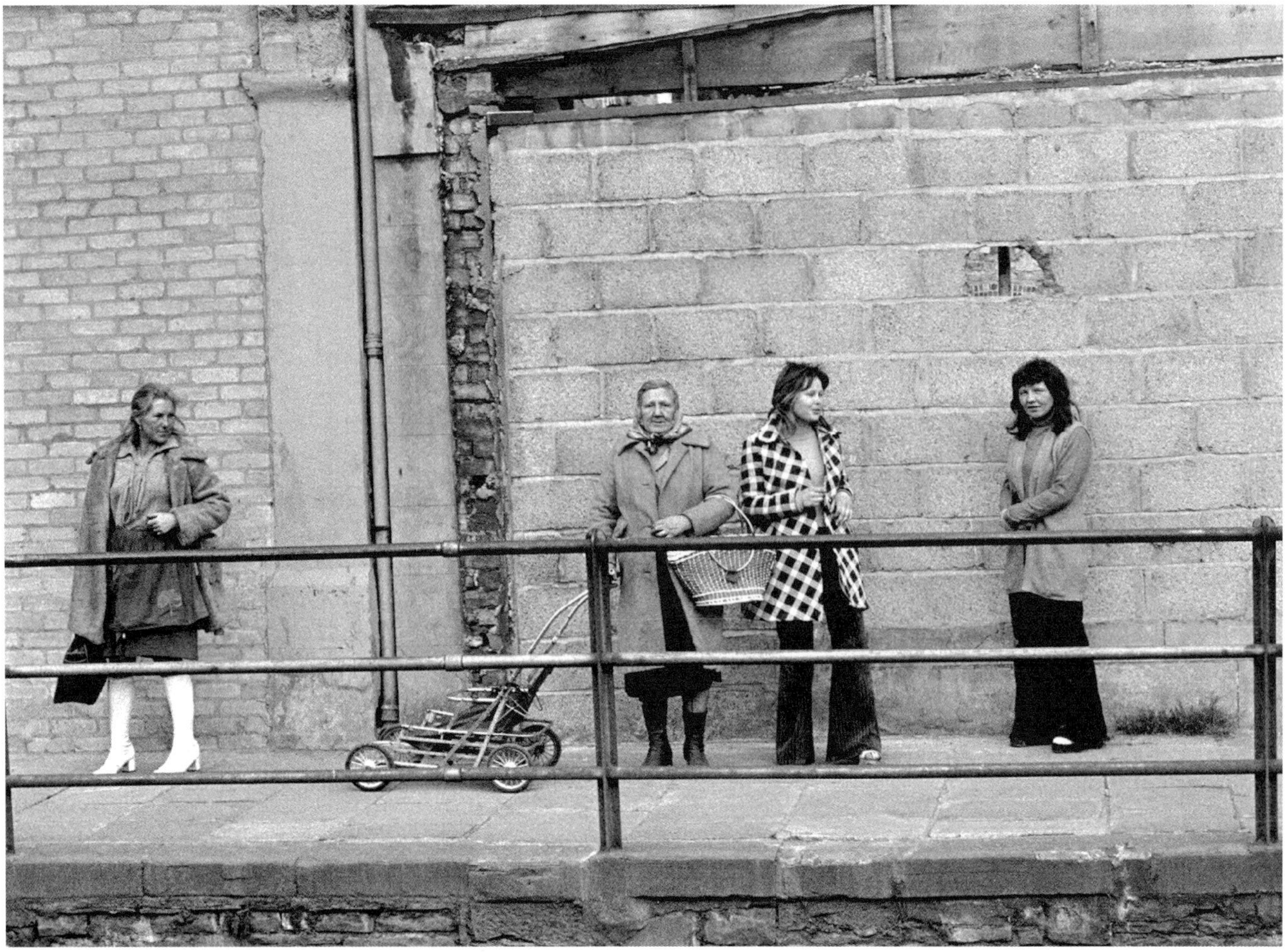

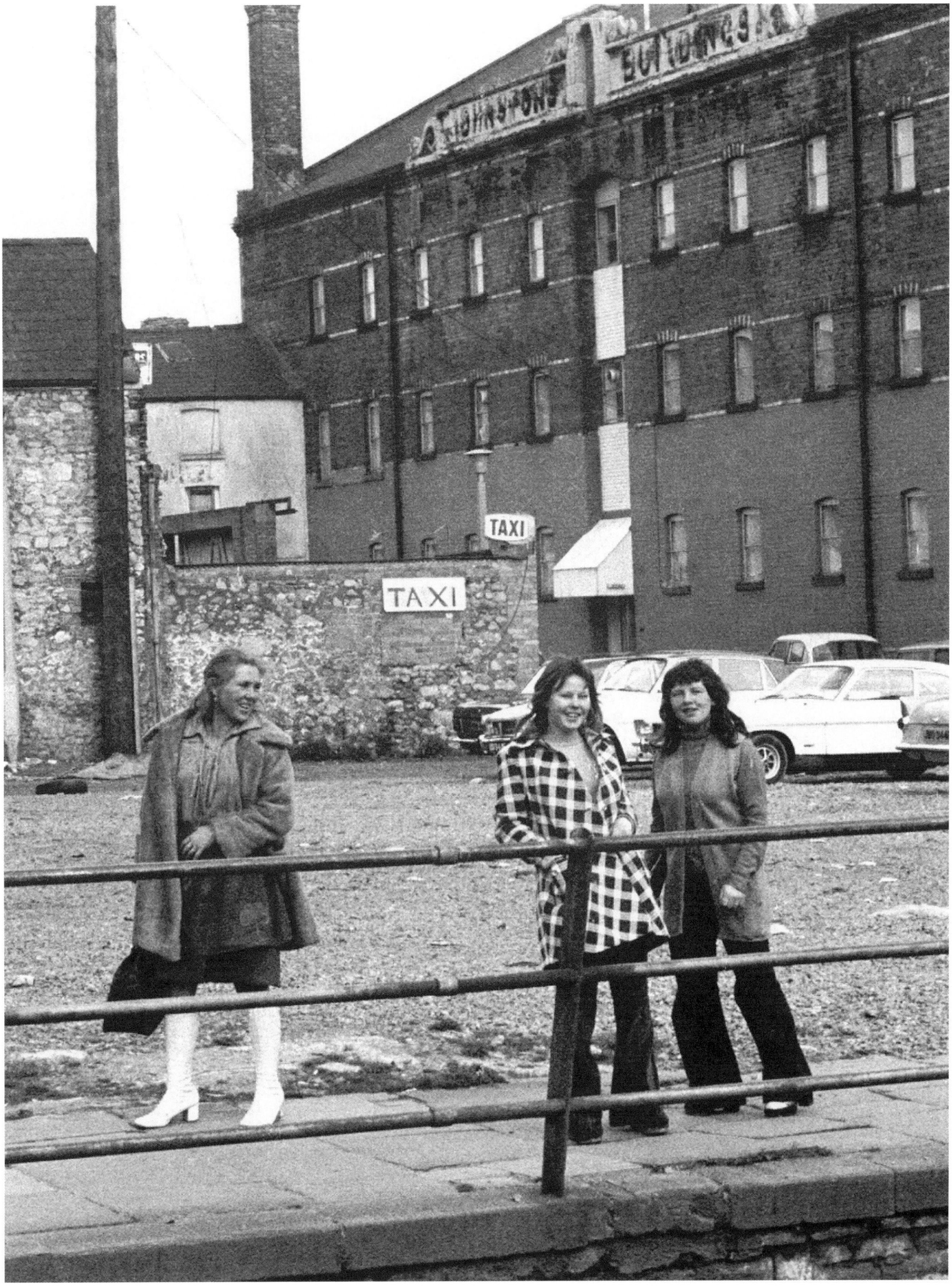
TAXI
TAXI

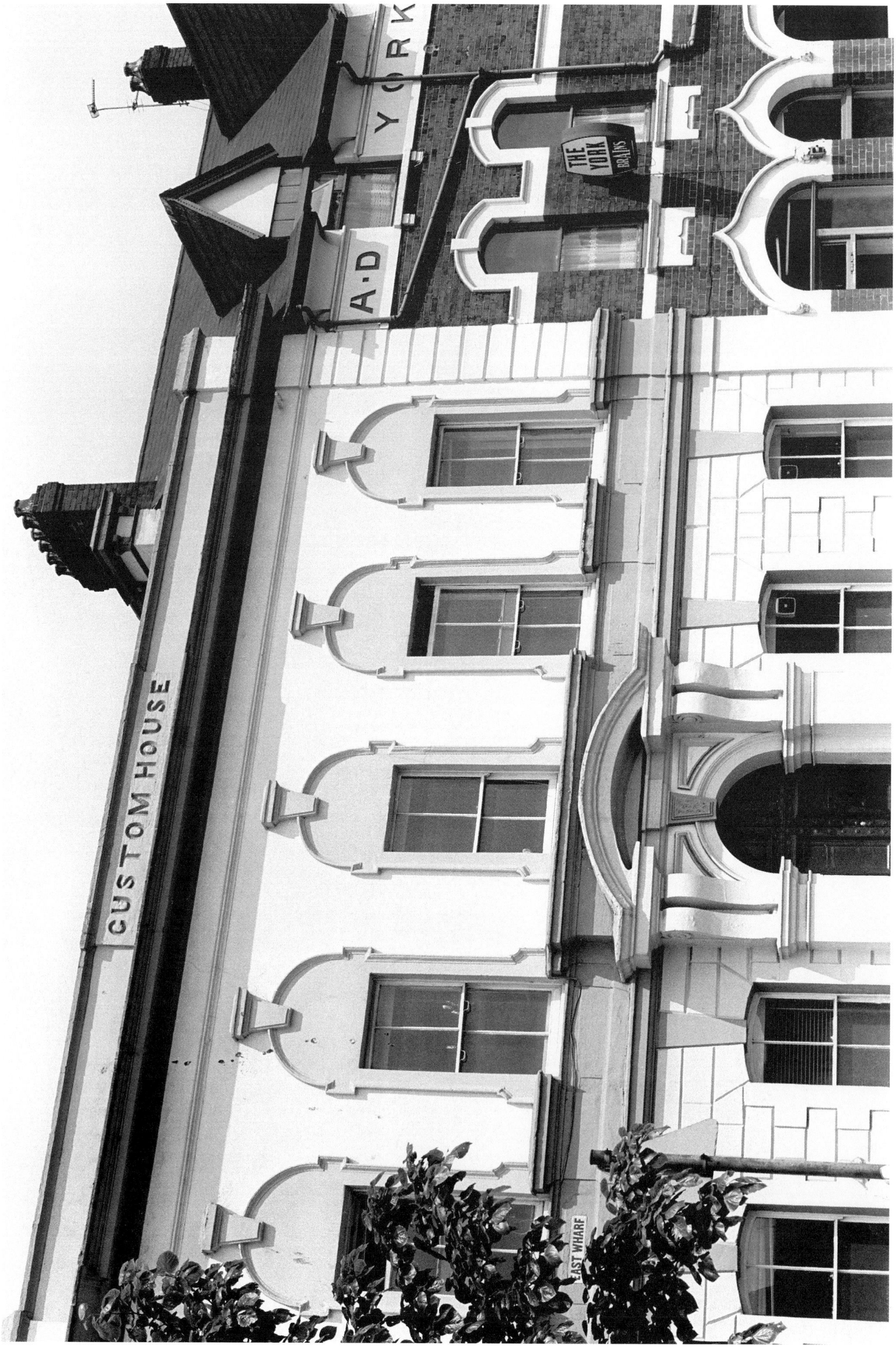
CUSTOM HOUSE
THE YORK
BRAINS

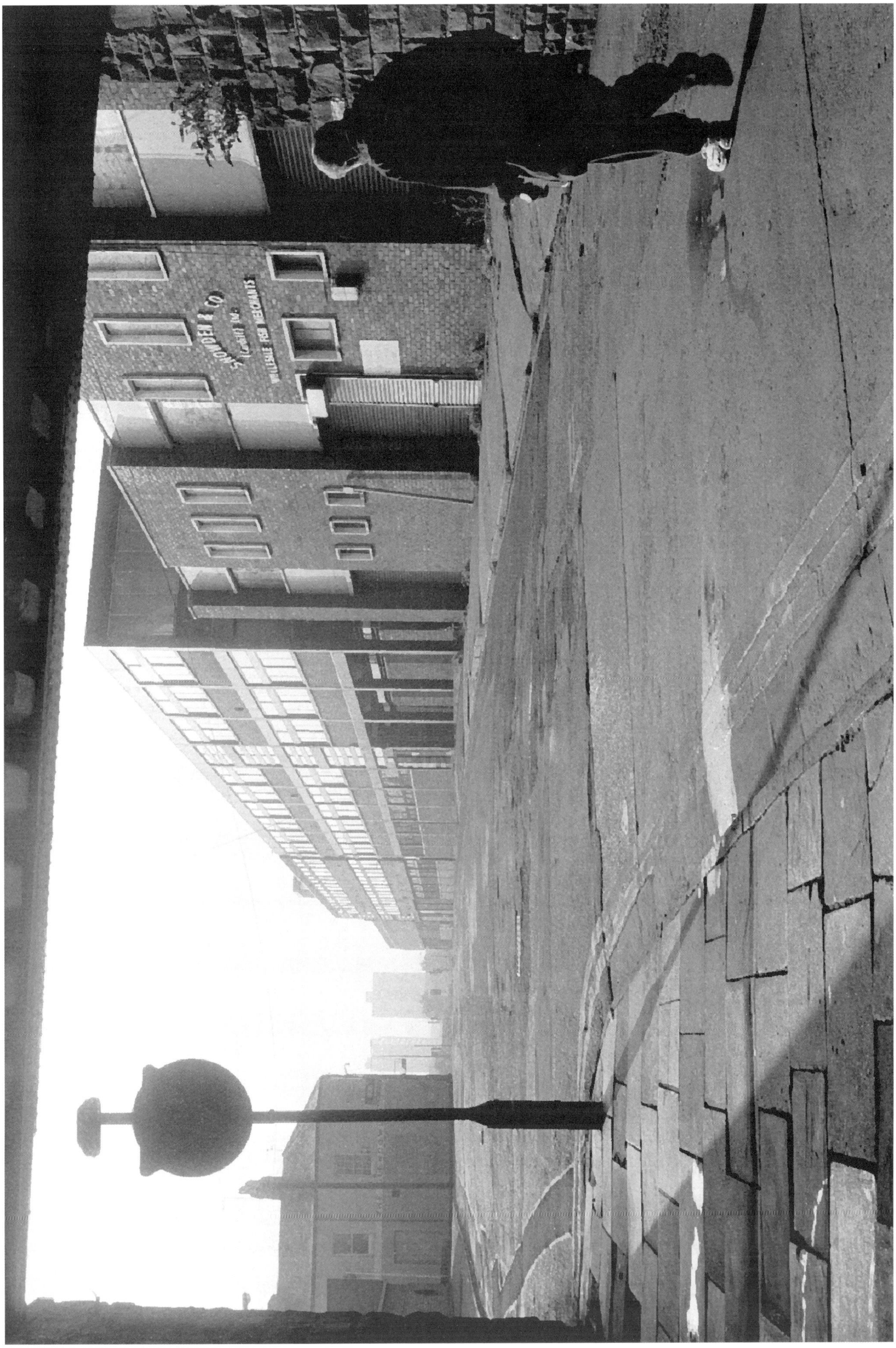
SNOWDEN & CO
WHOLESALE FEED MERCHANTS

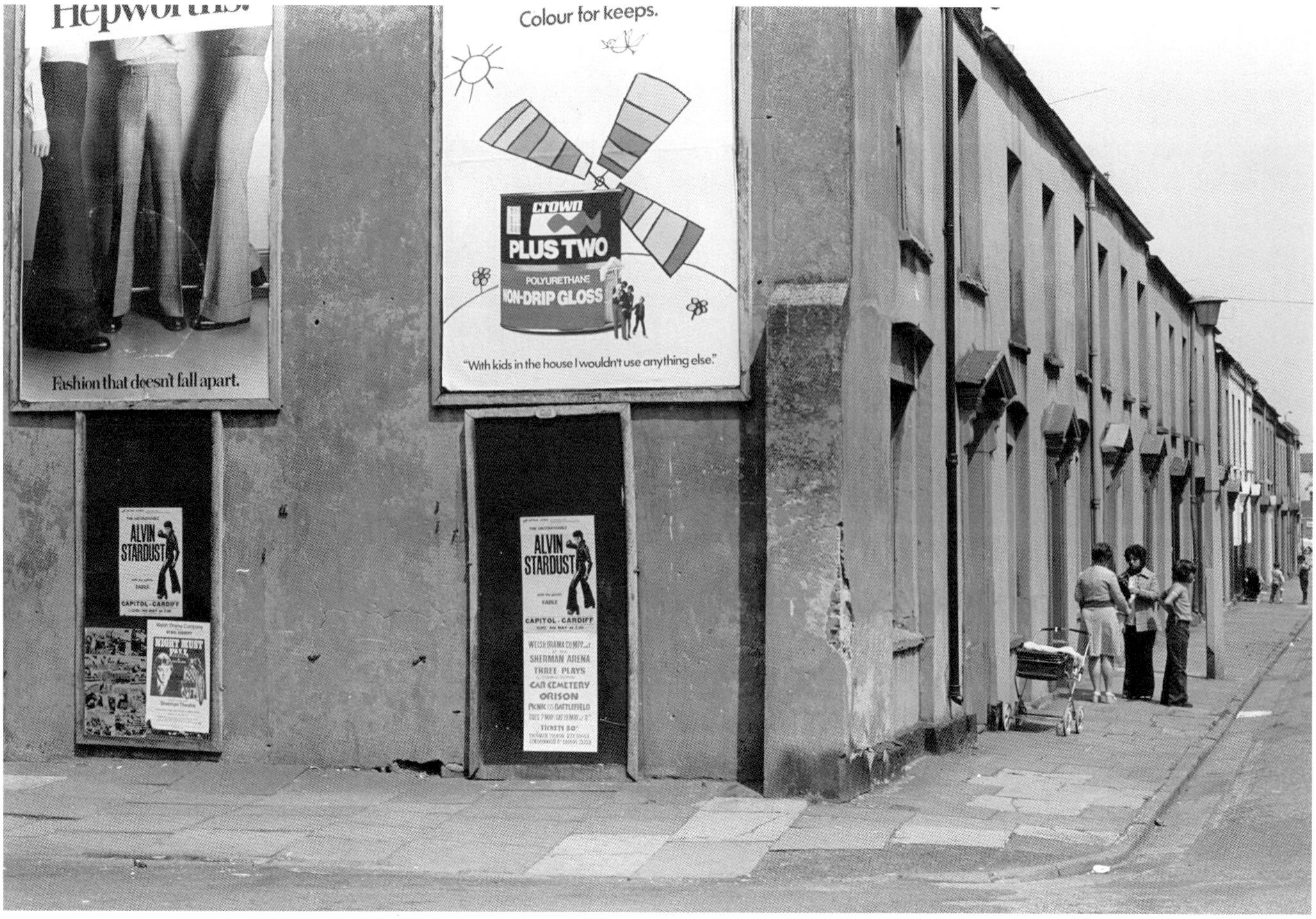
Fashion that doesn't fall apart.
Colour for keeps.
crown
PLUS TWO
POLYURETHANE
NON-DRIP GLOSS
"With kids in the house I wouldn't use anything else."
ALVIN STARDUST
CAPITOL · CARDIFF
ALVIN STARDUST
CAPITOL · CARDIFF
SHERMAN ARENA
THREE PLAYS
CAR CEMETERY
ORISON
PICNIC ON THE BATTLEFIELD

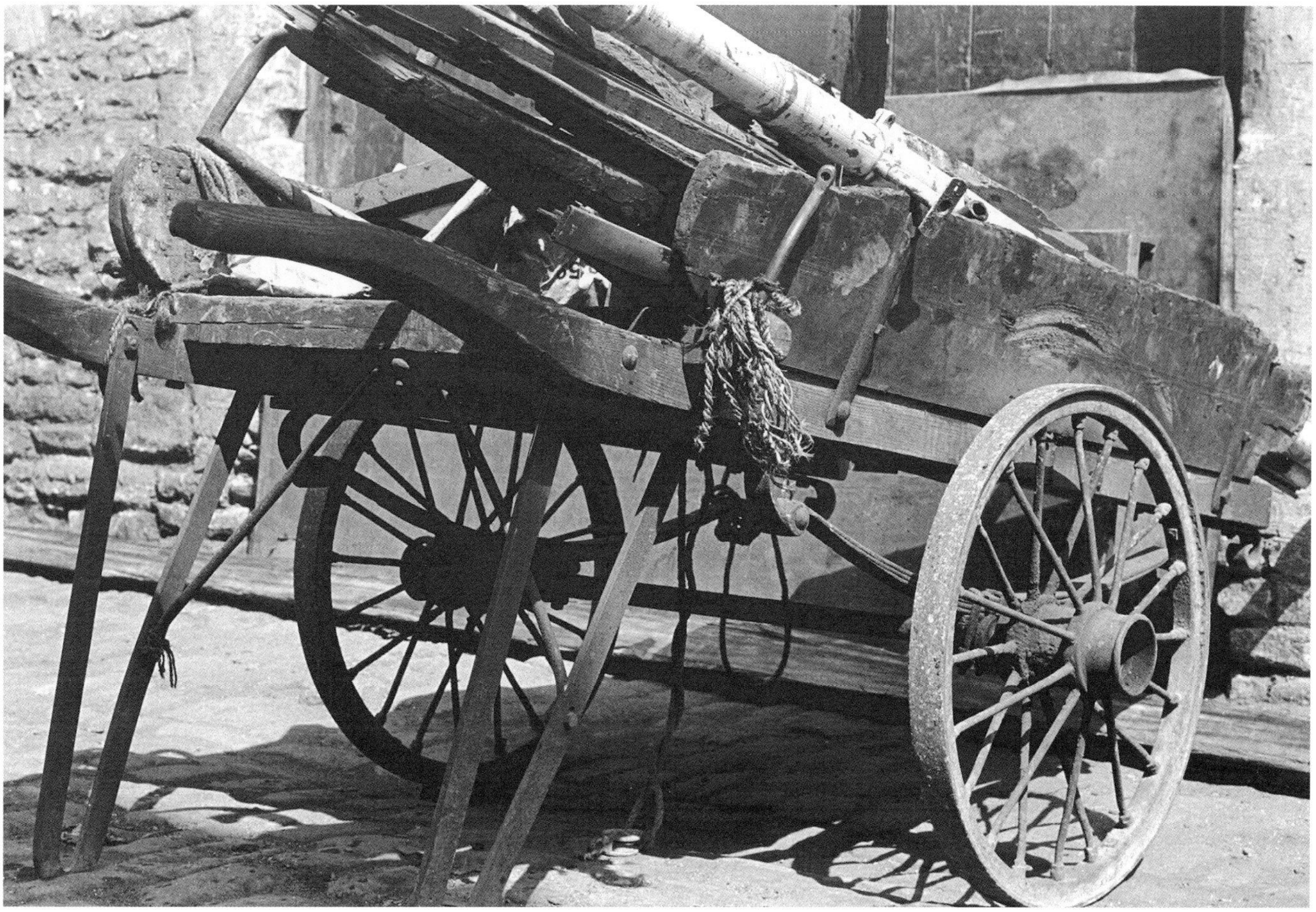

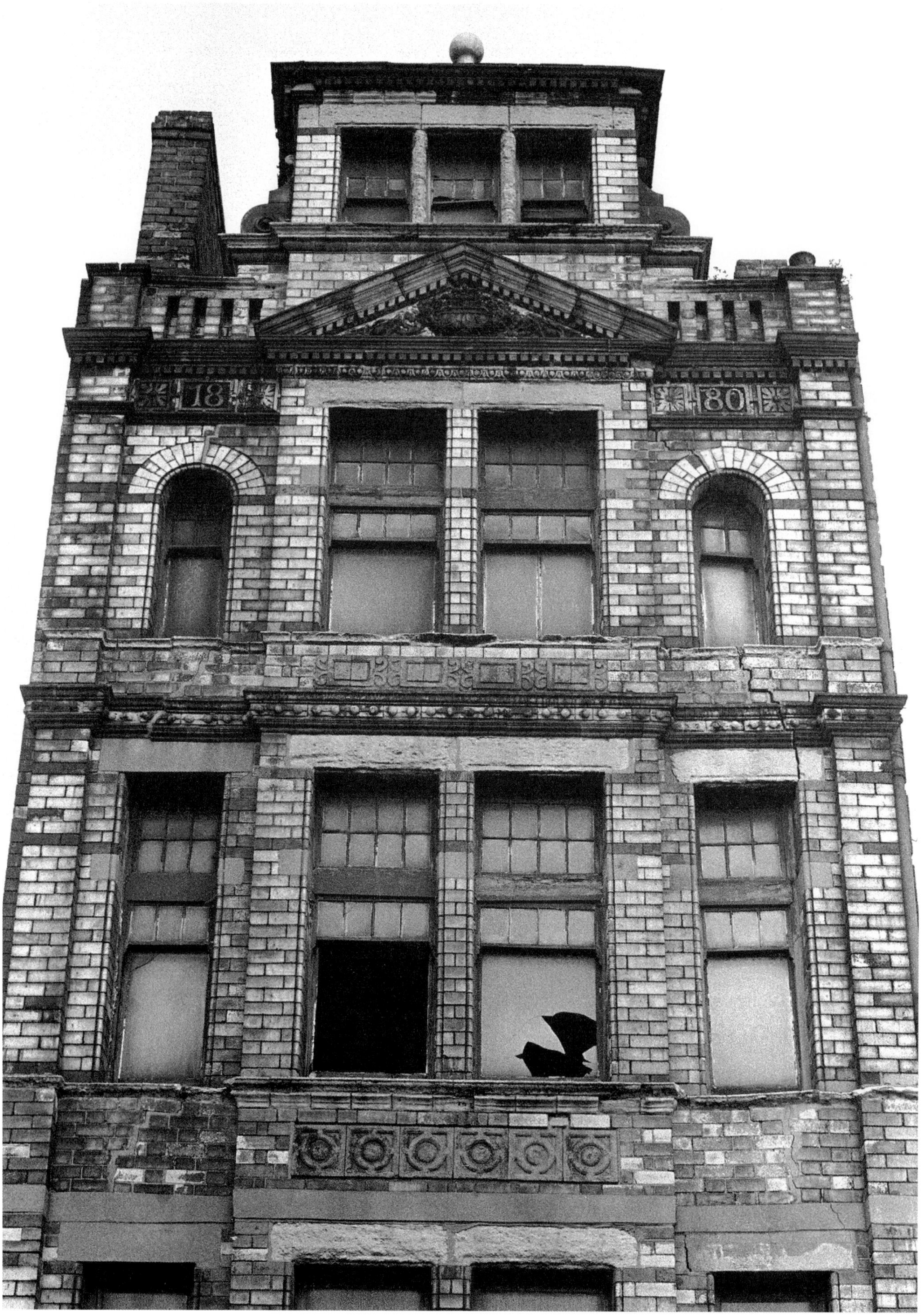
18
80

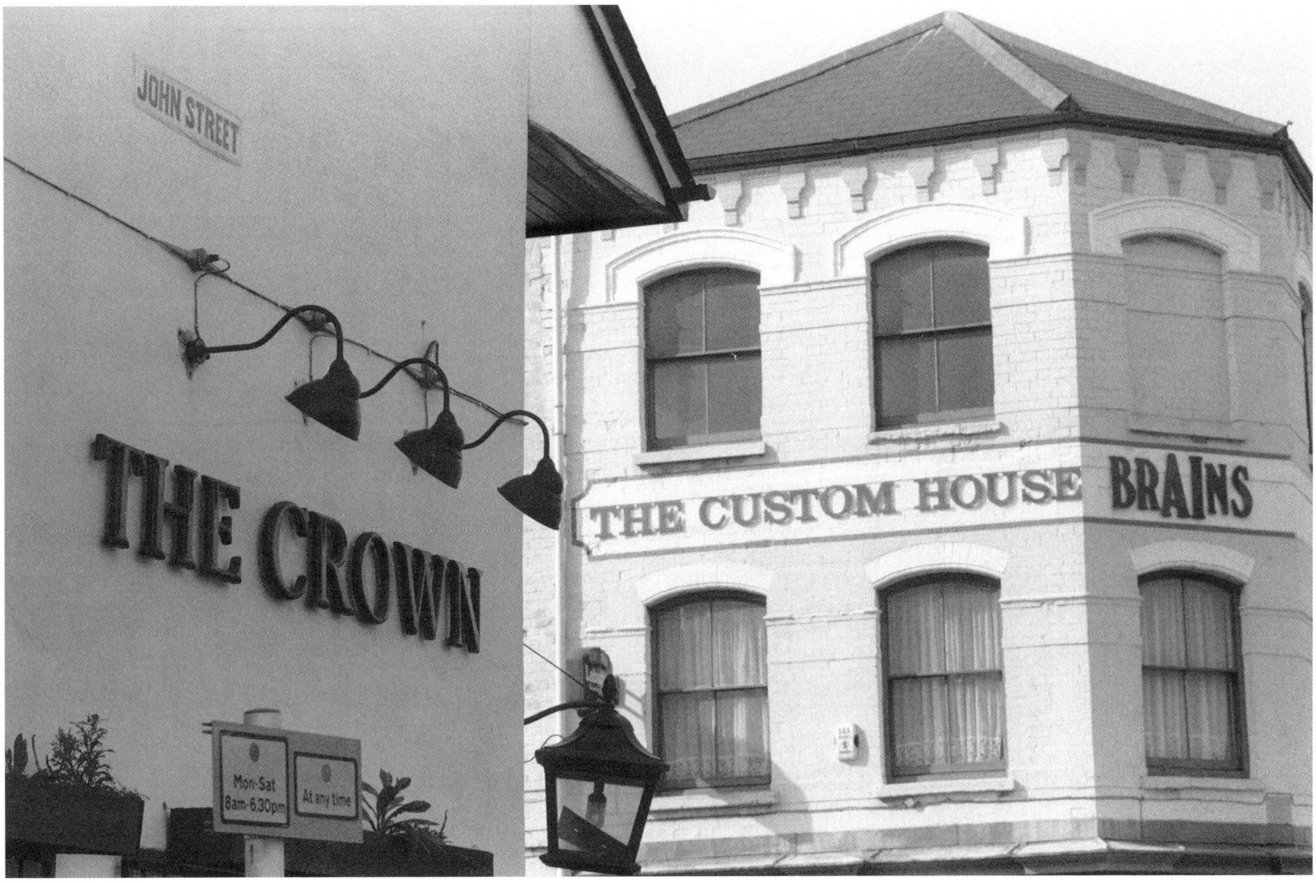
JOHN STREET
THE CROWN
THE CUSTOM HOUSE
BRAINS
Mon-Sat
8am-6.30pm
At any time

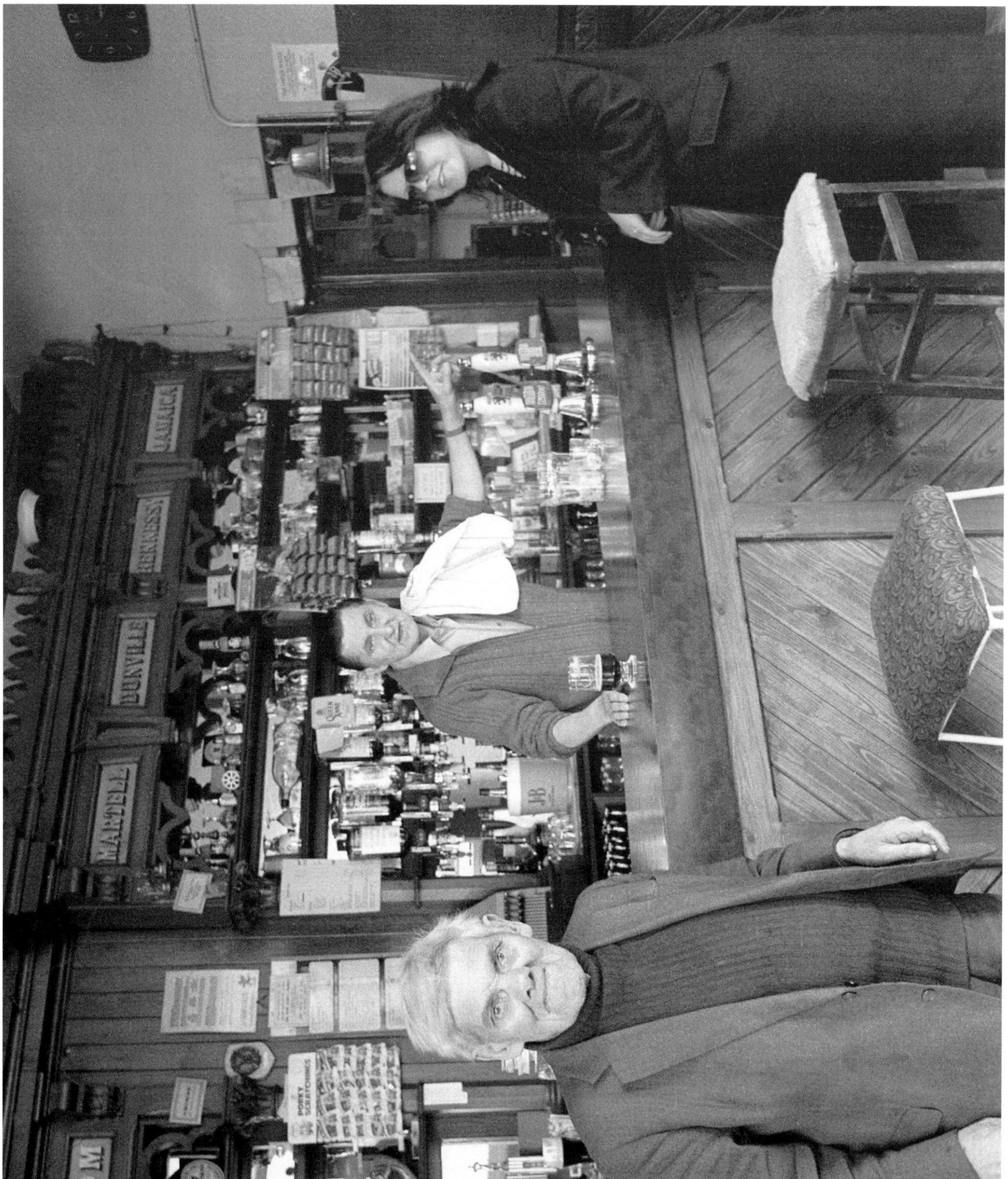
OLD TOM
MARTELL
DUNVILLE
HENNESSY
JAMAICA
PORKY SCRATCHINGS

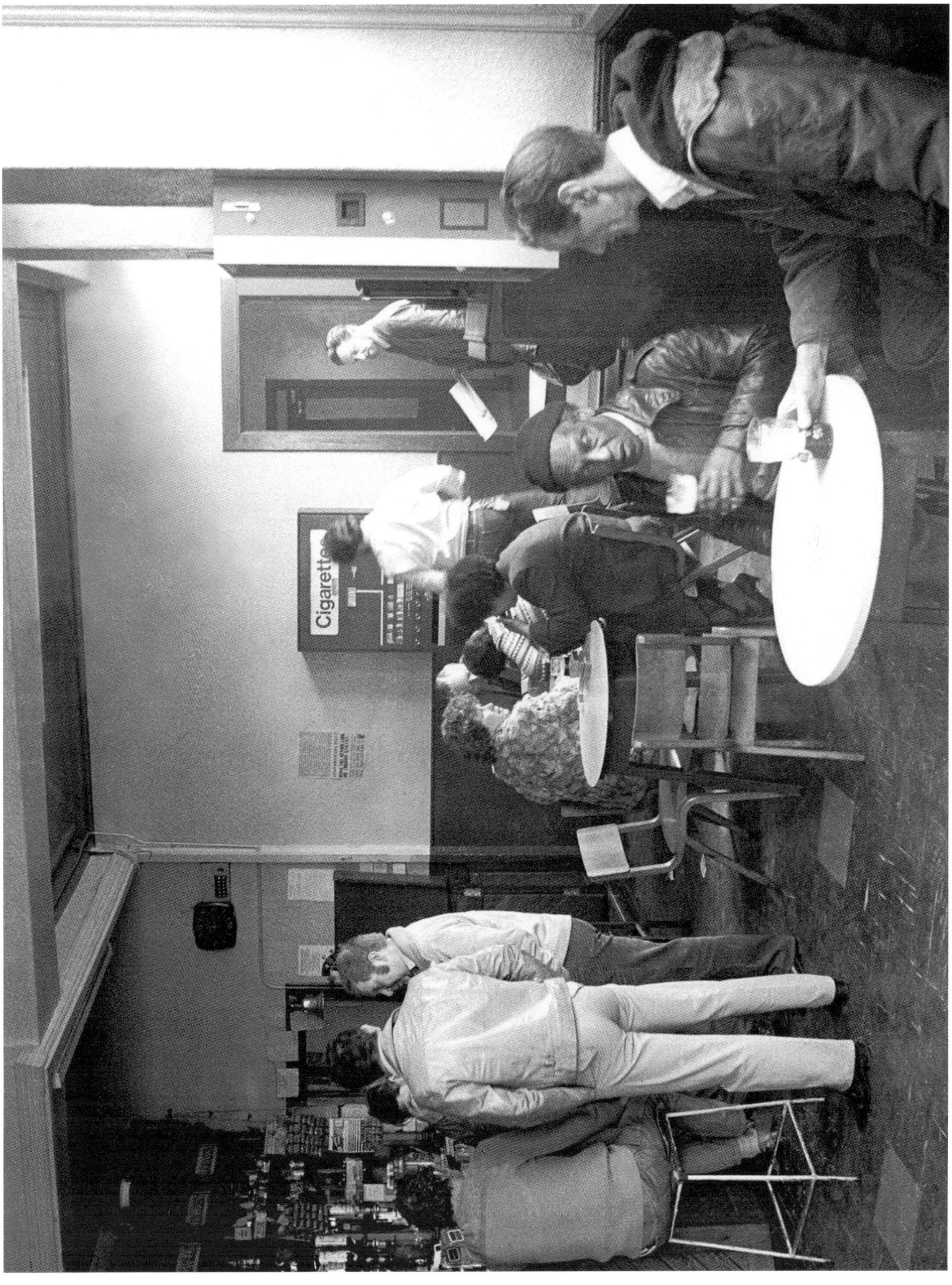
Cigarettes

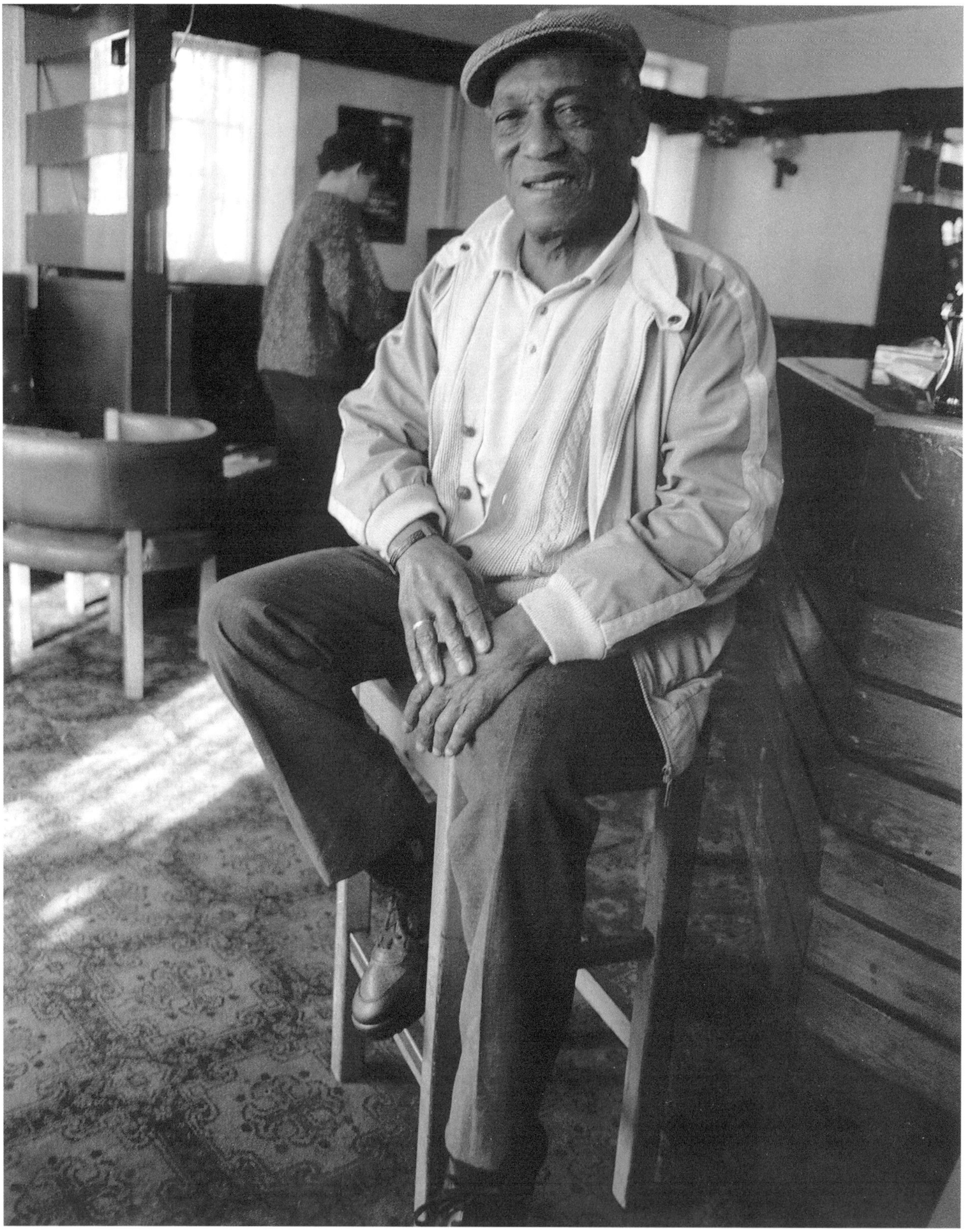

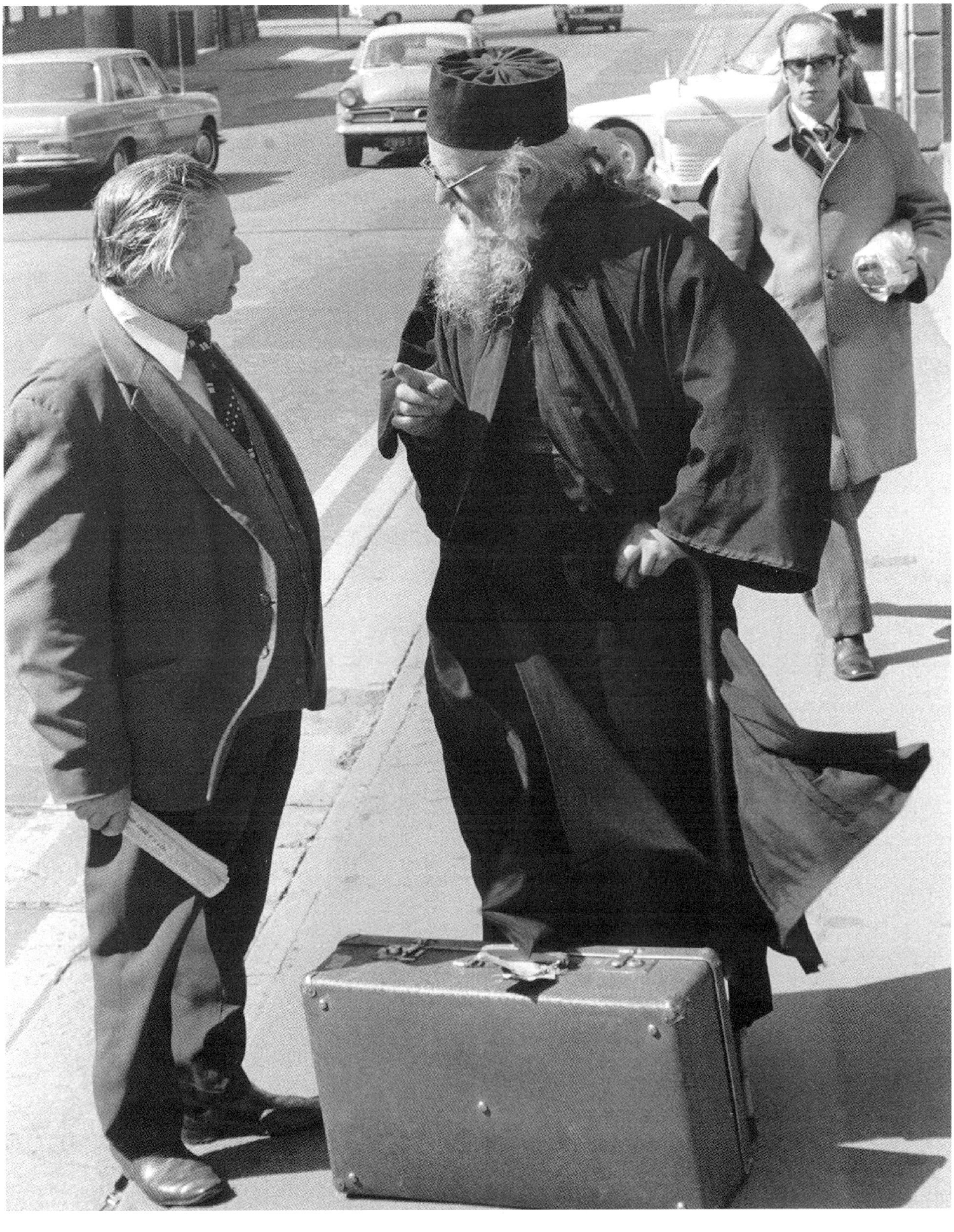

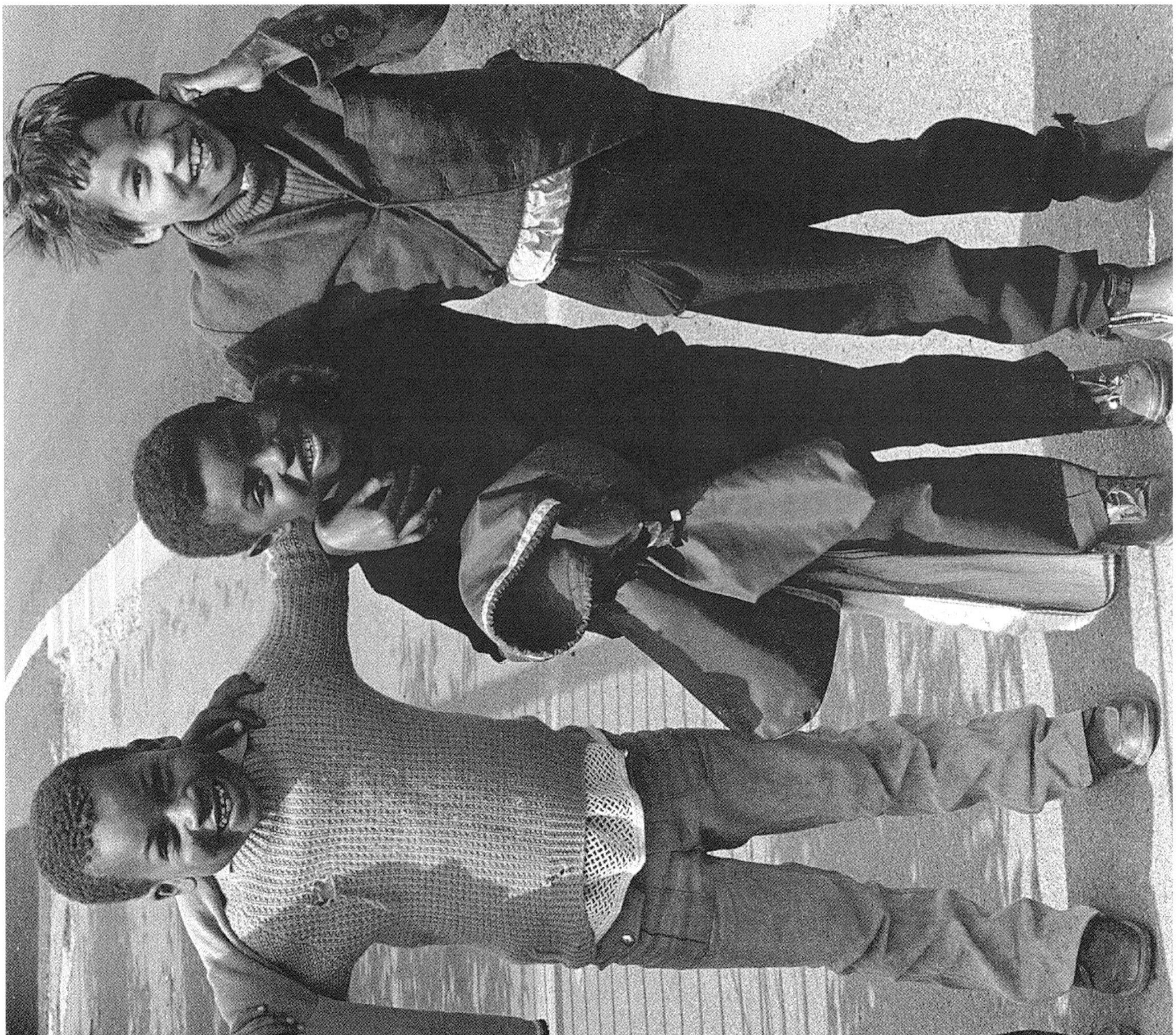

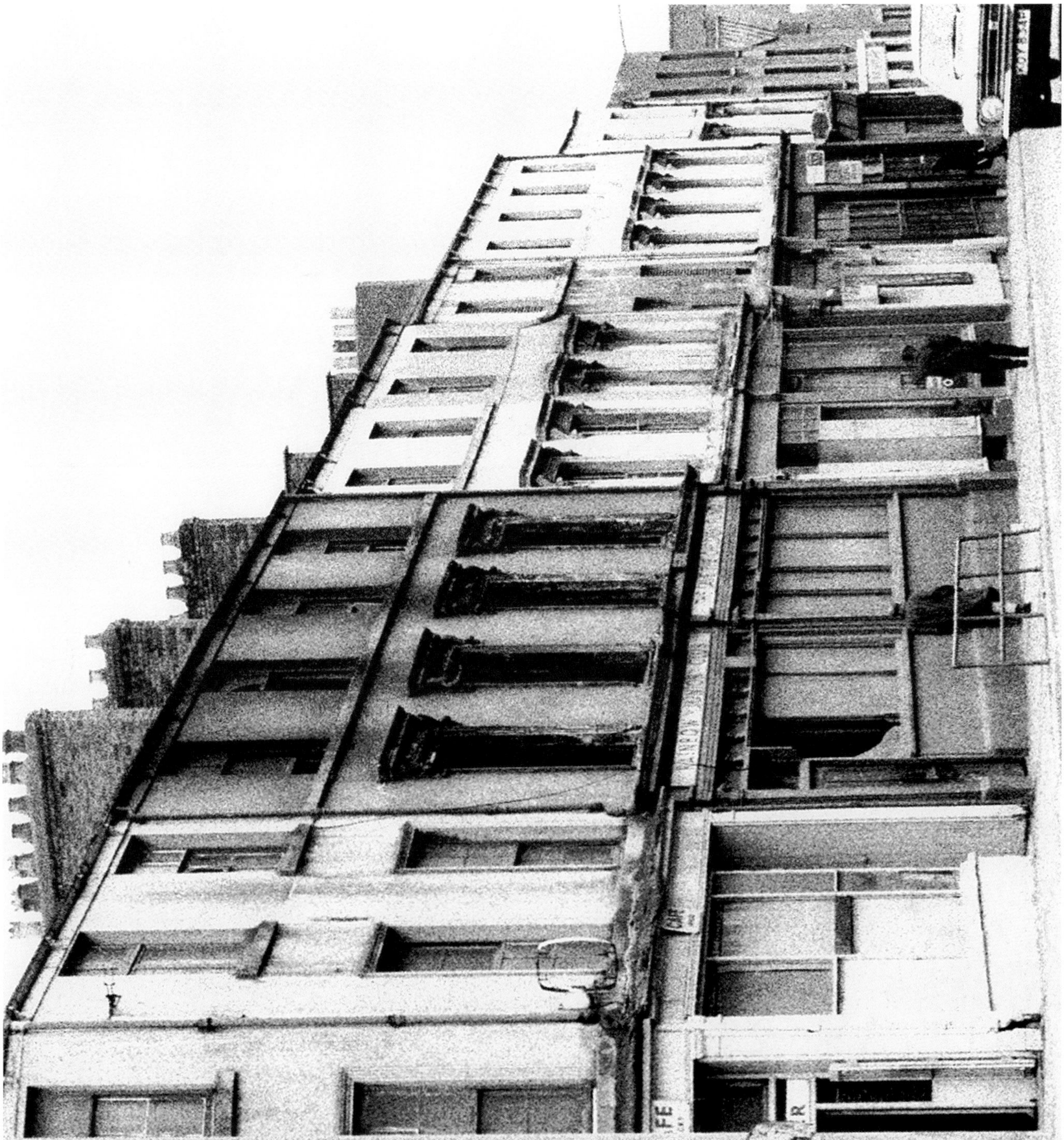

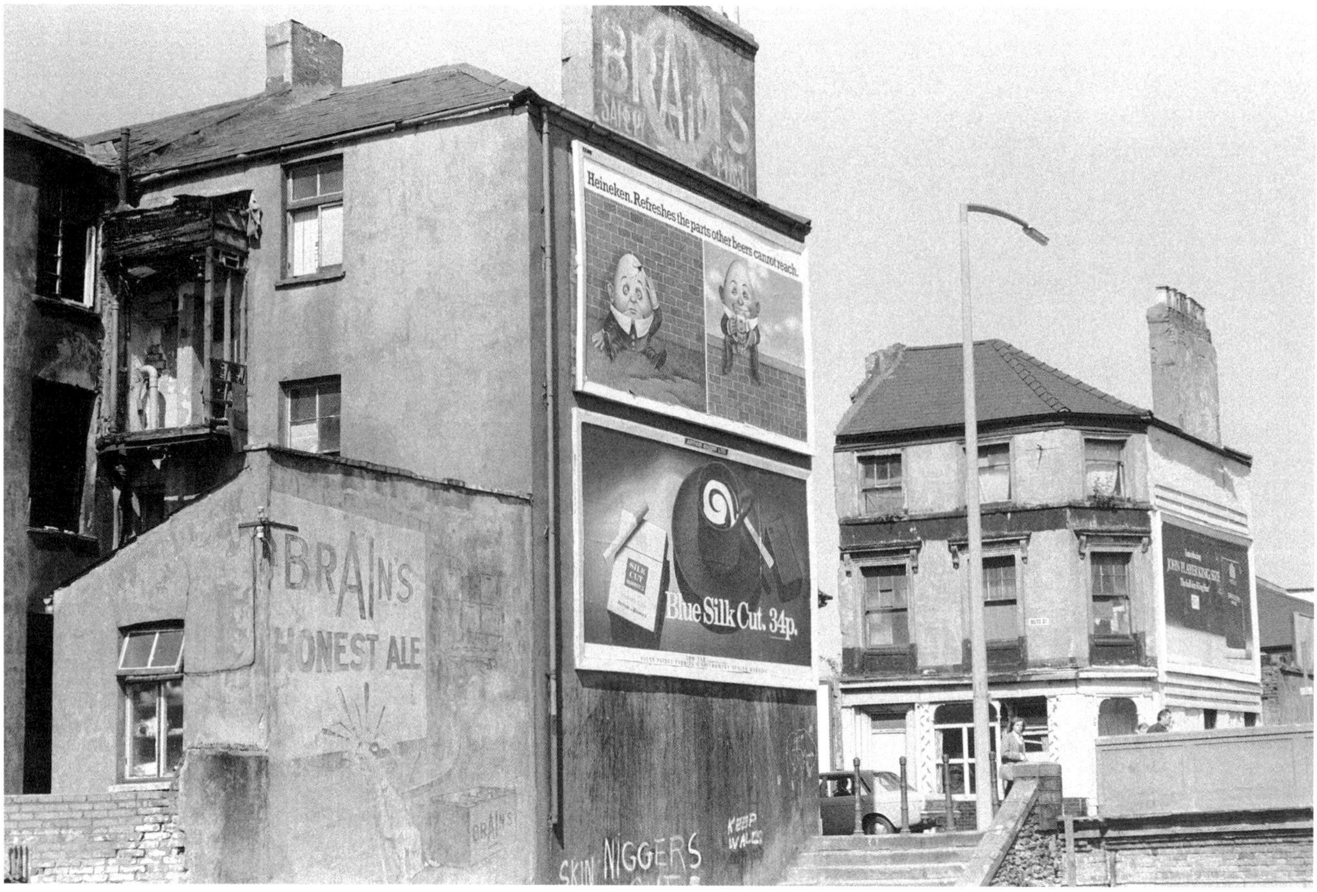
BRAIN'S
Heineken. Refreshes the parts other beers cannot reach.
BRAIN'S
HONEST ALE
Blue Silk Cut. 34p.
SKIN NIGGERS

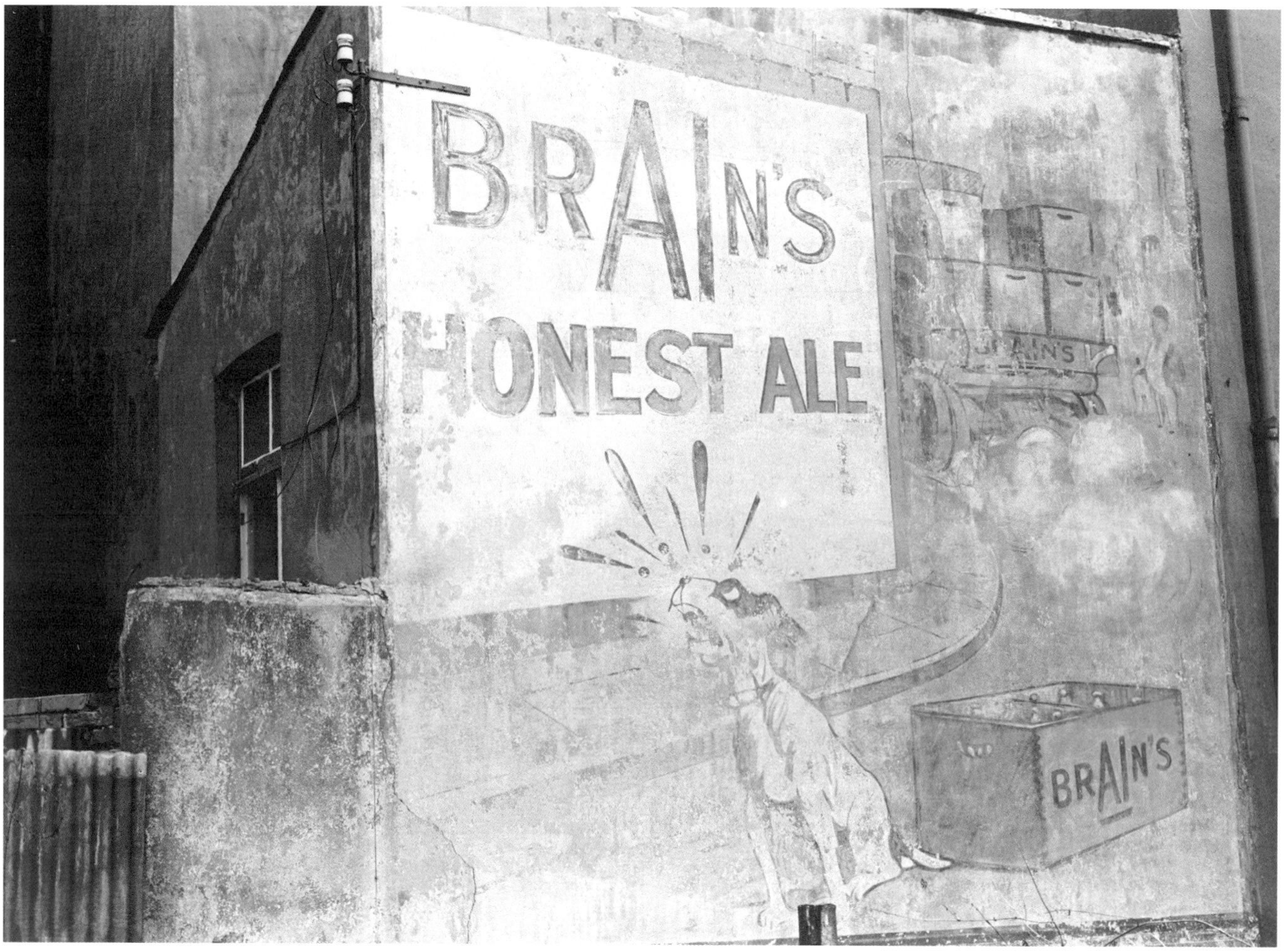
BRAIN'S
HONEST ALE
BRAIN'S

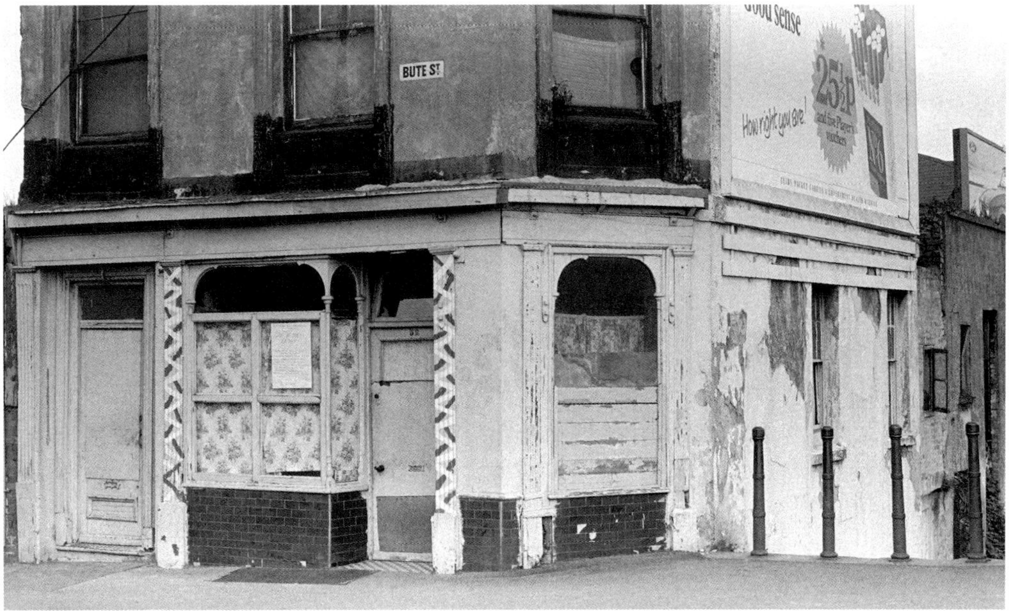
BUTE ST
sense
How right you are!
25½p

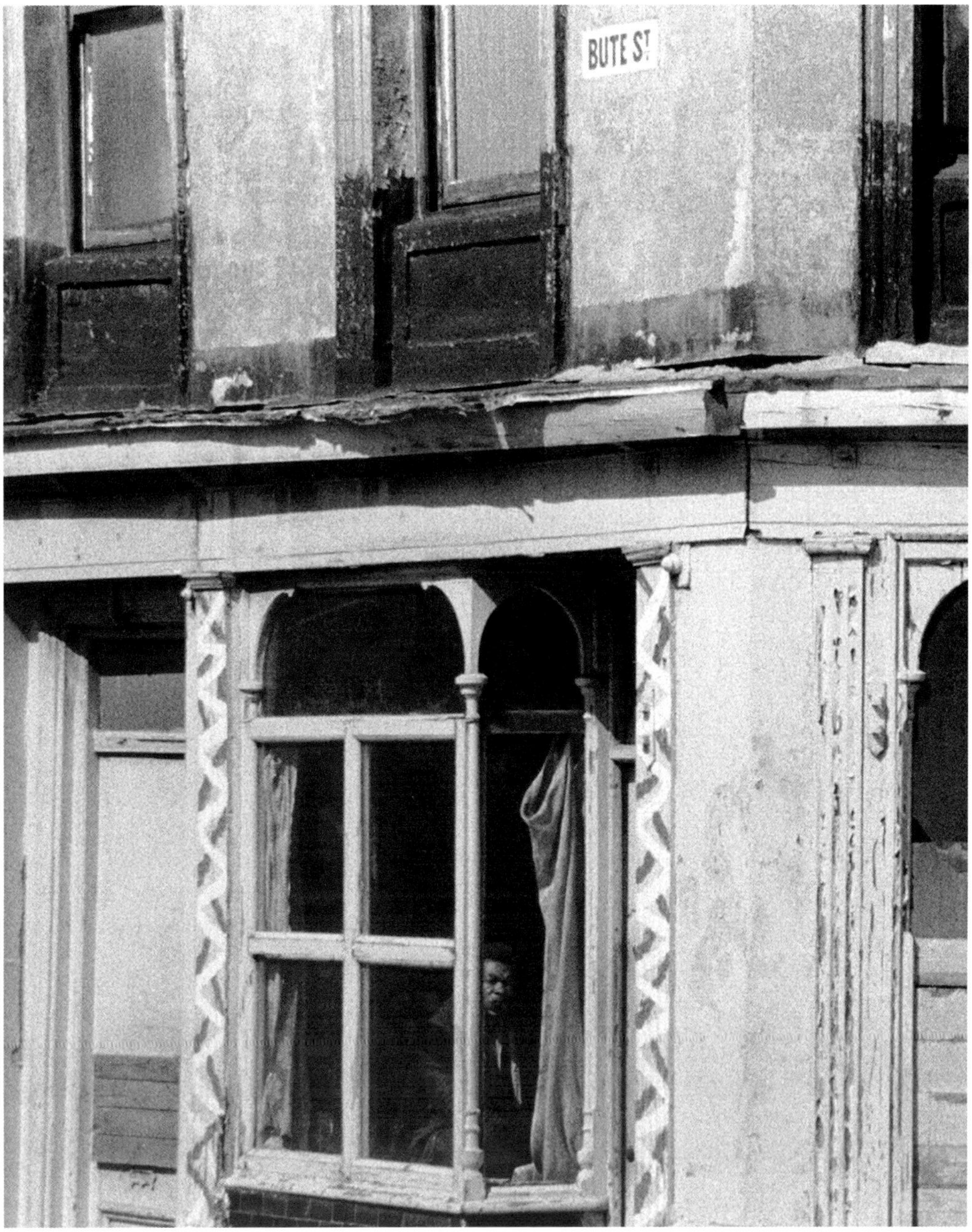
BUTE ST

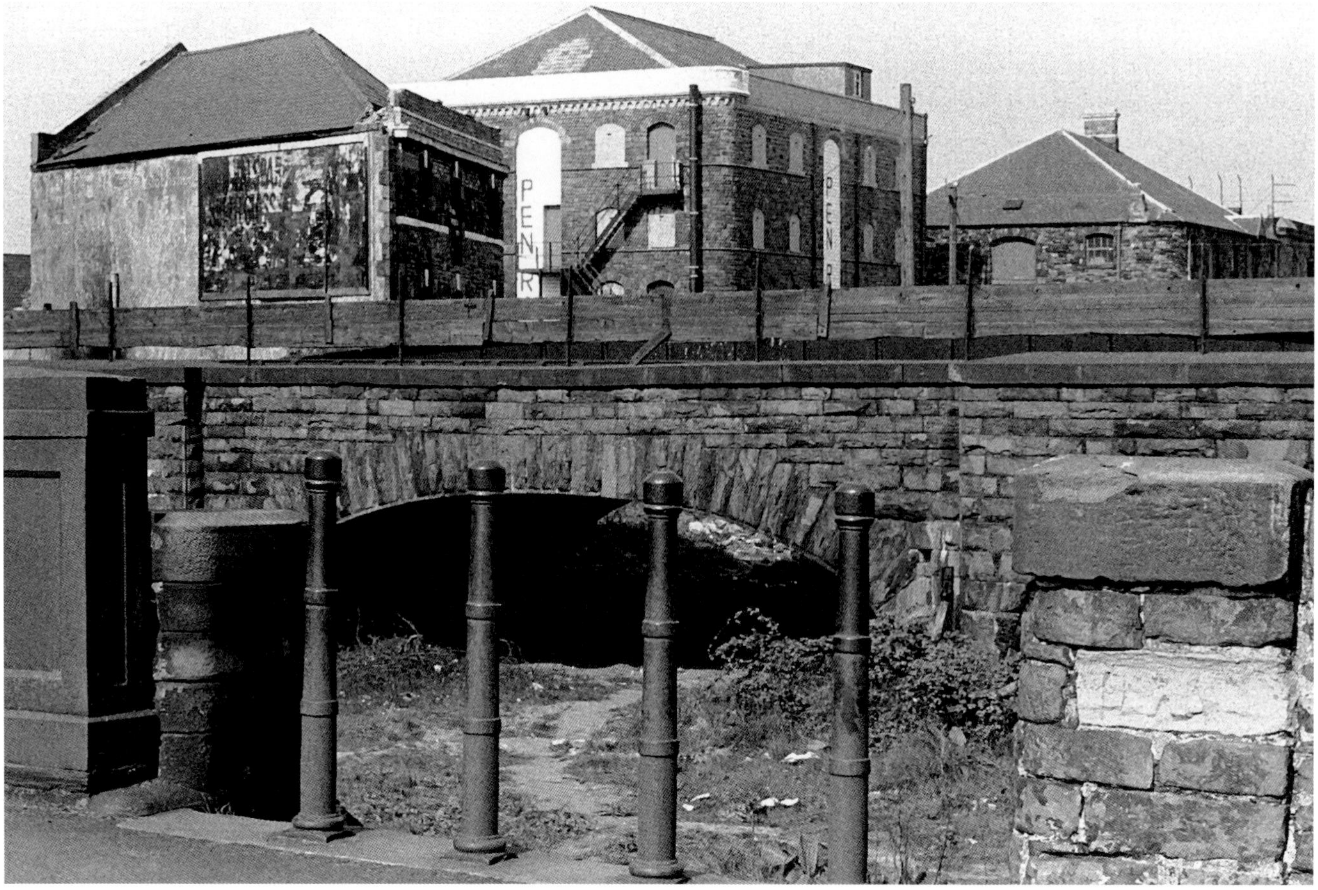
PENIR

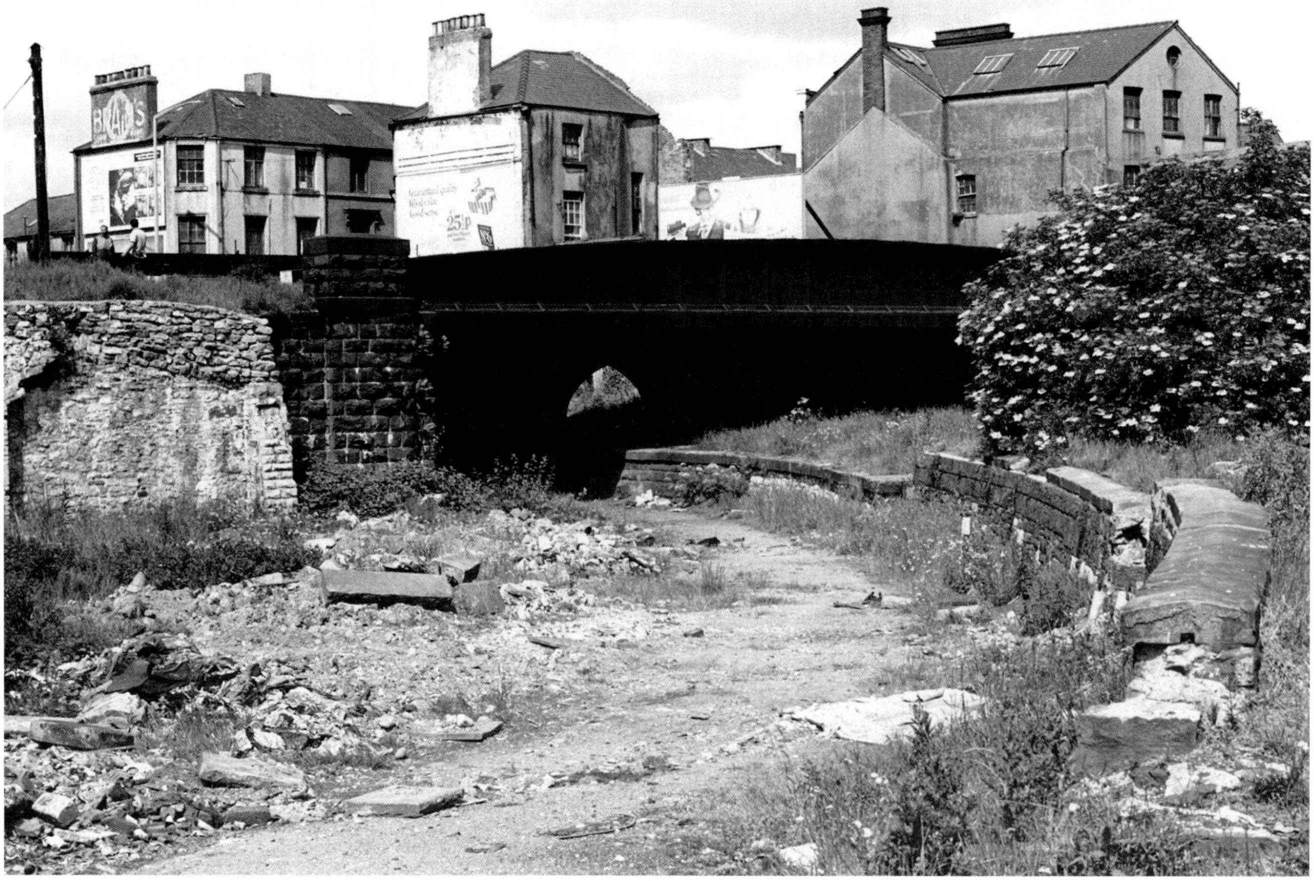
25p

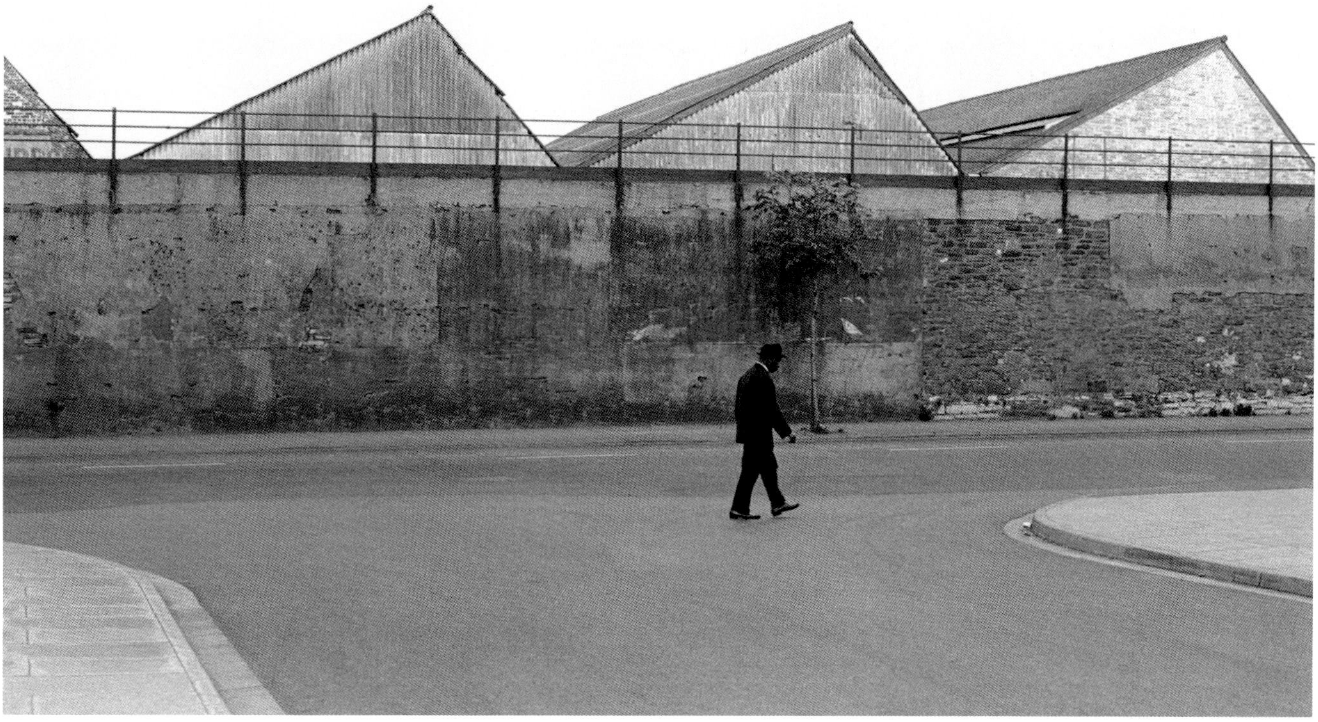

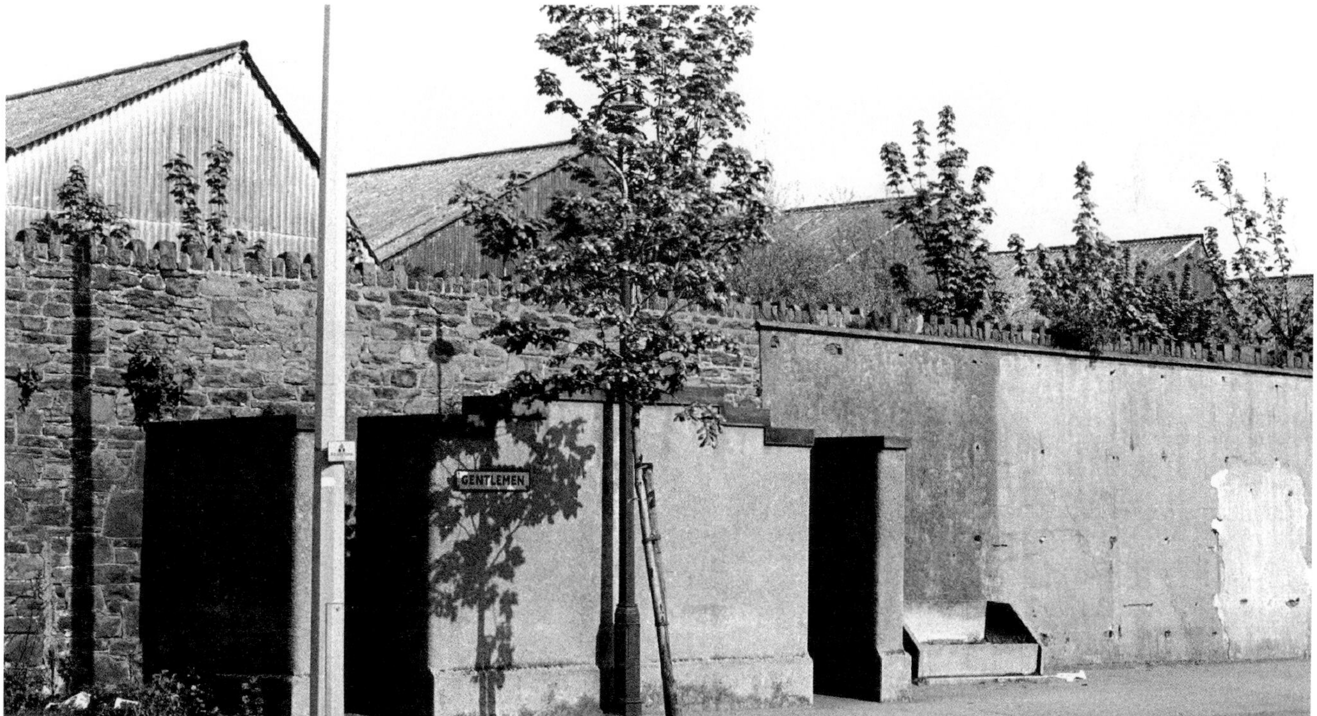
GENTLEMEN

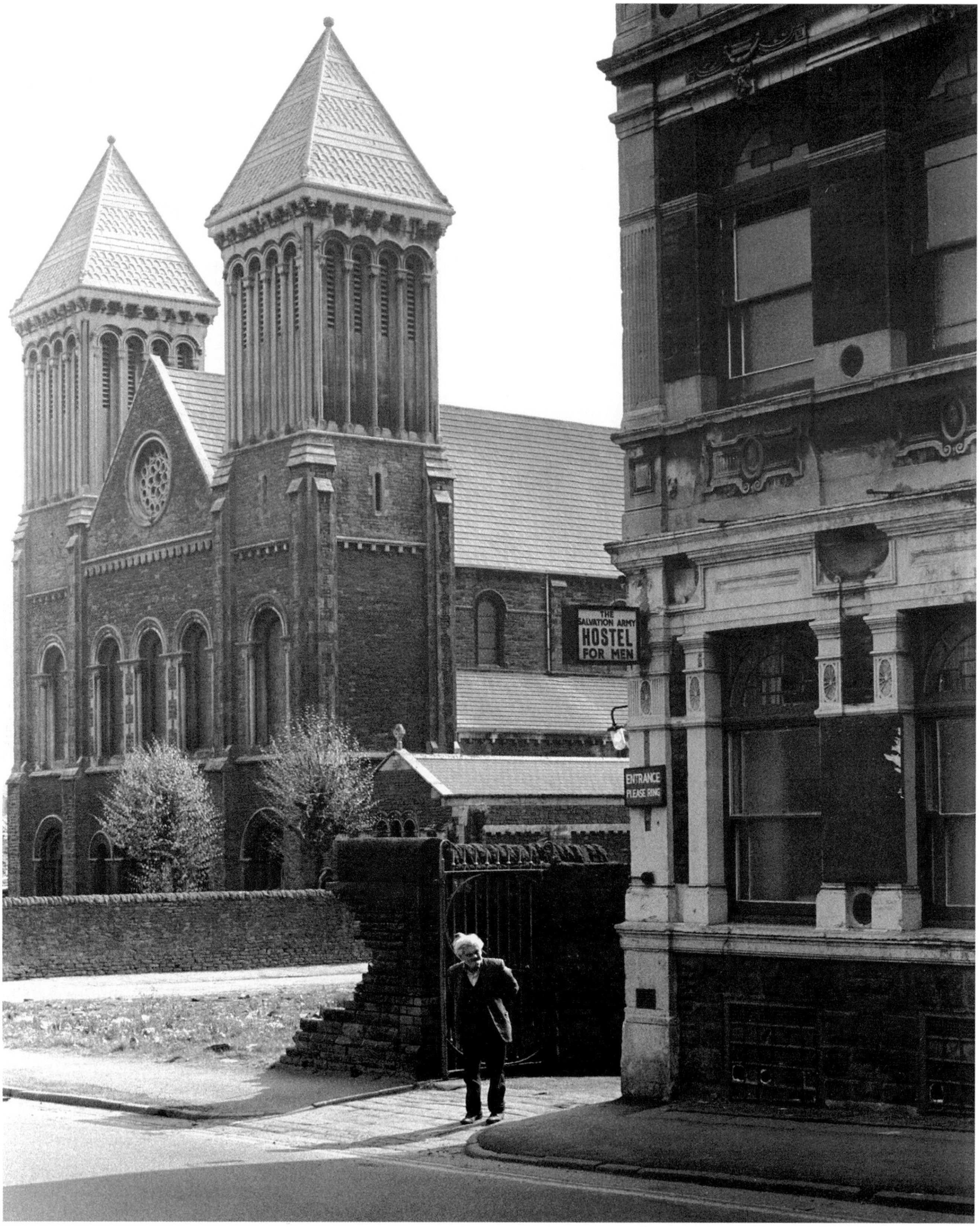
THE
SALVATION ARMY
HOSTEL
FOR MEN
ENTRANCE
PLEASE RING

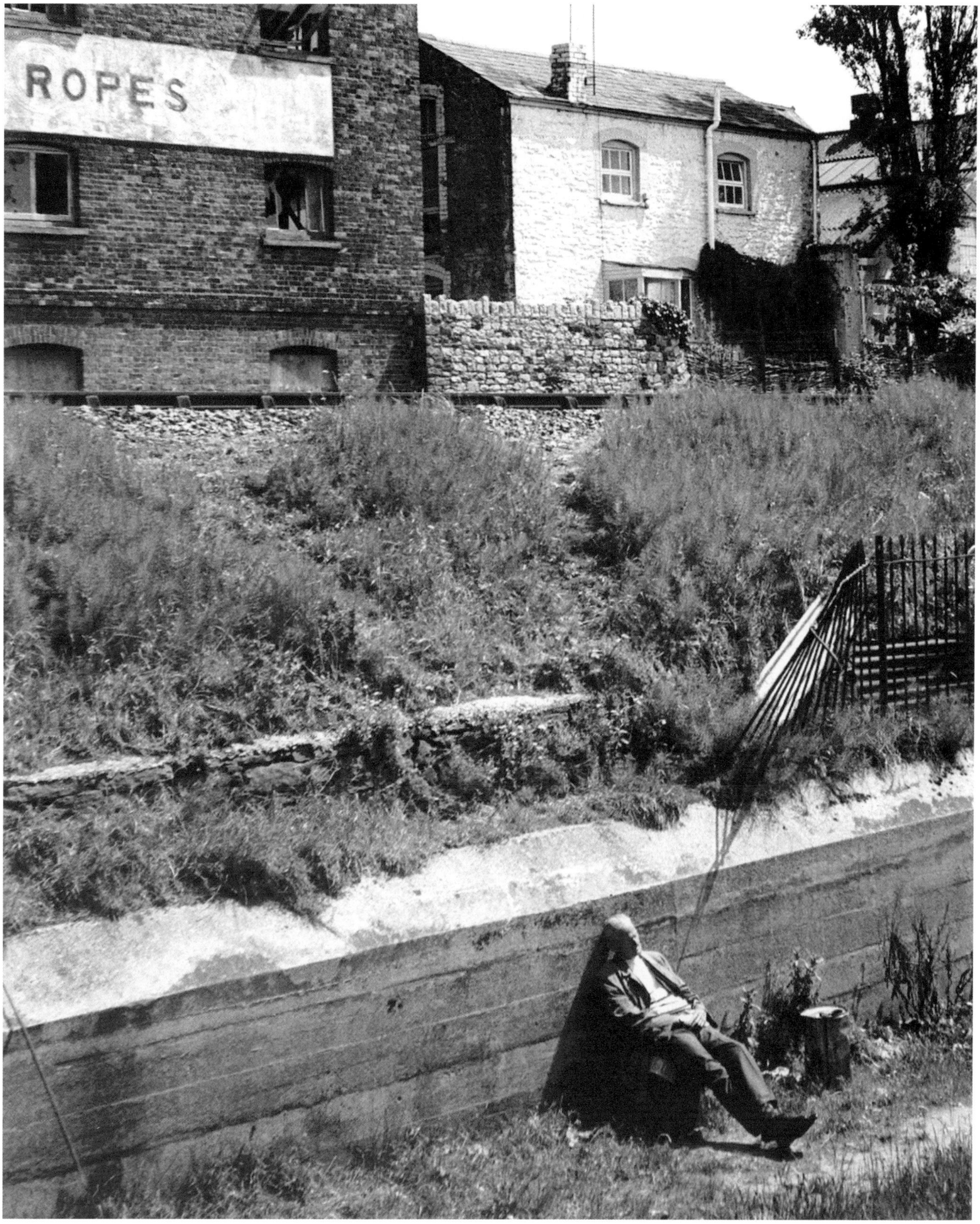
ROPES

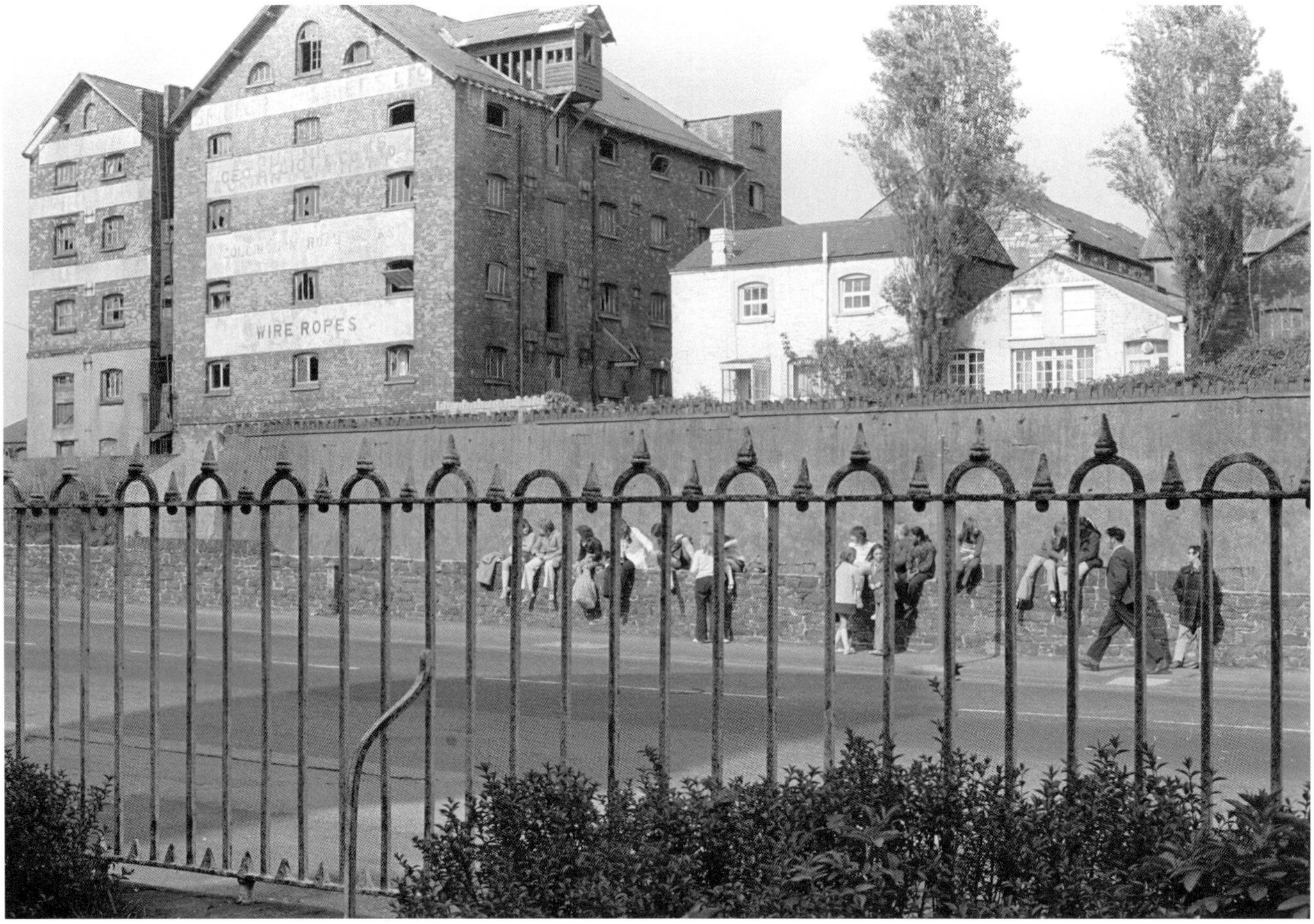
WIRE ROPES

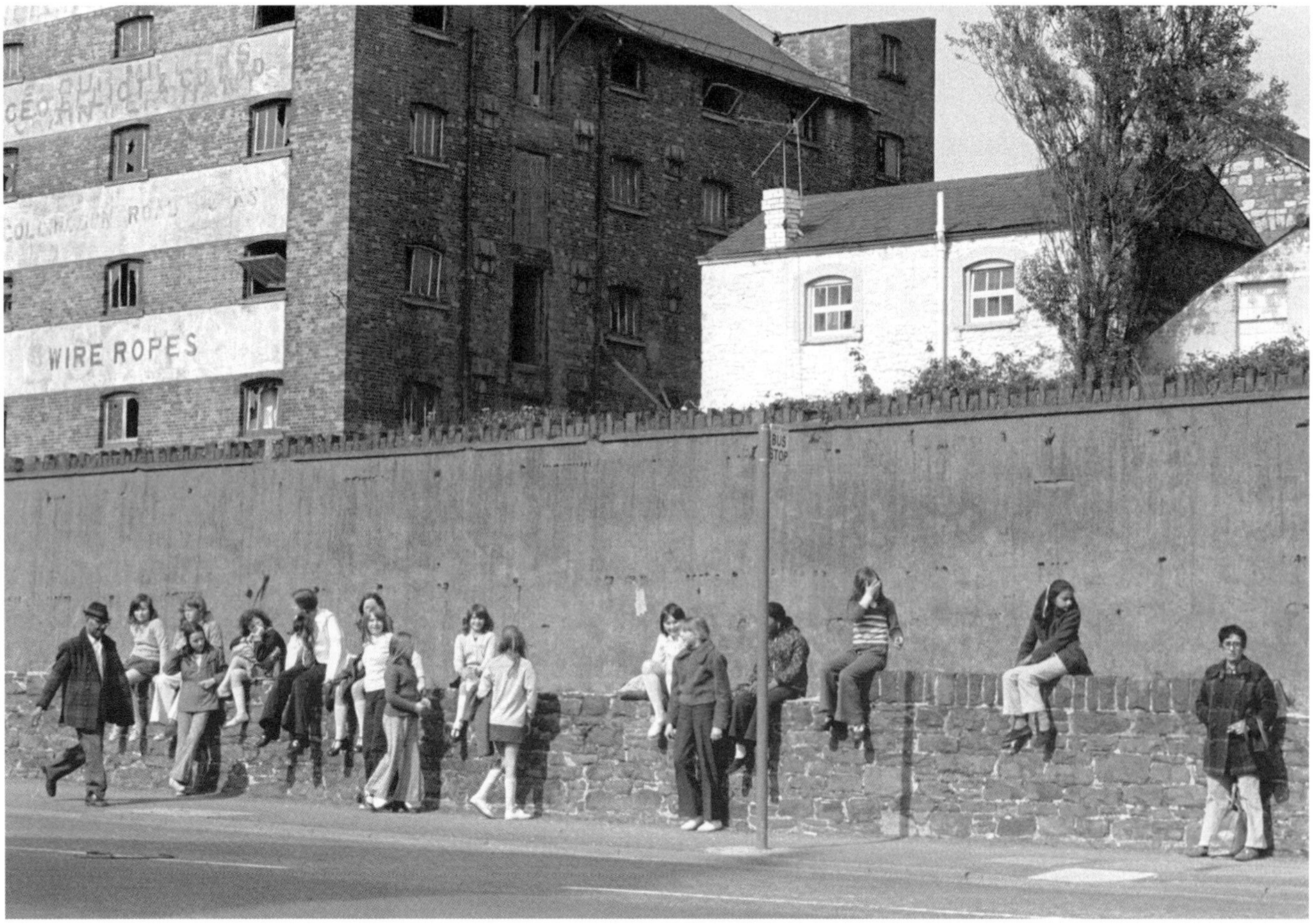
WIRE ROPES

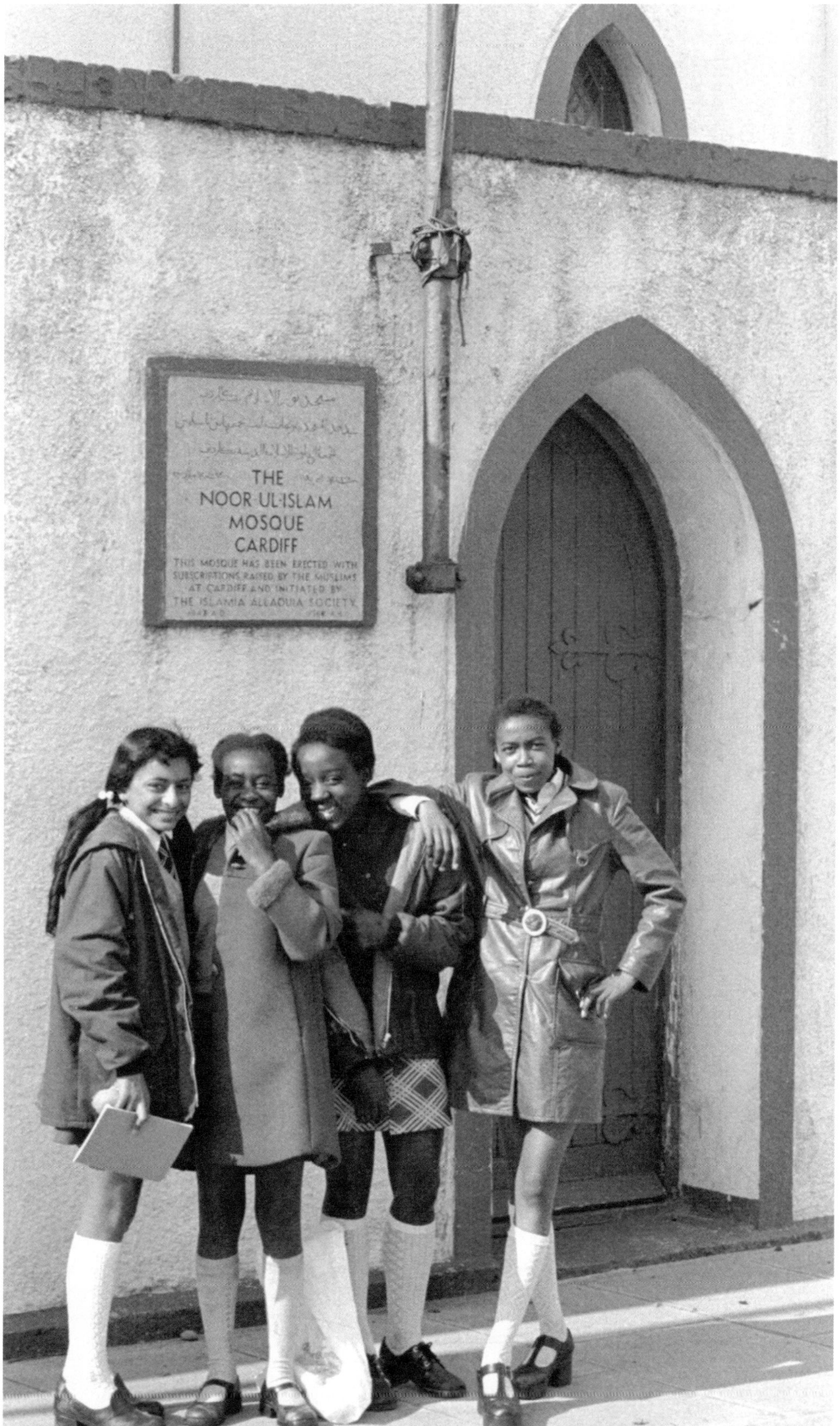
THE
NOOR-UL-ISLAM
MOSQUE
CARDIFF
THIS MOSQUE HAS BEEN ERECTED WITH
SUBSCRIPTIONS RAISED BY THE MUSLIMS
AT CARDIFF AND INITIATED BY
THE ISLAMIA ALLAOUIA SOCIETY

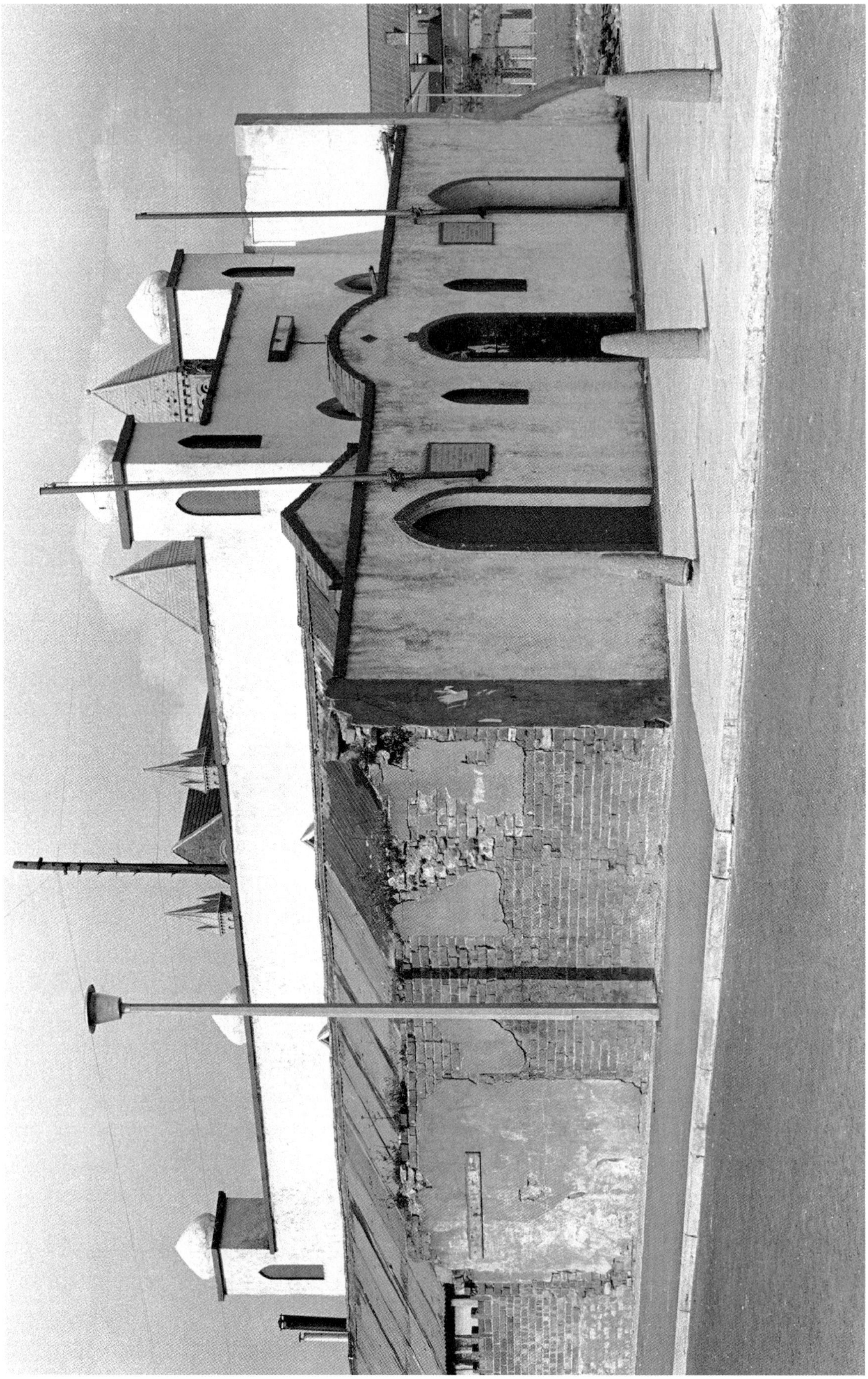

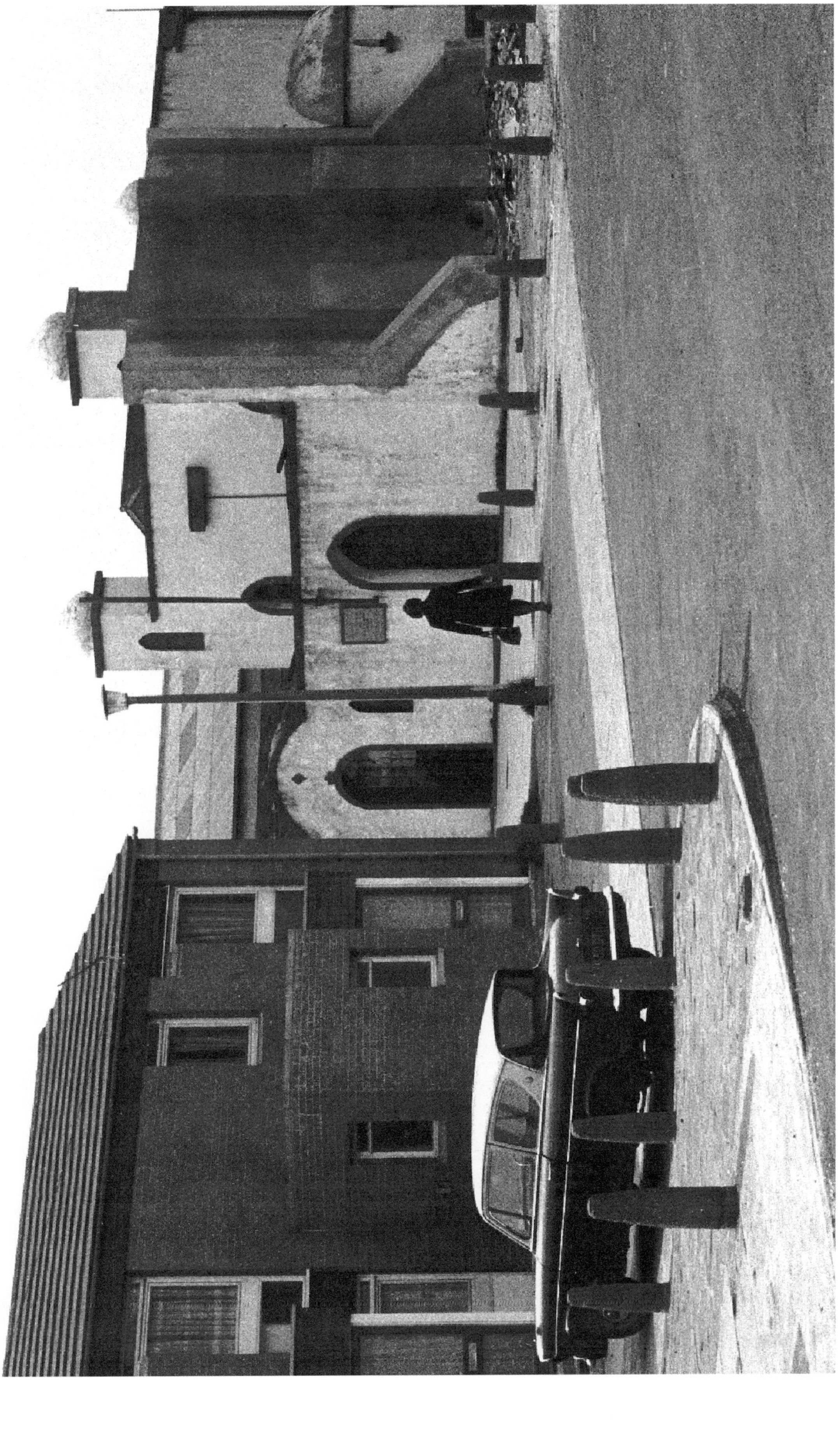

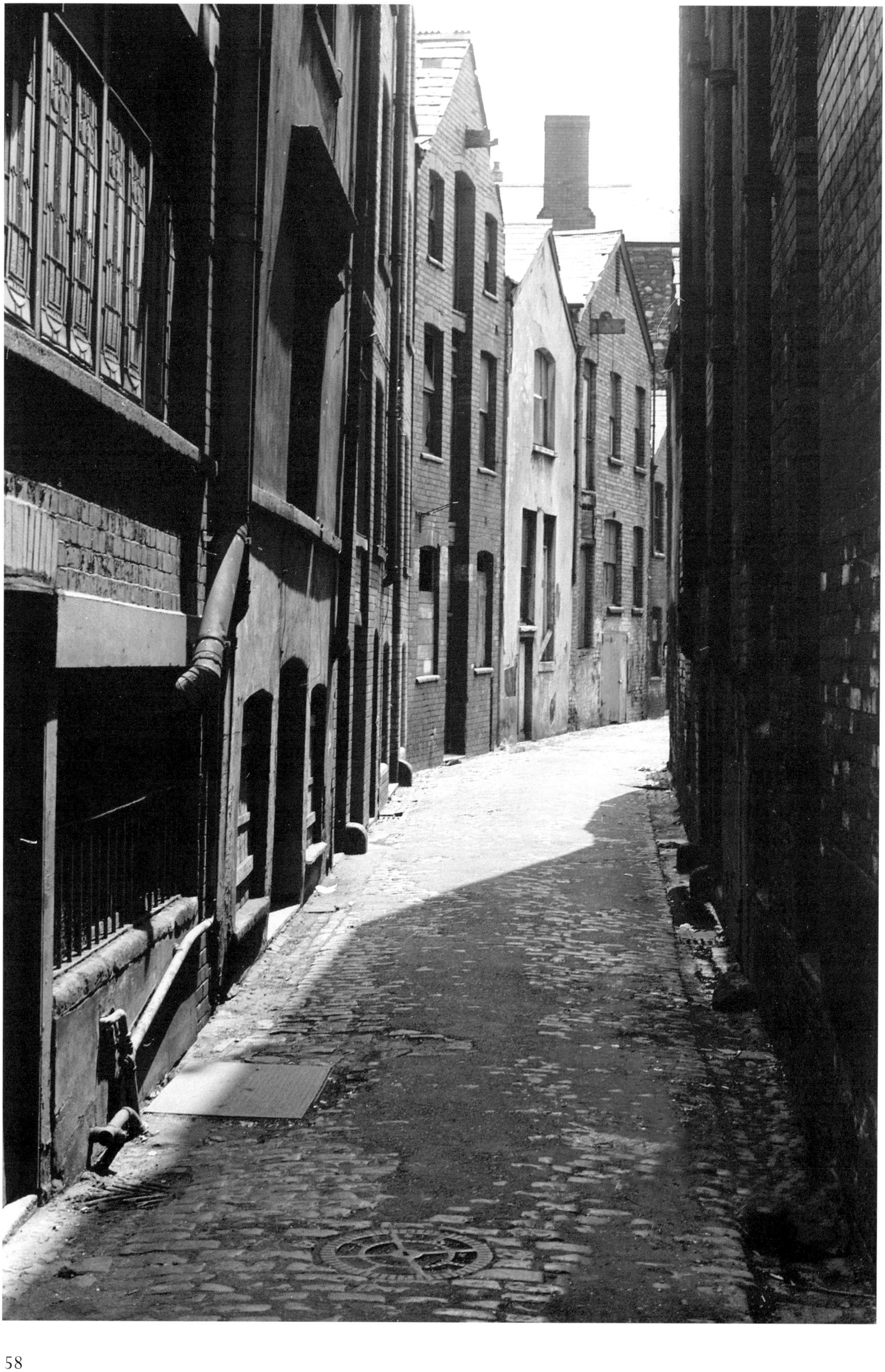

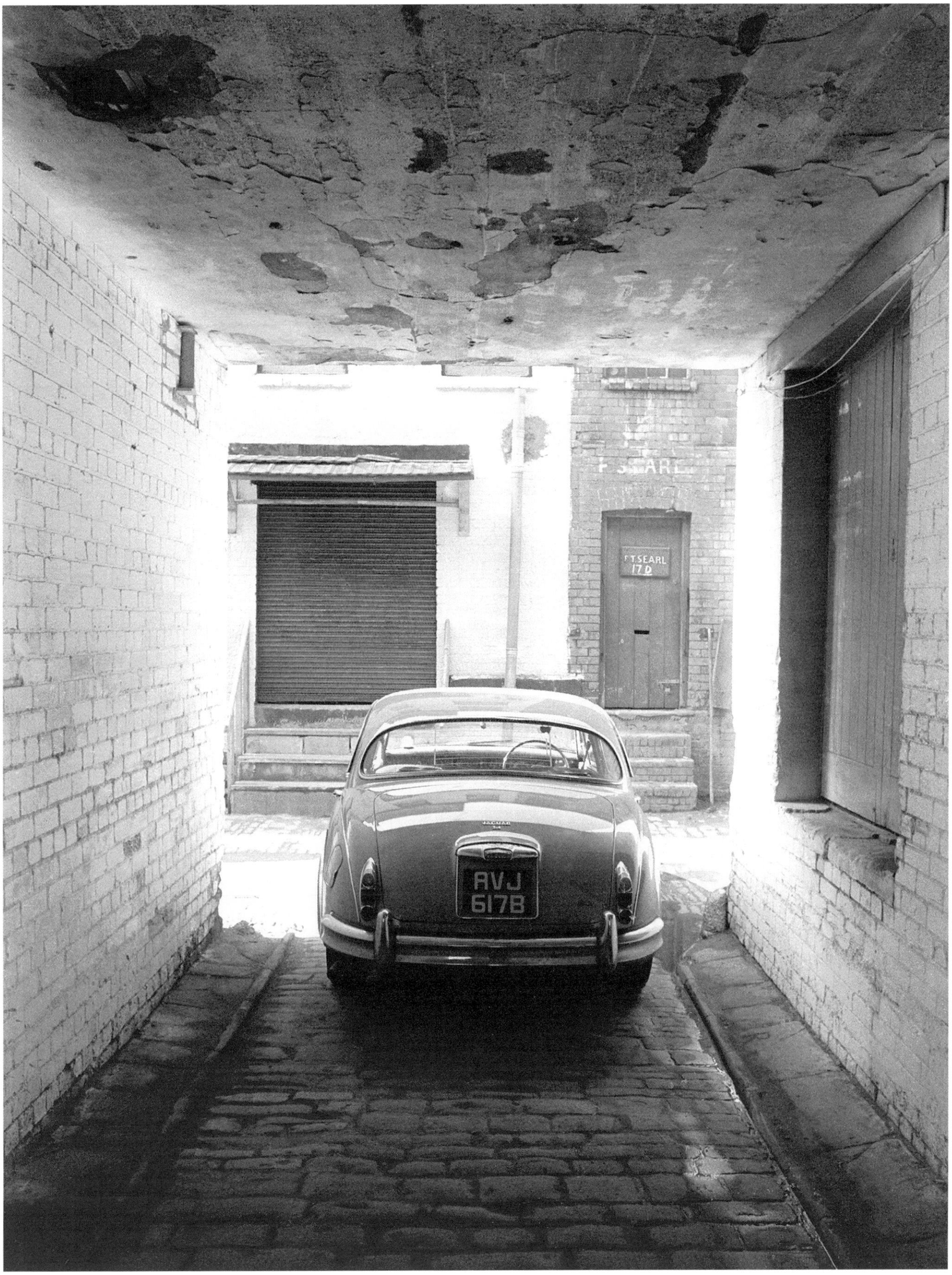
F.T.SEARL
17D
AVJ
617B

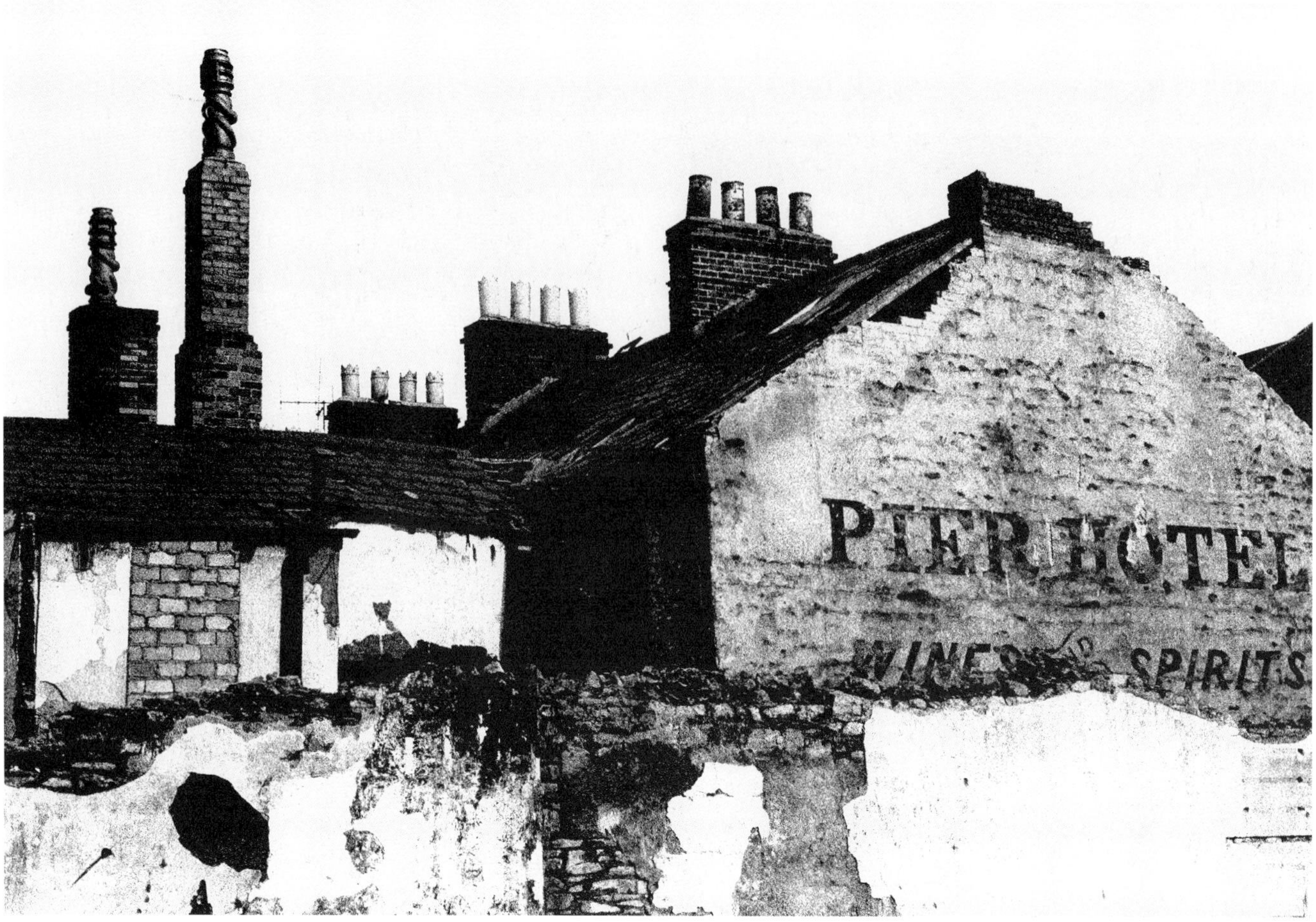
PIER HOTEL
WINES SPIRITS

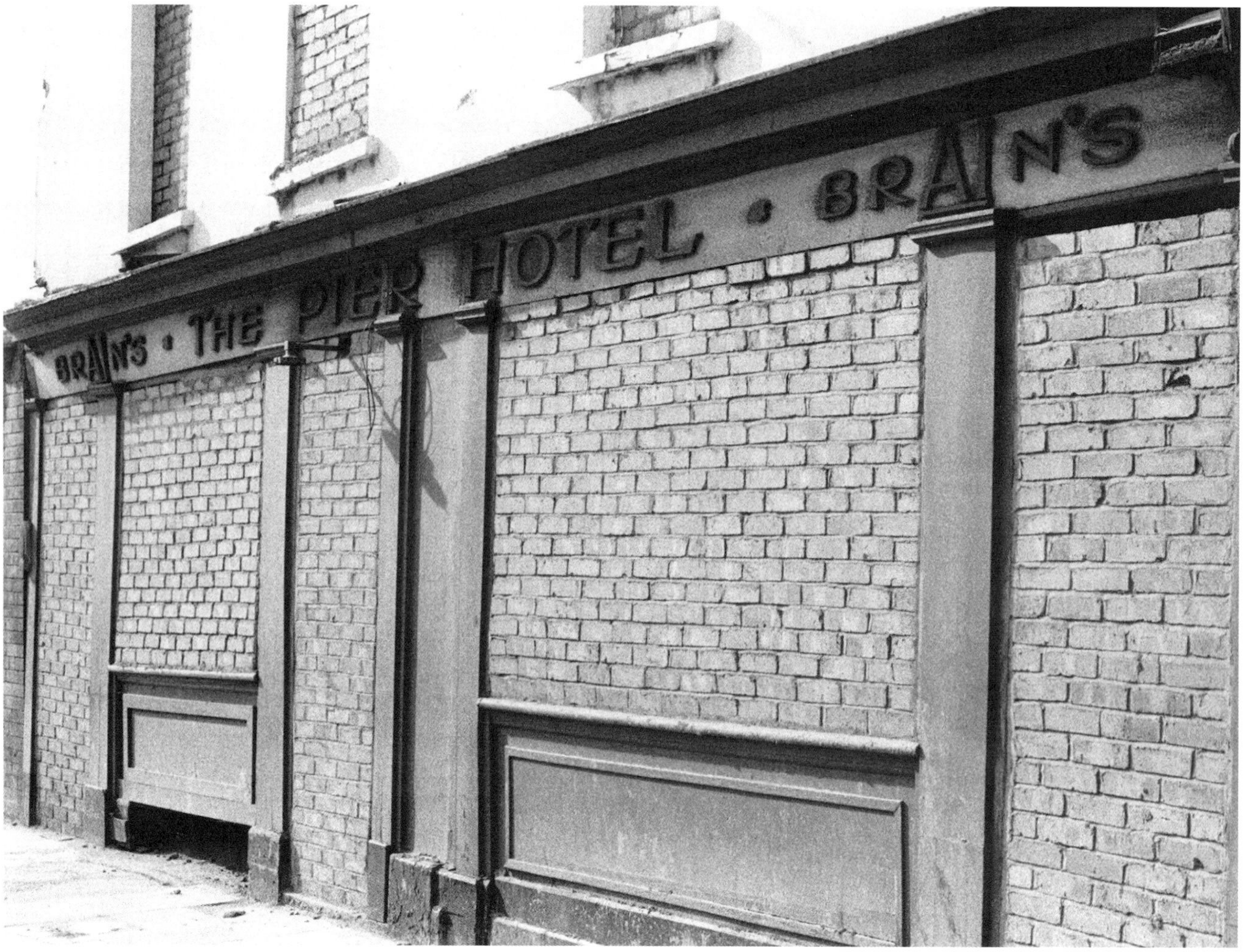
BRAIN'S • THE PIER HOTEL • BRAIN'S

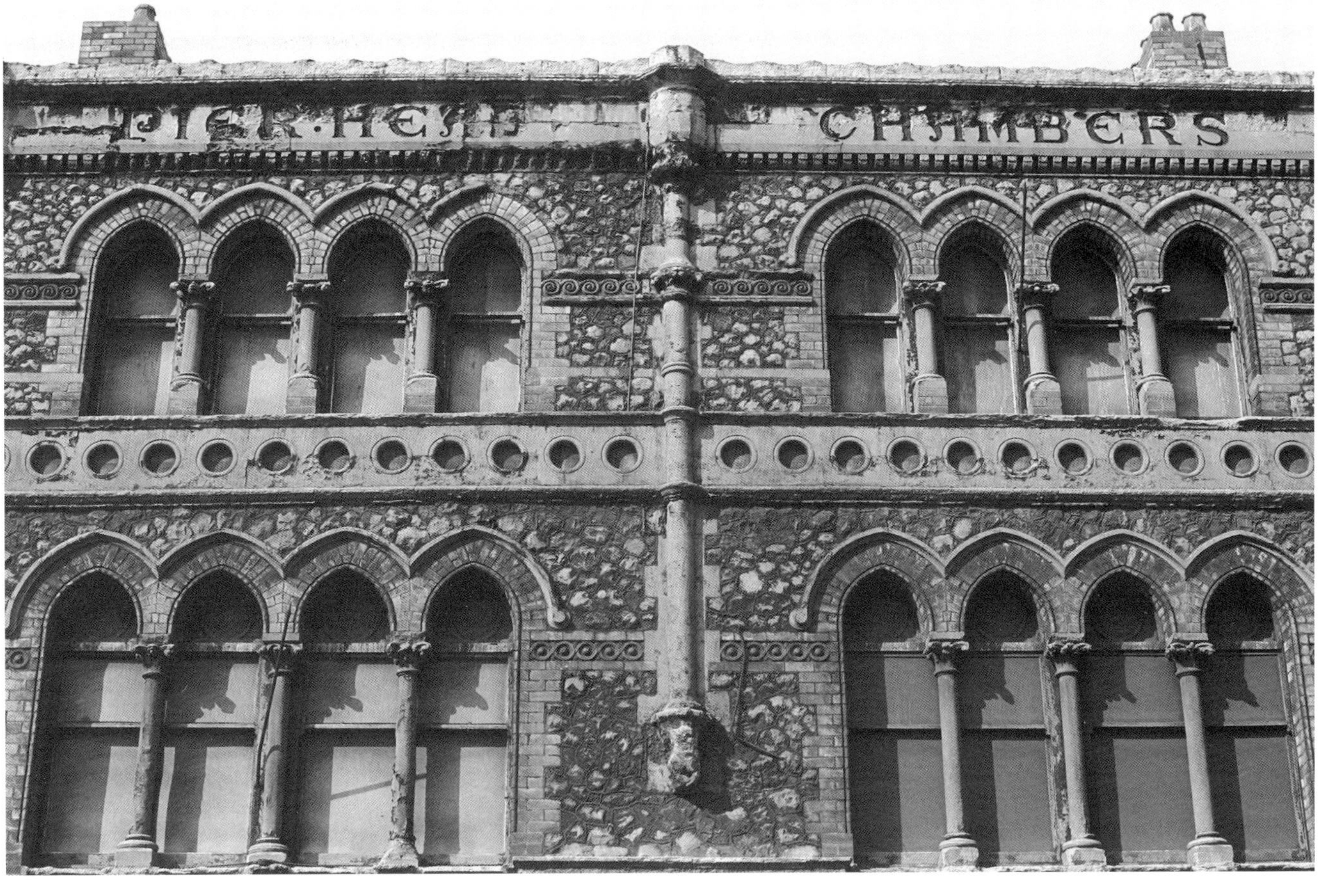
CHAMBERS

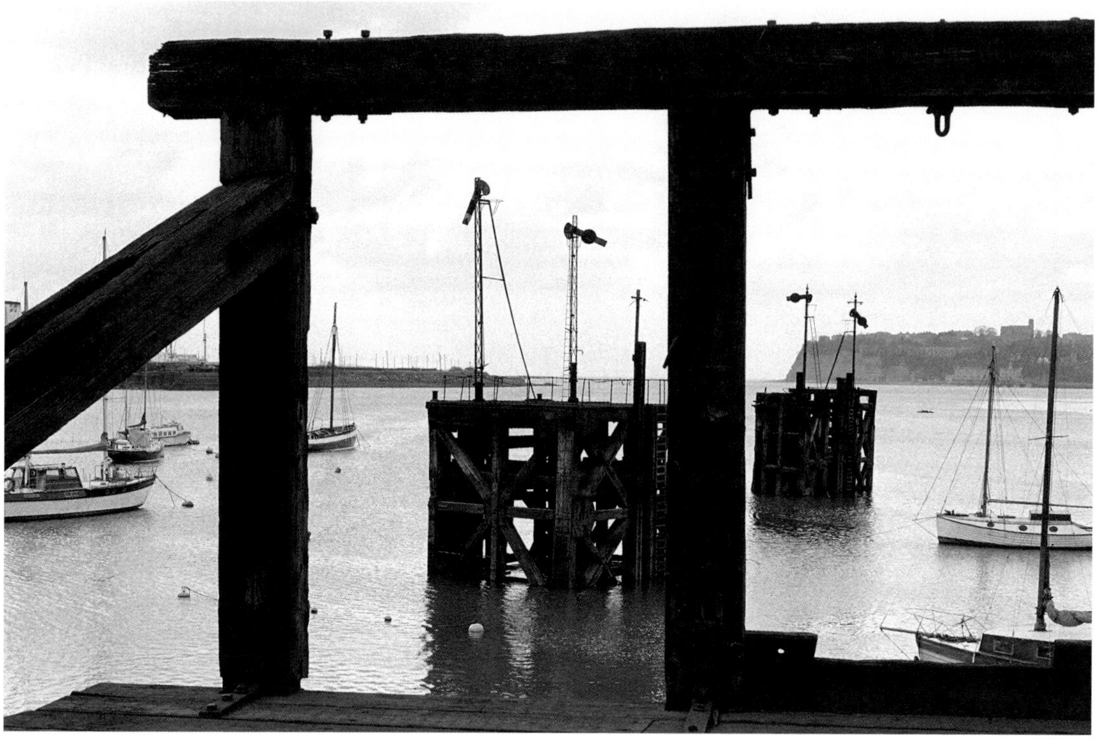

Two

The Docks

Following Bute Street past Tiger Bay and to its conclusion in the Lower Docks ultimately brought one to the Pier Head where the sea beckoned – if the tide was in, that is. If it was not, the arresting sight of the harbour at low tide with its vast mudflats, small craft resting on the muddy sand, docks' cranes in the background and 'dolphins', the ancient wooden mooring towers, in the foreground, provided focal points for anyone who happened along, be they local resident, occasional visitor or American stranger with a camera. But it is most fitting to start this section with the centrepiece of it all, the Pier Head Building, before dwelling on other singular features of Cardiff's Inner Harbour, as it used to be known prior to being permanently swamped by the waters of Cardiff Bay. The photographic itinerary then covers the old Cardiff Docks proper, showing what remained in the mid-1970s of Bute's original excavations, what there was of past and present industry around them and finally what maritime activity was taking place during that decade in some other parts of the port.

The first Bute Dock, originally known as the Bute Ship Canal and then later as the Bute West Dock, opened in 1839. It was a long and narrow affair running north for half a mile from what is now the wood-slatted Oval Basin (originally the West Dock Basin) to the Edward England potato warehouses just before Herbert Street. The Atlantic Leisure Village complex and new apartments extending along Lloyd George Avenue (what did he have to do with the docks?) have lately been built on the site of this historic waterway, filled in the late 1960s.

Not having a historical perspective on the building of the Bute Docks when I came to Cardiff in 1974, I am sure that I first walked through the large, empty expanse of the West Dock landfill site without knowing what it was. Shortly after, I paid 35p in a shop for what I assumed to be an up-to-date street map of Cardiff and found that it had an area of water on it, in blue just like the River Taff, the other docks and the Inner Harbour, called Bute West Dock. I went back to find it but couldn't. After a little research I also discovered that I was in fact about ten years too late to see the waters of the West Dock. Nonetheless I kept the map and have found it useful, at long last, for showing the location of Bute West Dock in relation to certain photographs in this section. What I did find at the time were a few derelict but fine examples of historic buildings which had stood along or near the Second Marquess of Bute's maritime masterstroke. The first was the Norwegian Church, originally sited right at the dock's eastern edge and now rebuilt near the Roath Basin lock. The second was the Junction Dry-dock with its dependent buildings, fine Victorian industrial buildings which were enthralling to photograph and to which I returned a number of times. I will always maintain that these would have been a fine setting for a maritime museum, which regrettably has never been a part of the new Cardiff Bay redevelopment scheme.

The second Bute Dock, the East Dock, opened in 1855. Much wider than the first and just as long, it was built, like the three others which followed it, to cater for the world's growing and eventually insatiable appetite for Welsh coal up to the end of the First World War. At one time ships crammed into it as tightly as possible, its quays were packed with dozens of coal hoists fed by a huge network of railway sidings and thousands of railway wagons. Then, years after the boom days of coal and after the other industries along its wharves such as milling, warehousing and ship-repair had become redundant it too closed, like the West Dock, in the late 1960s. In contrast to the West Dock, however, the East Dock was retained as a water feature for the new post-1970s developments installed along its old wharves. These include the new pagoda-like South Glamorgan County Hall, the Victorian-styled Wharf pub and restaurant and the waterside Holiday Inn. The old Junction Canal at its north end has also been impressively restored and it now flows past new houses, apartments and converted Victorian industrial structures.

For me as photographer in the 1970s the area was a fascinating study in dereliction. So much had simply been left behind that the abandoned structures and other relics formed a world of their own. And so complete and so extensive was the spectacle of industrial desolation, and all of it wrapped in a serene and death-like stillness, that, like any other perfect landscape, it could only fill one with a sense of wonder. The grimy industries that survived at the dockside – a trolley-bus scrapyard, a car crusher and scrap merchant's, small engineering and repair shops, a tyre dump, East Moors Steelworks – only reinforced the nether-worldly aspect of the surroundings. It all seemed to conspire first to amaze and then to evoke in me the desire to capture it photographically. And in the midst of all this were fine buildings – derelict monuments to Victorian architectural style crying out for attention and I was only too happy to oblige with my camera. Some of these buildings, I am relieved to report, eventually attracted the eye of the constructive developer as well as that of me, the photographer. The transformation of the Spiller's biscuit works into an apartment building, the No 11 Bonded Warehouse into office accommodation and the LNWR Goods Warehouse into a quality hotel give a good idea of what is possible when architectural heritage is preserved then adapted for the use and appreciation of future generations. I fervently wish that this had been a much greater priority during docklands redevelopment, that new structures had been designed with many more echoes of the past and that those Victorian buildings still standing derelict in Butetown may one day be restored to their former glory.

With access to the docks not being an obstacle to the occasional photographer, I also sought out whatever shipping activity I might find beyond the old Bute Docks. Nearer the active parts of the Port of Cardiff, – the Roath and Queen Alexandra Docks – lay the almost empty waters of Roath Basin, first opened in 1874. Here the one or two wharves still in use were home to a single 'sander' as it disgorged its muddy sludge sucked up from the bottom of the Bristol Channel. At the Bute Dry-dock, in one of only two active bays from the remaining nine in the port, the lone ship berthed there gave the impression of being marooned more than of undergoing restoration to maritime fitness. The derelict pleasure steamer *King George V*, having met its fate in a disastrous fire while being refitted in the dry-dock as a floating restaurant, waited to be towed to the shipbreaker's and rested in peace along what is now a newly refurbished Britannia Quay. The mangled wreckage of its decks yielded such fantastic shapes that dozens of strong images were possible. I wish there was room in this book for more of them. Further on in the Roath Dock, telling images of dwindling activity could be seen and captured in the form of a hulking Gulf tanker from Belgium discharging oil at a terminal where coal hoists had once been and, in a scene soon to disappear, a bulk carrier from Singapore unloading its iron ore at the East Moors Steelworks wharf.

Who else, other than someone with a camera's warped eye, might be drawn to such scenarios? No-one but a solitary fisherman waiting quietly for a catch by the still waters of the East Dock.

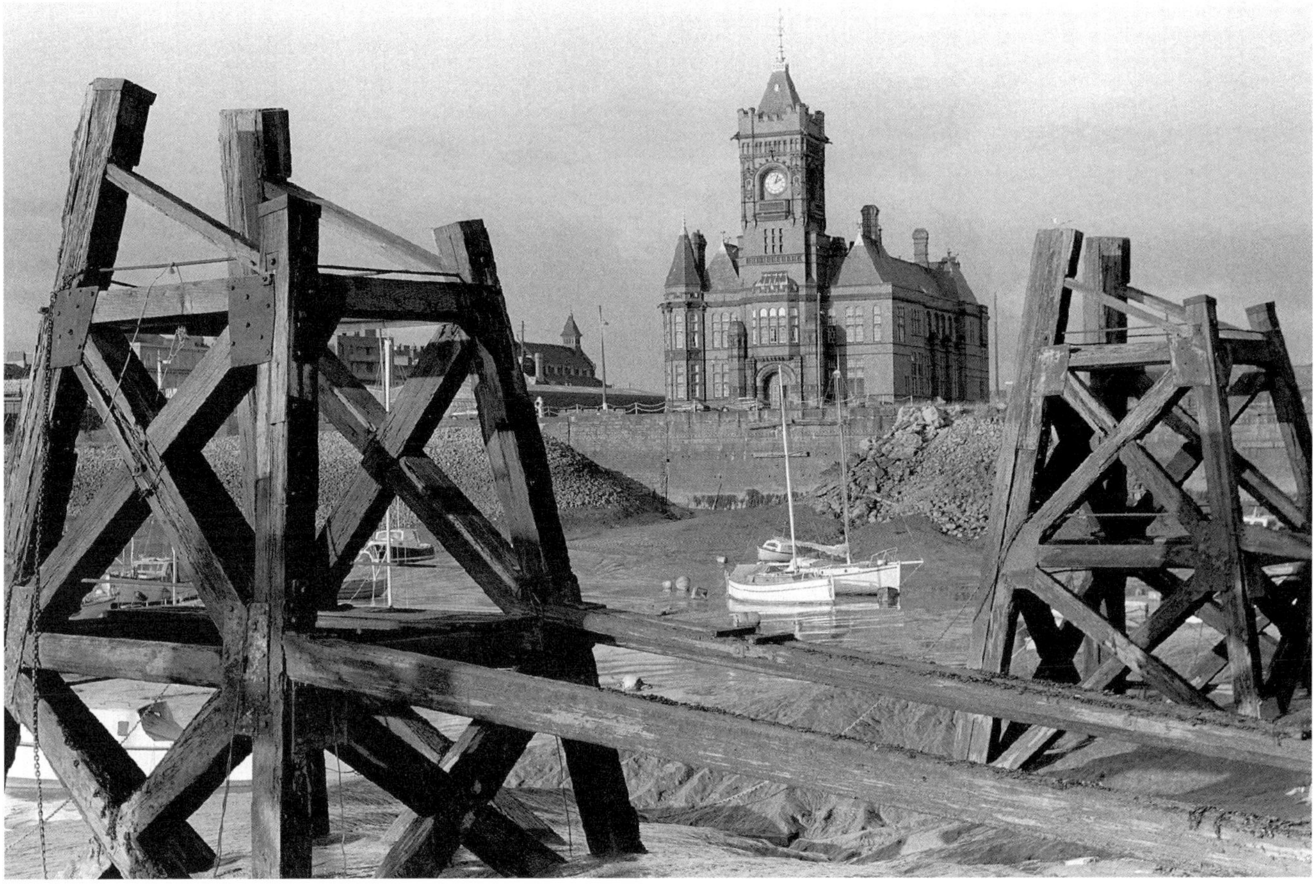

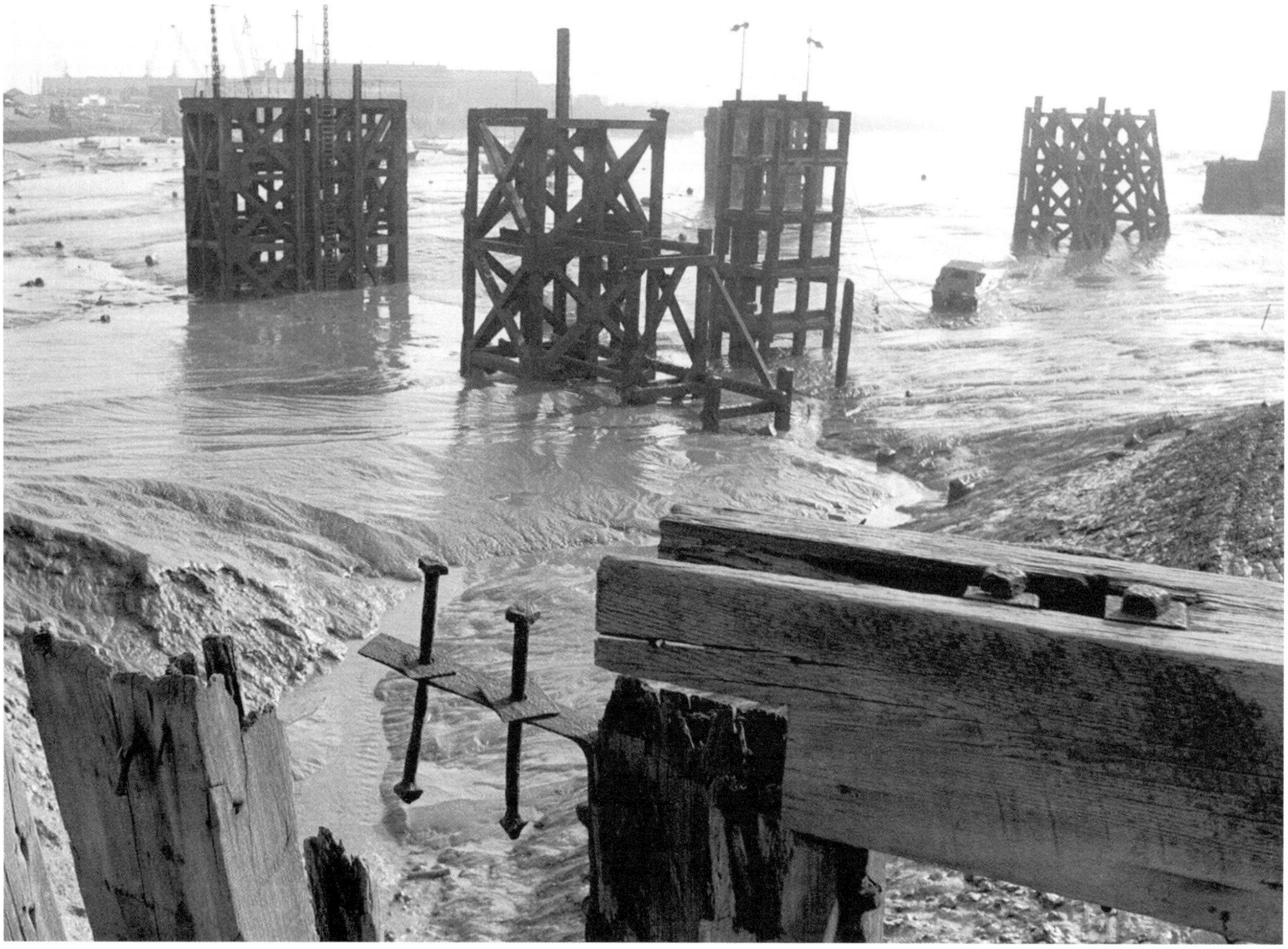

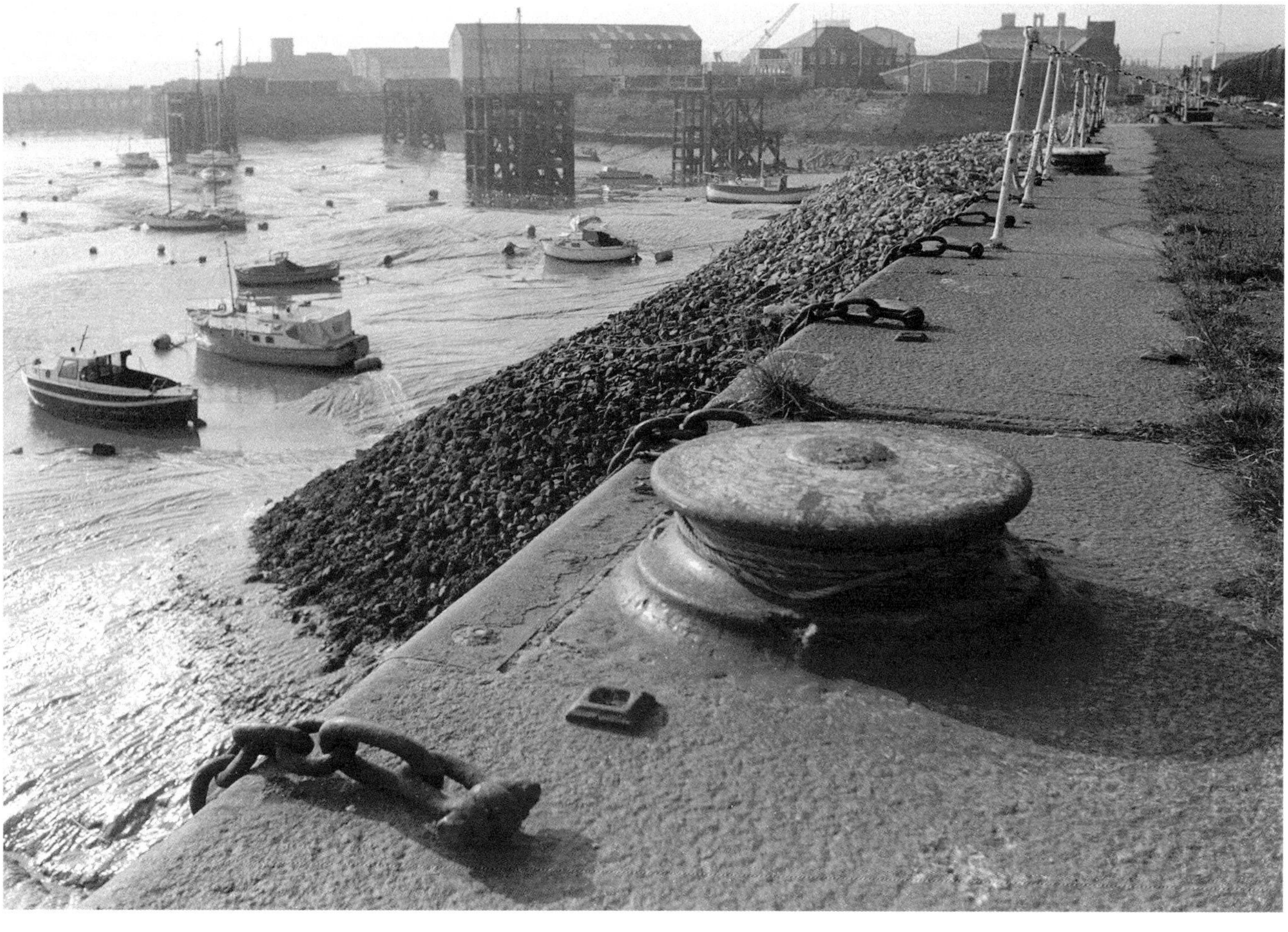

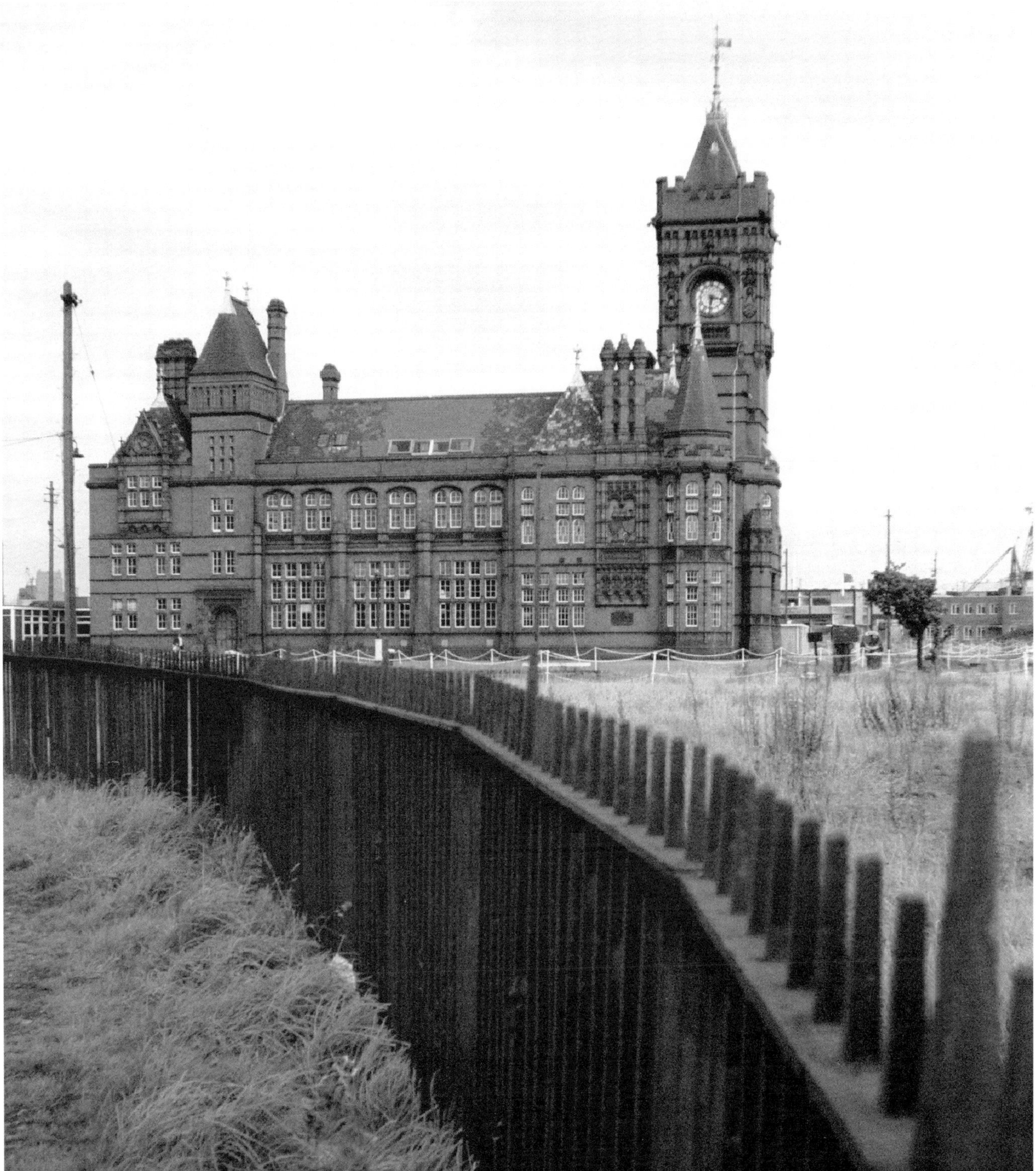

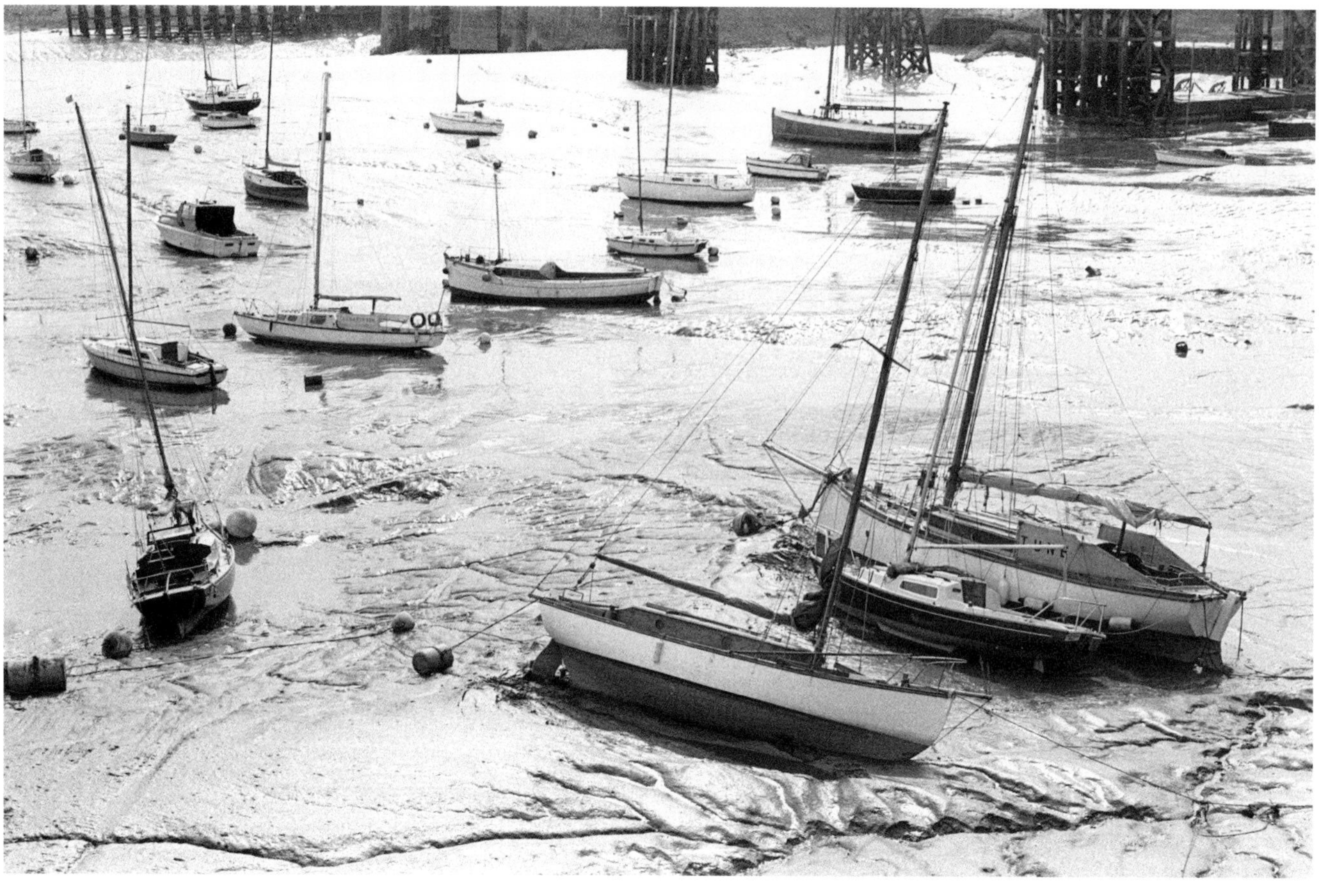

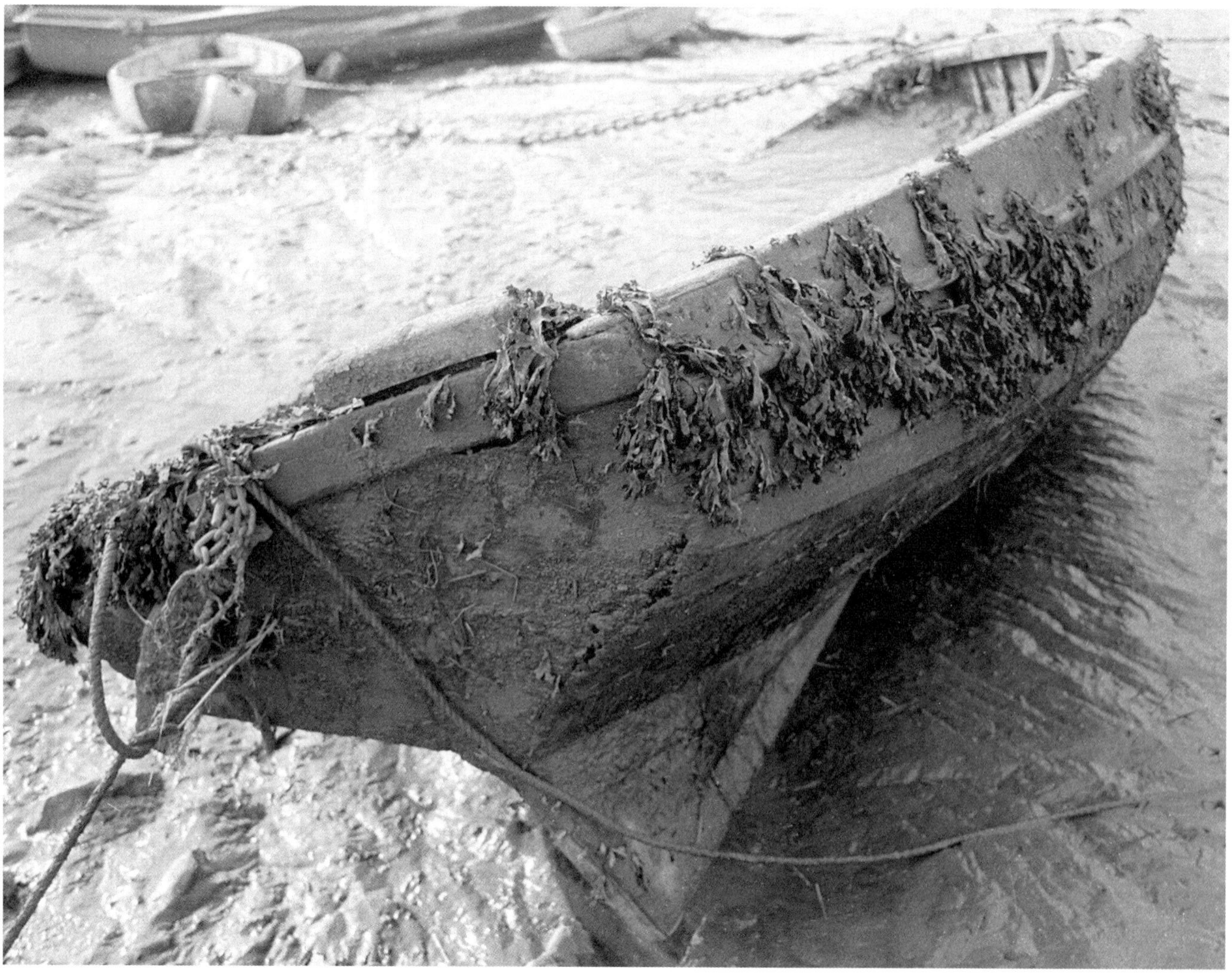

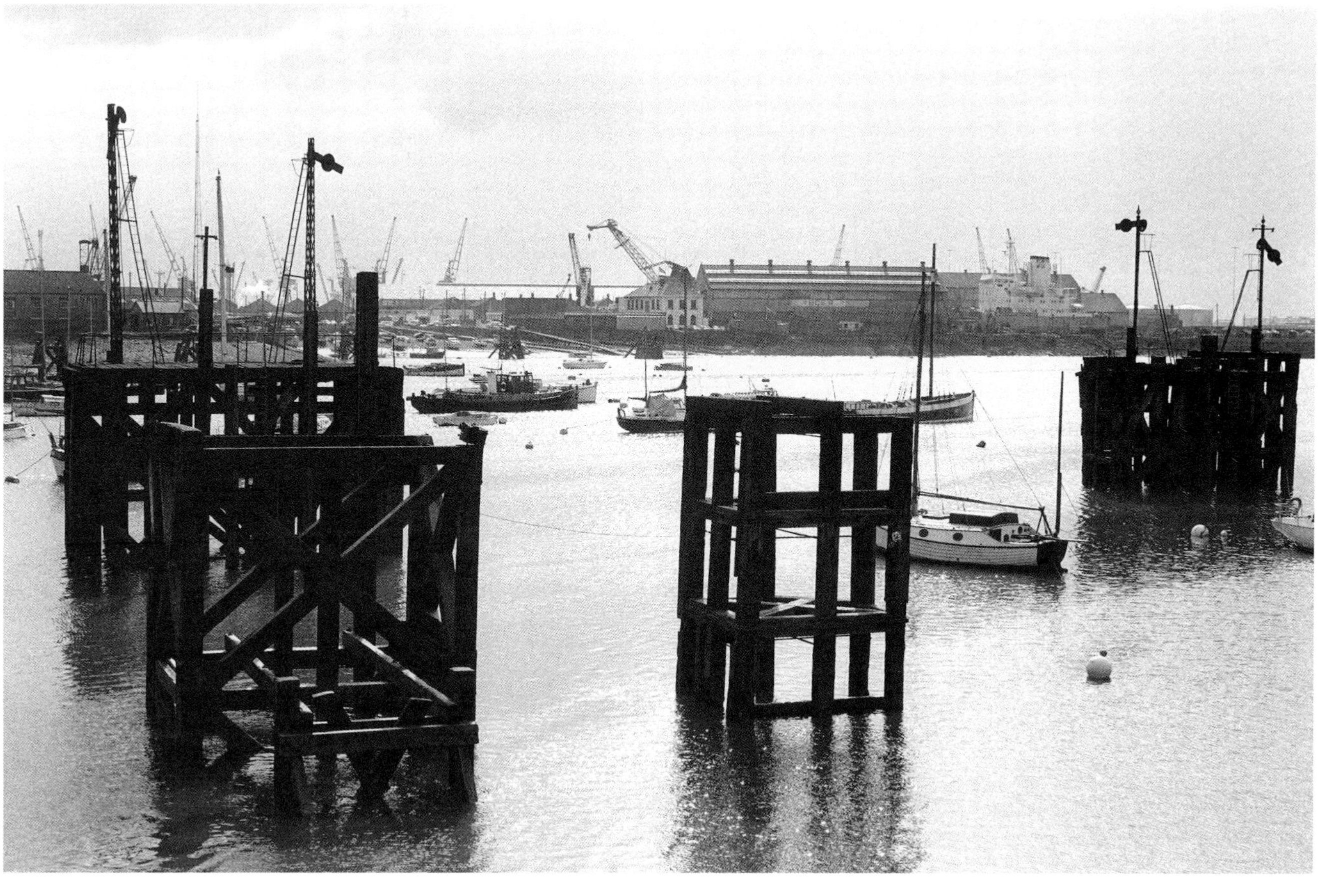

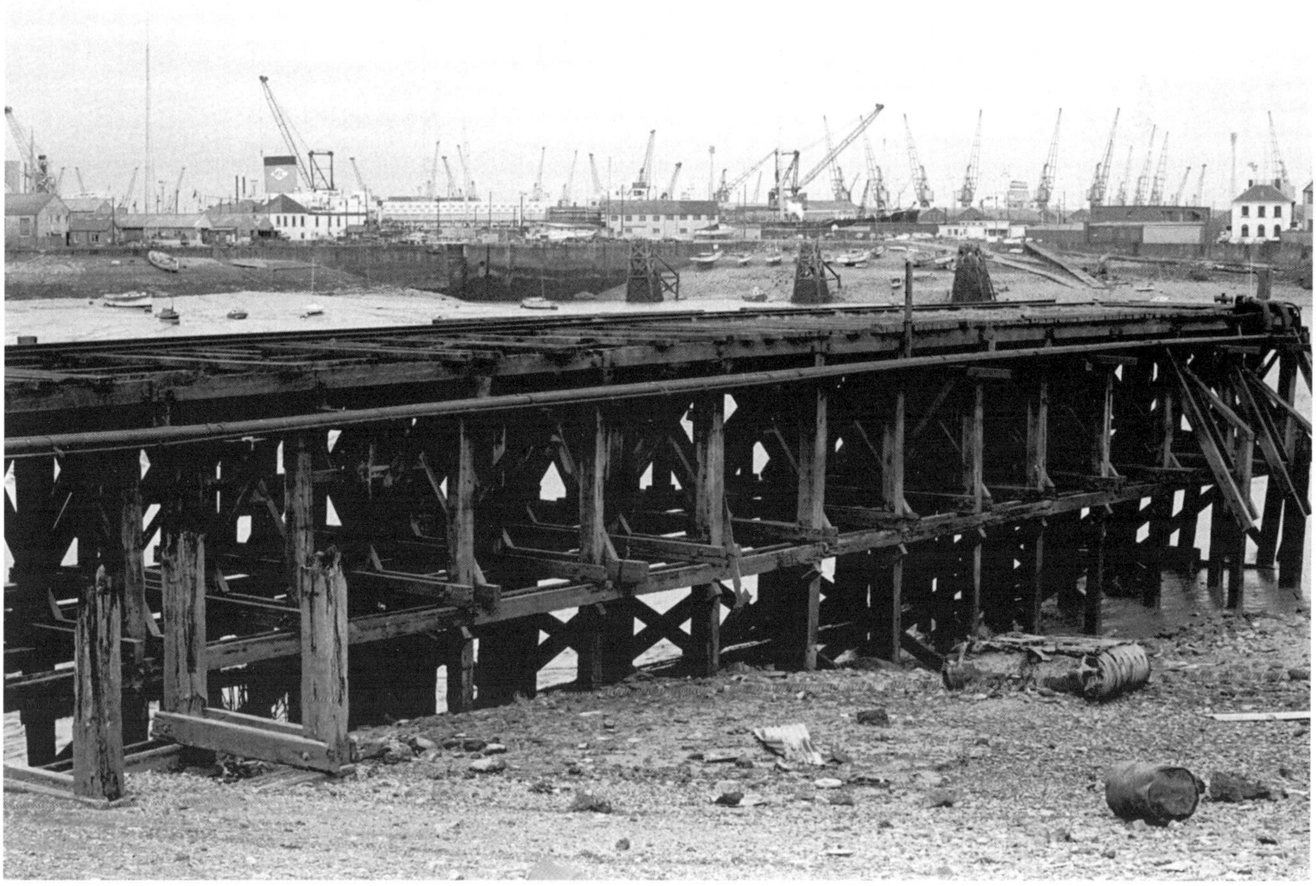

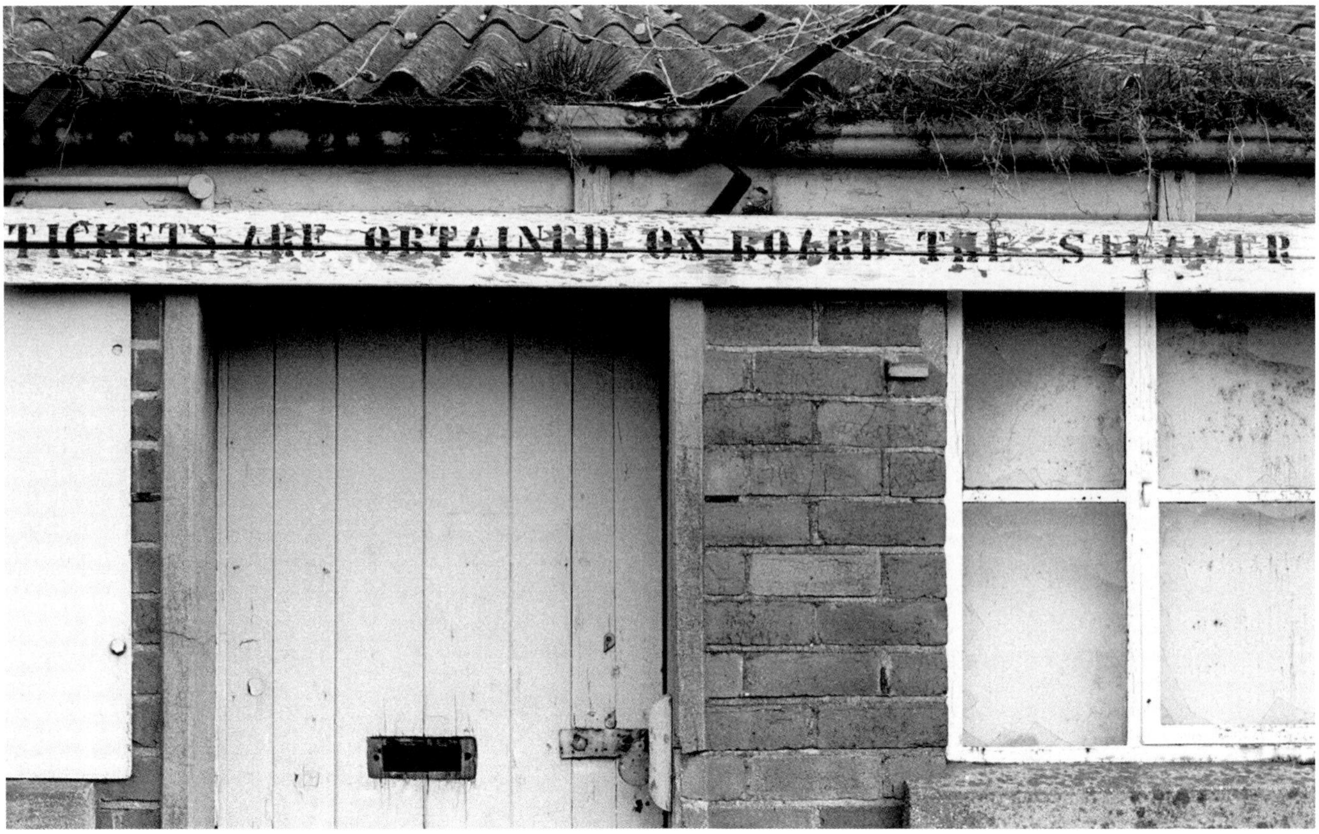
TICKETS ARE OBTAINED ON BOARD THE STEAMER

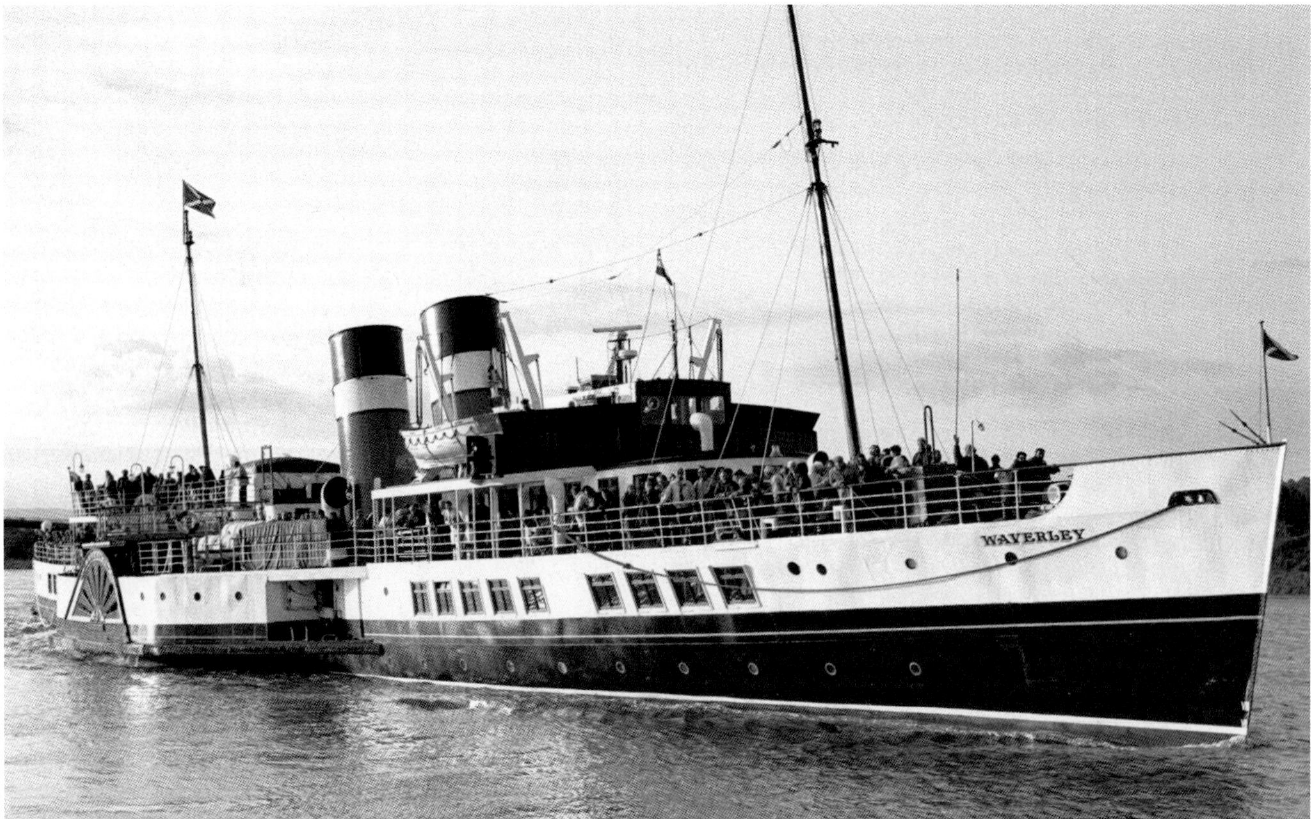
WAVERLEY

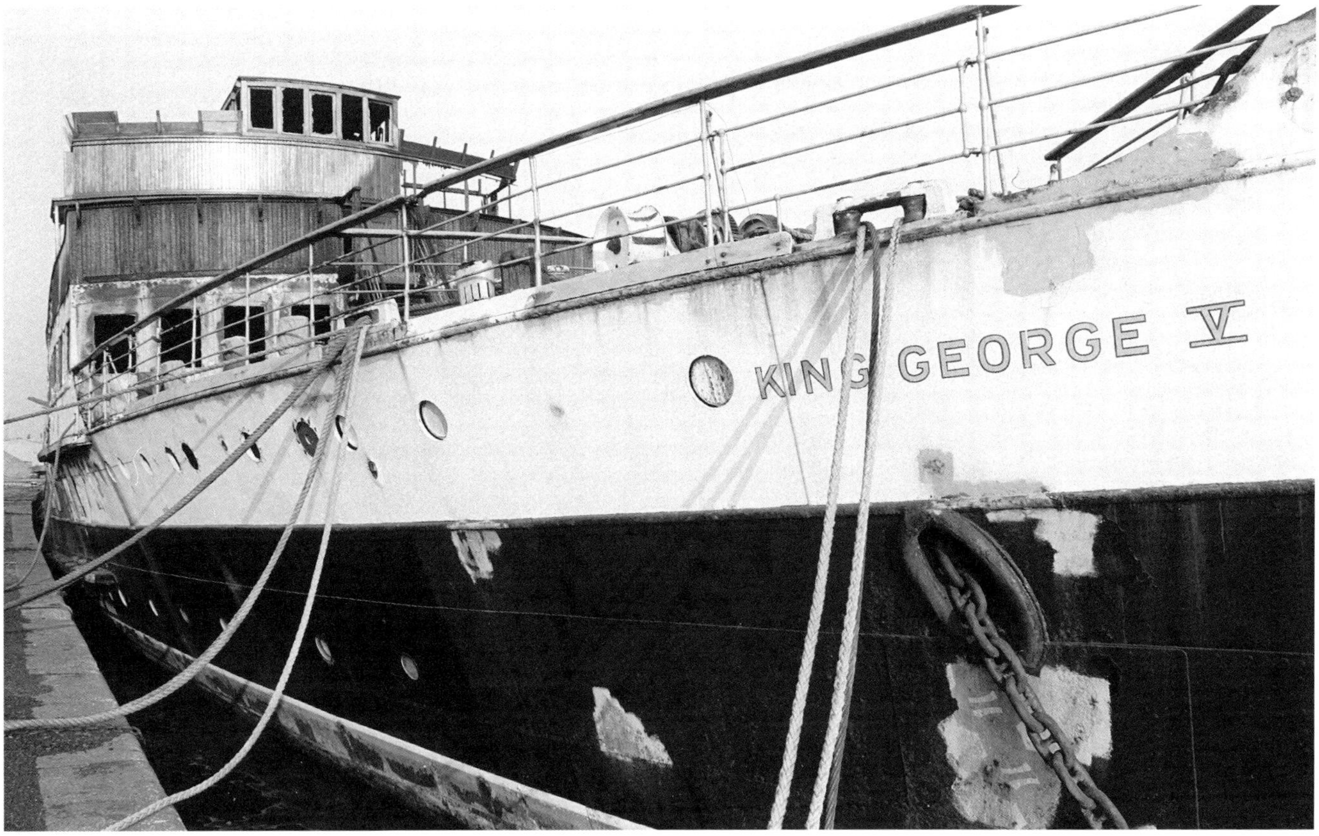
KING GEORGE V

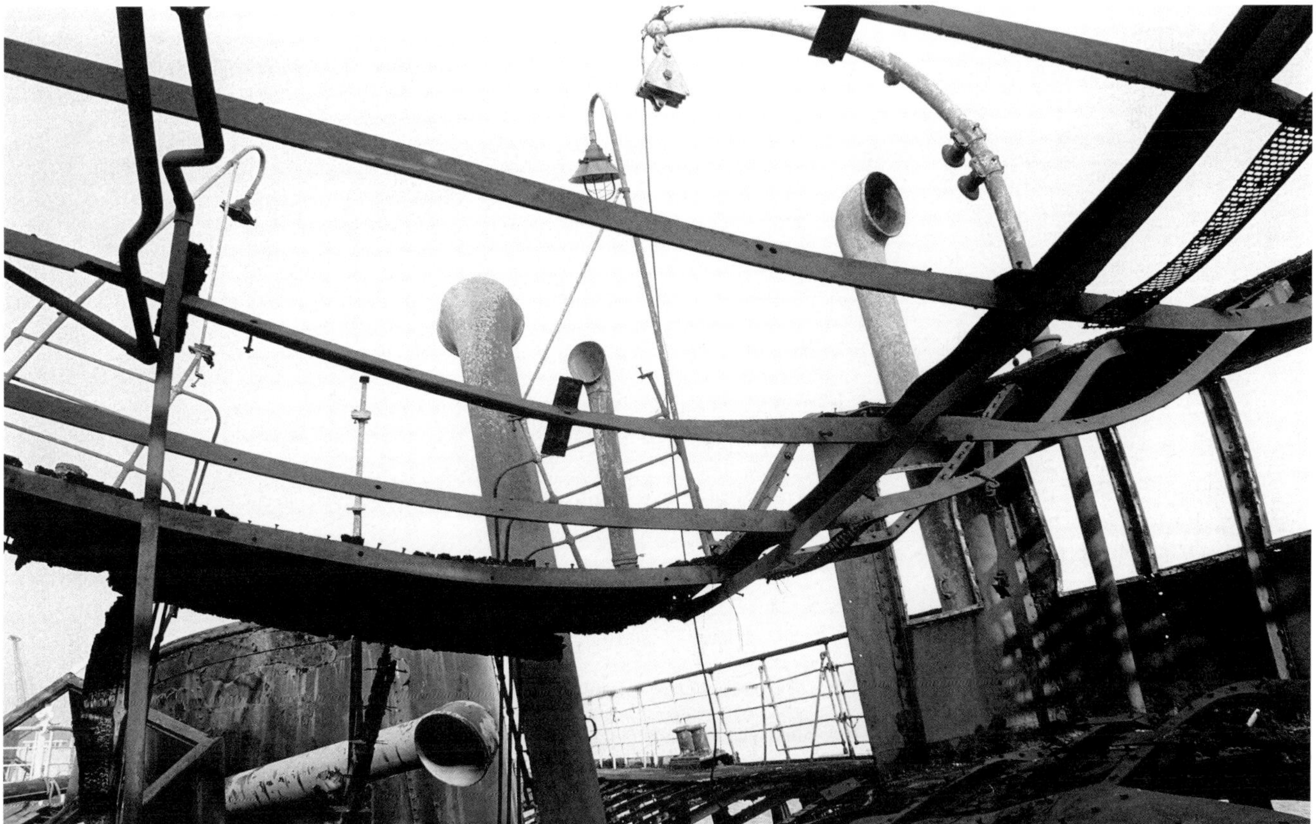

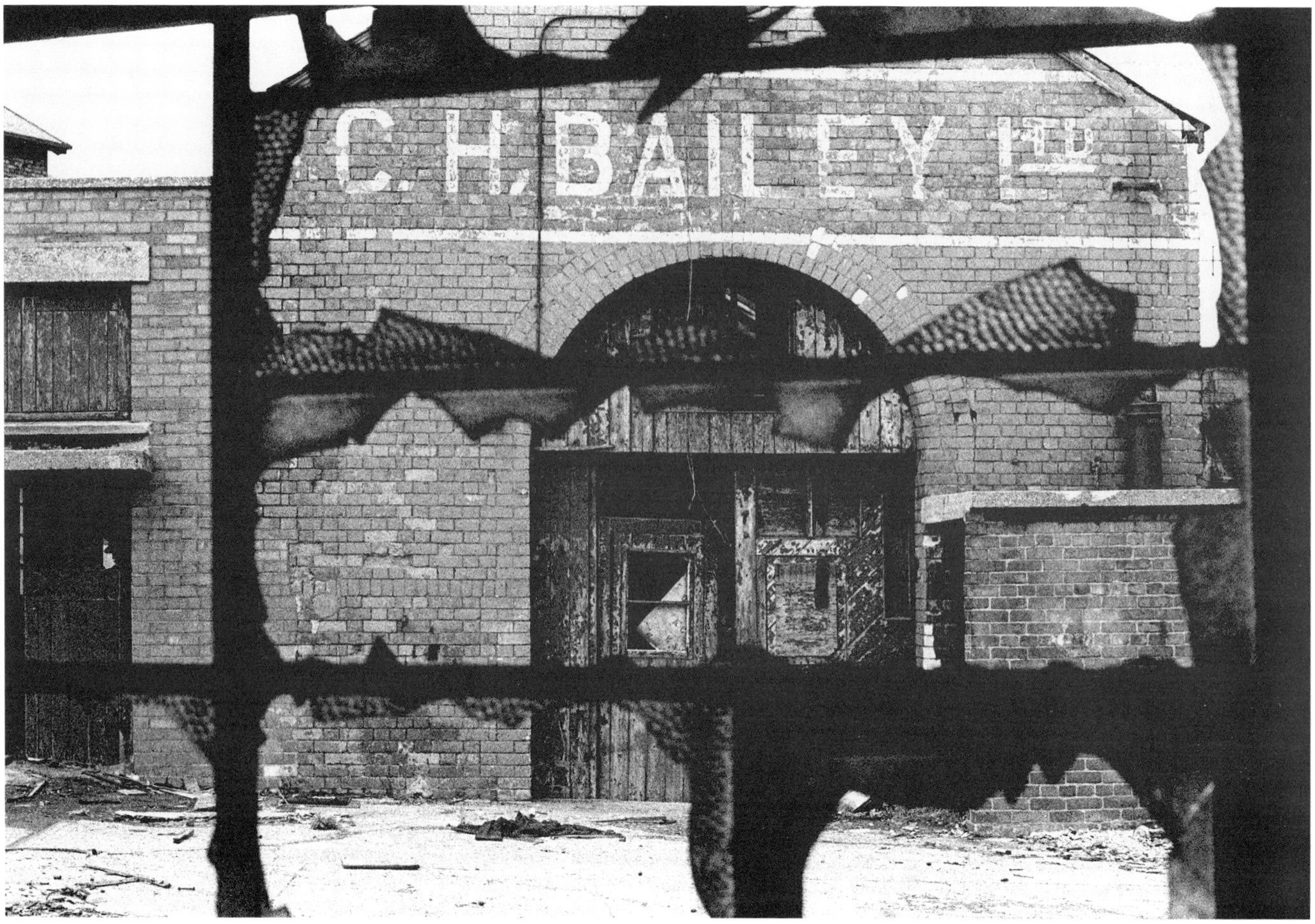
C.H. BAILEY Ltd

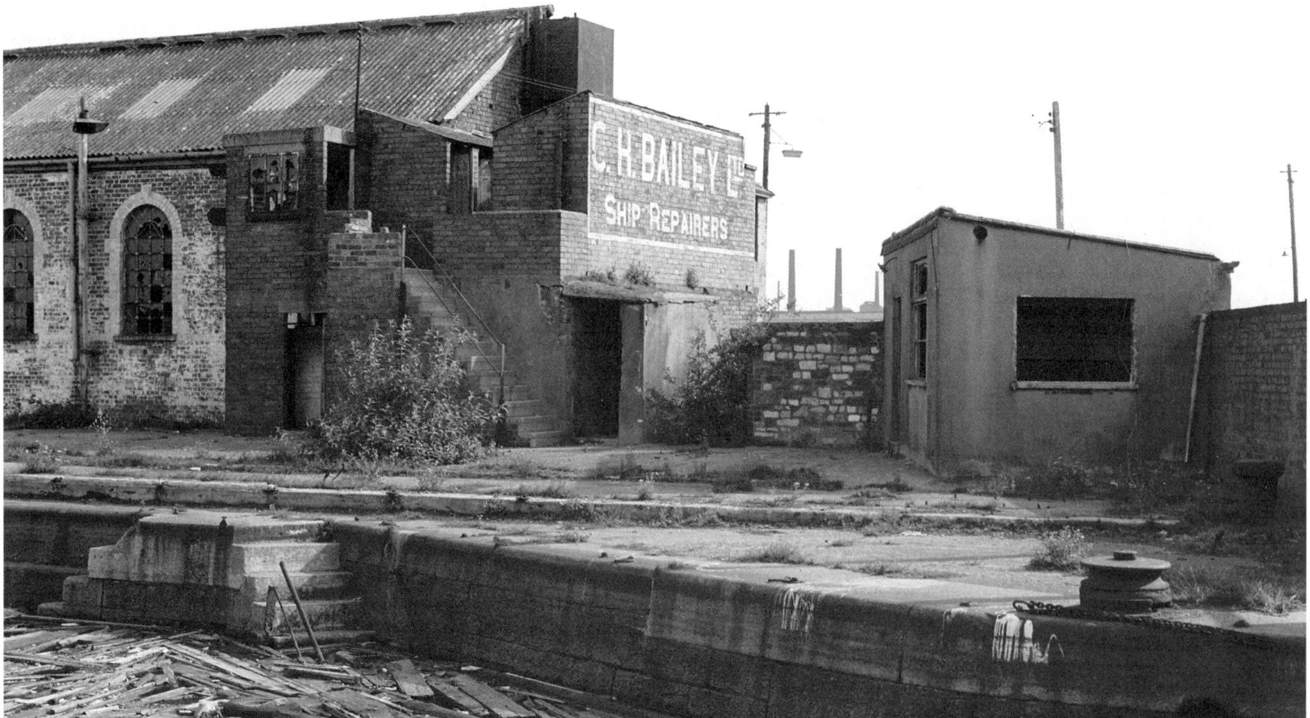
C.H. BAILEY Ld
SHIP REPAIRERS

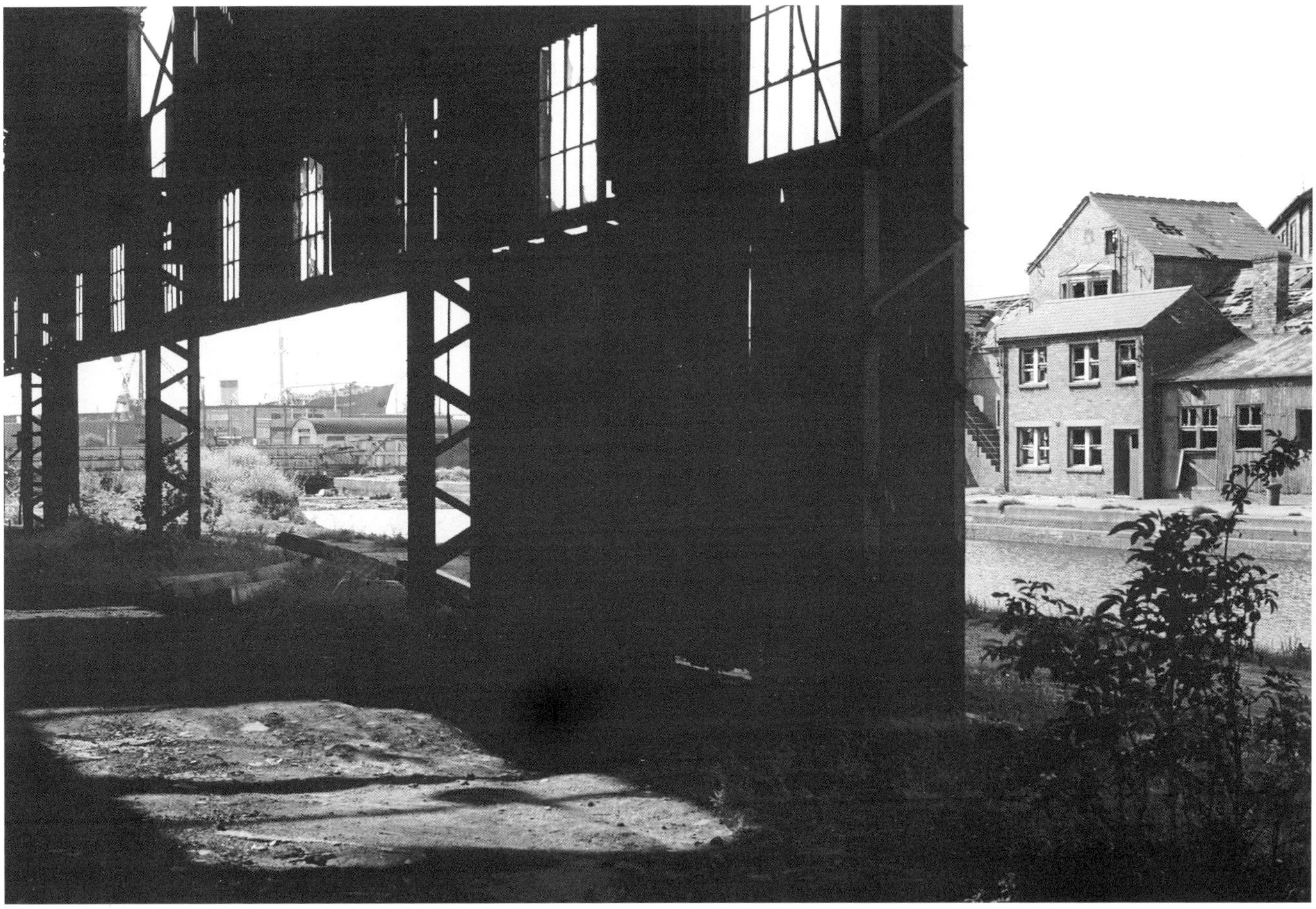

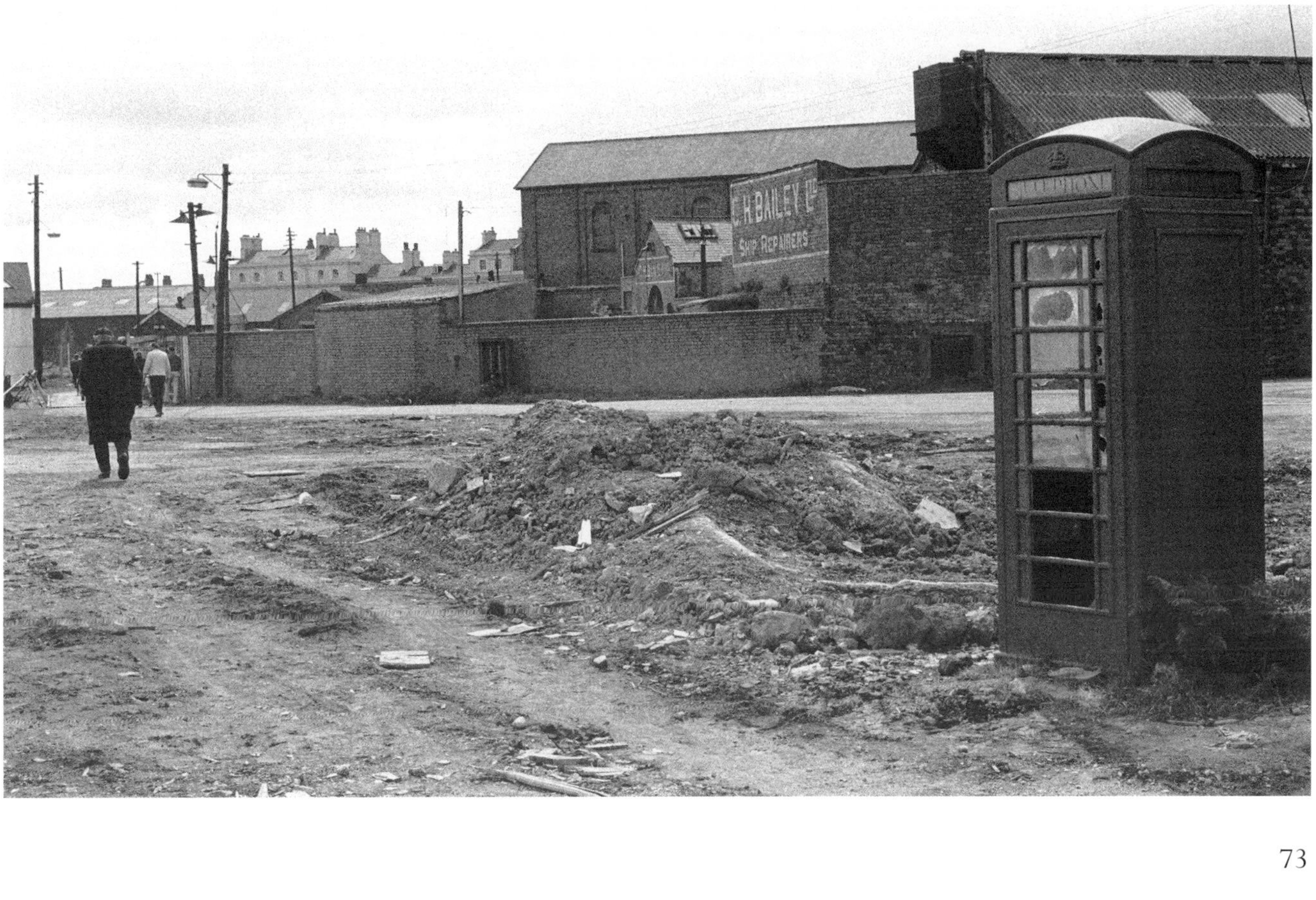
C. H. BAILEY Ltd
SHIP REPAIRERS

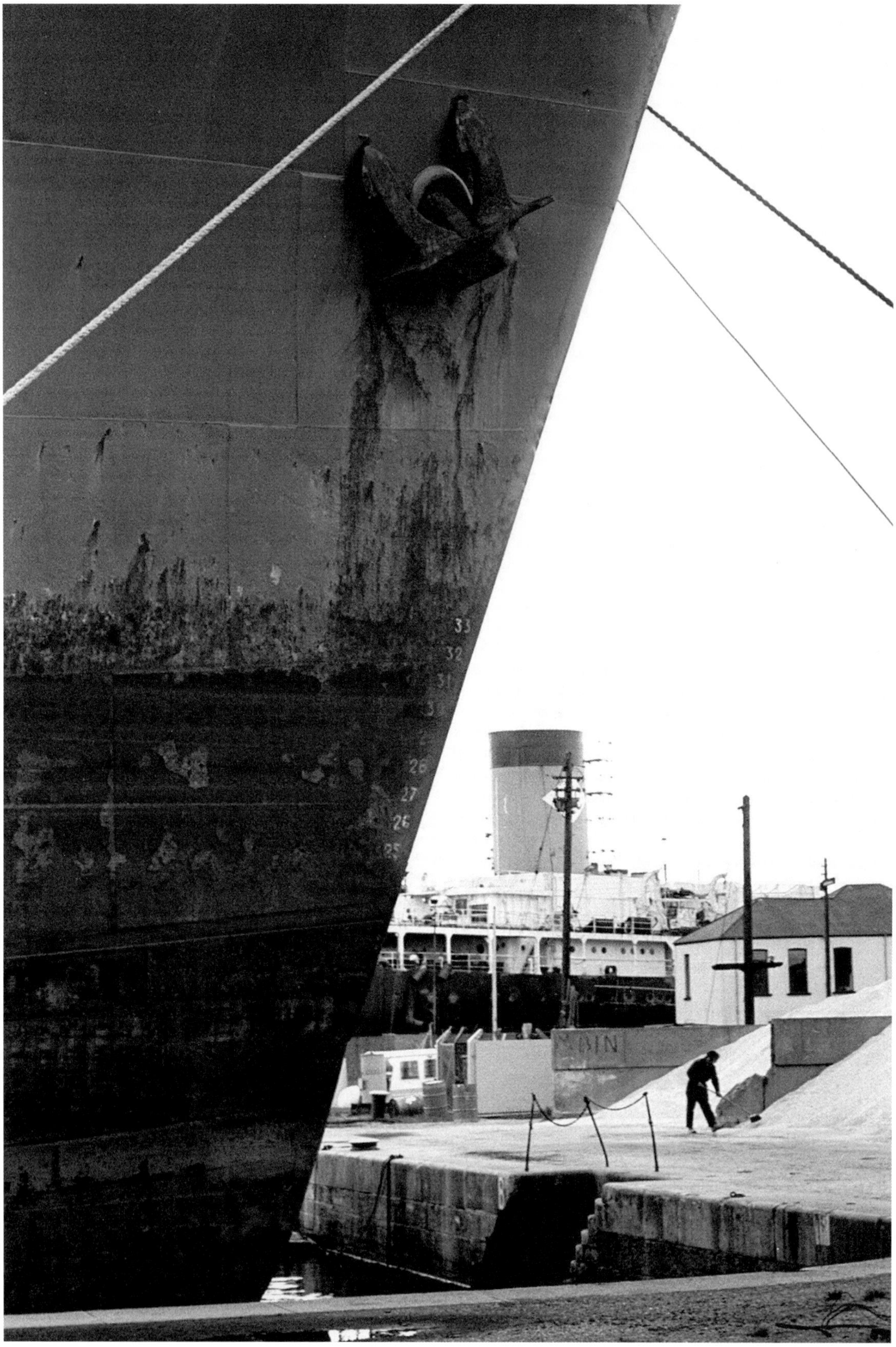

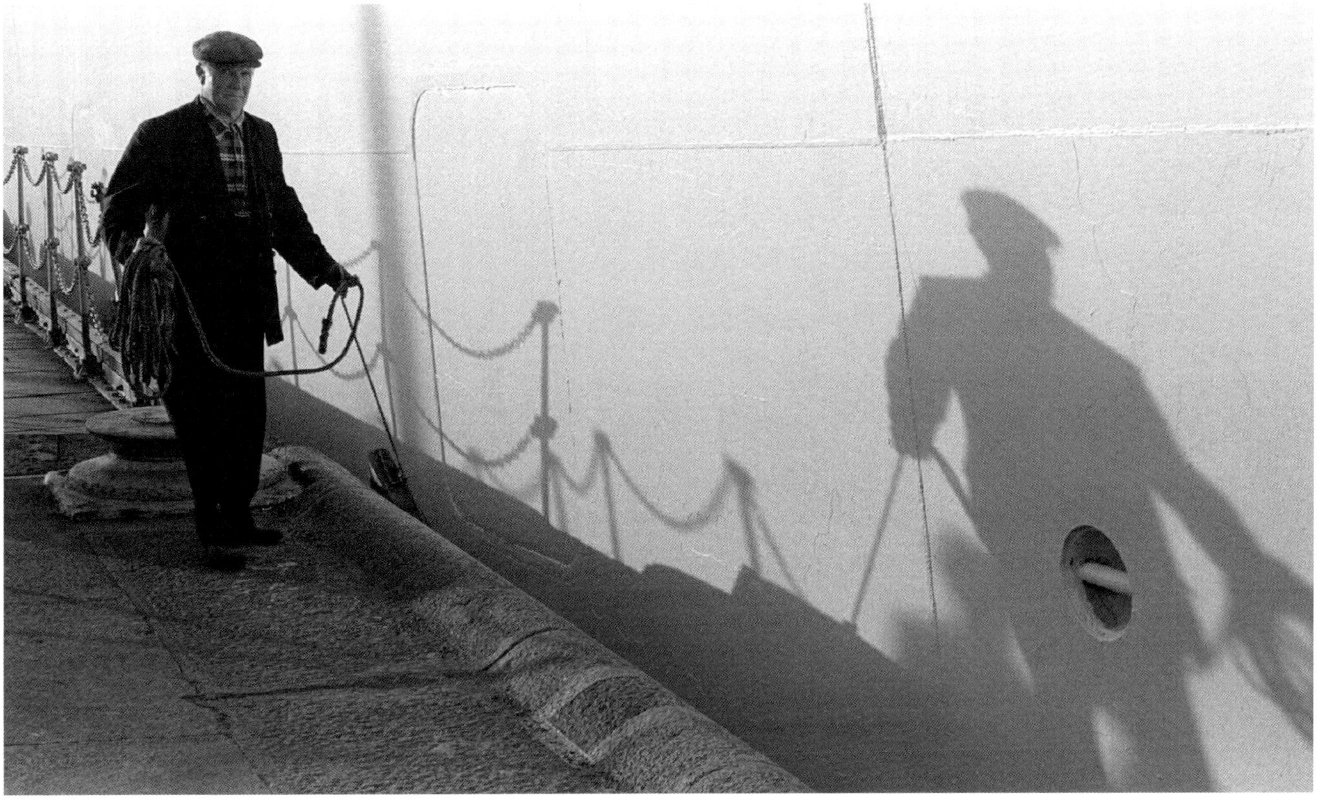

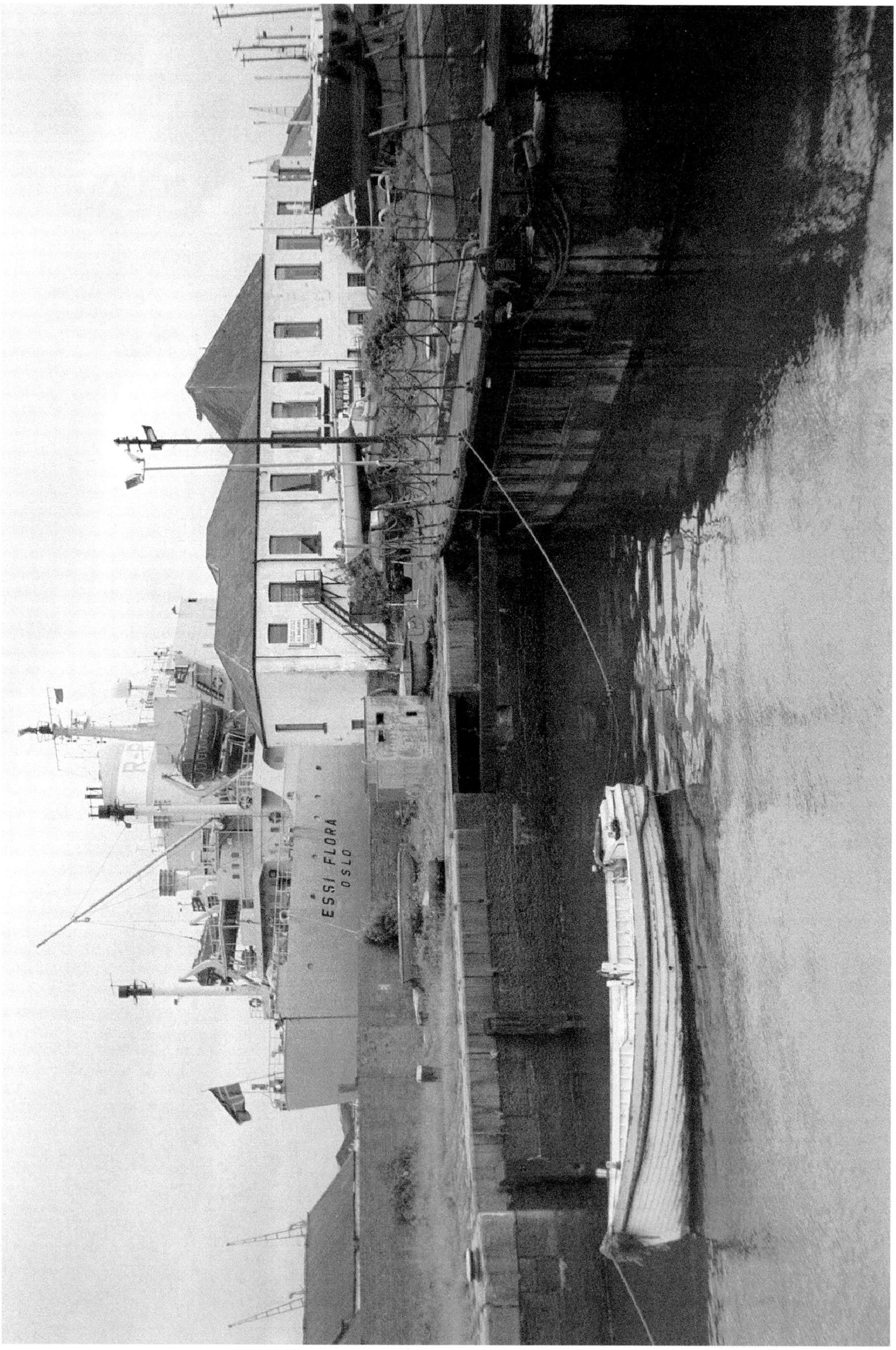
ESSI FLORA
OSLO

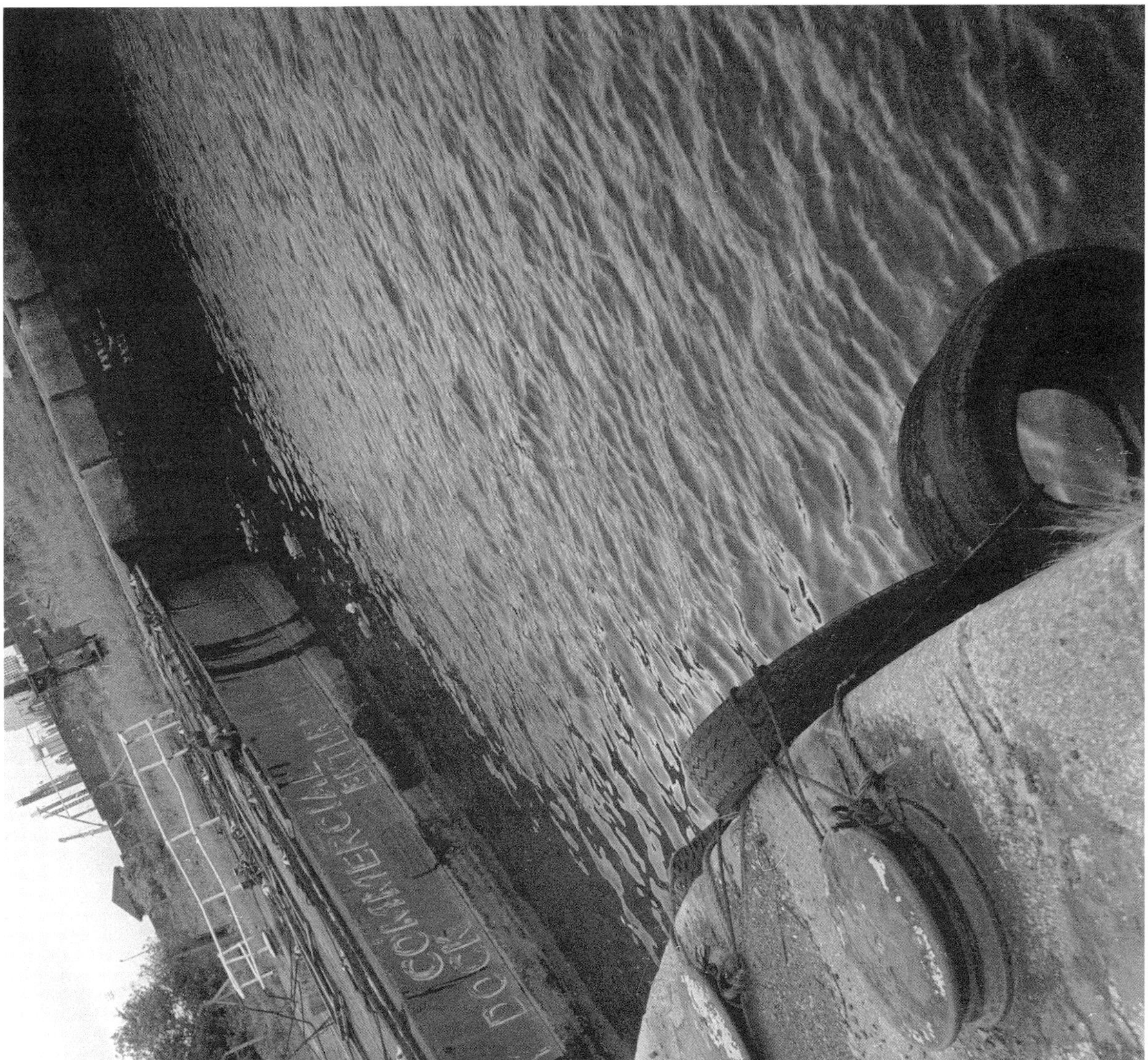
COMMERCIAL
DOCK

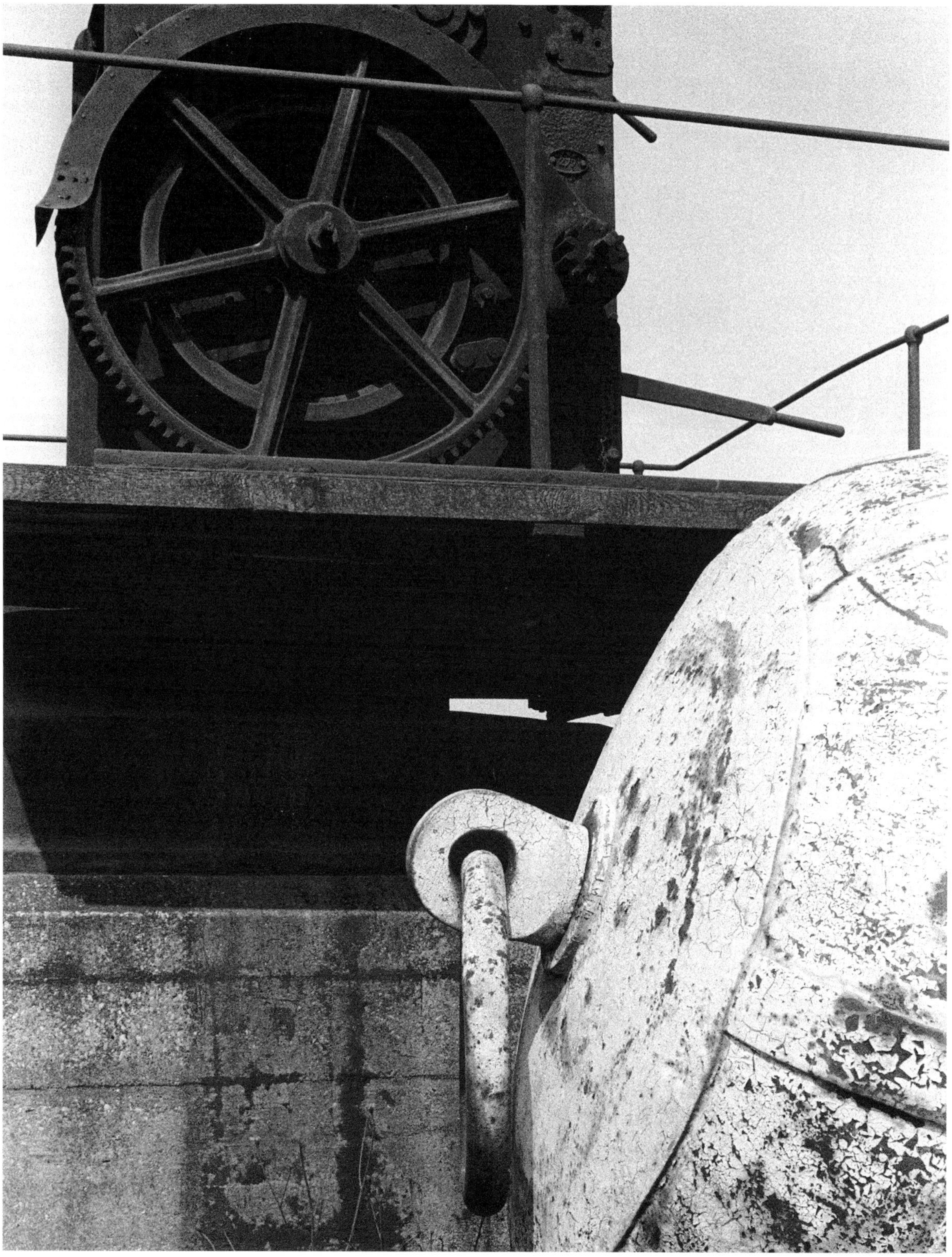

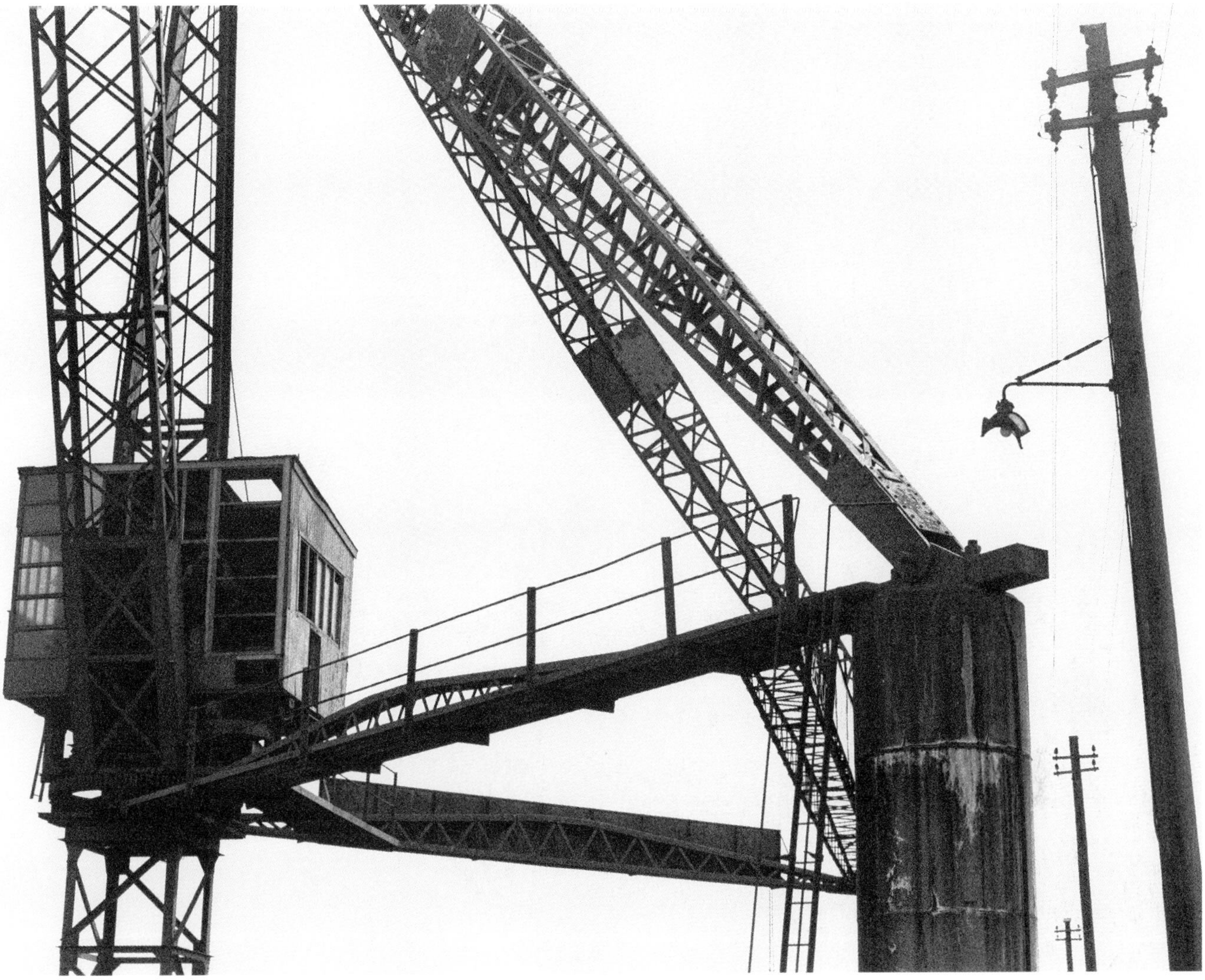

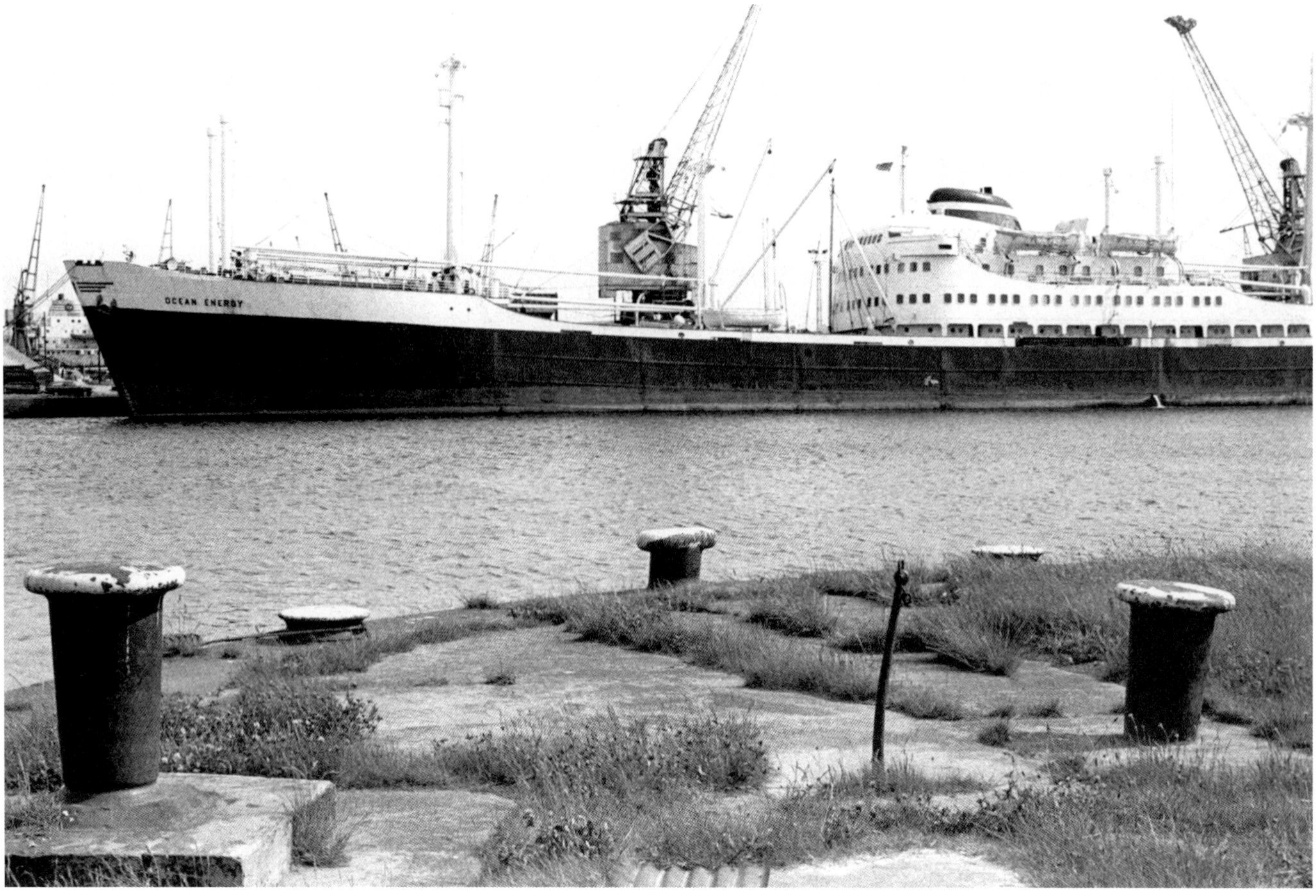
OCEAN ENERGY

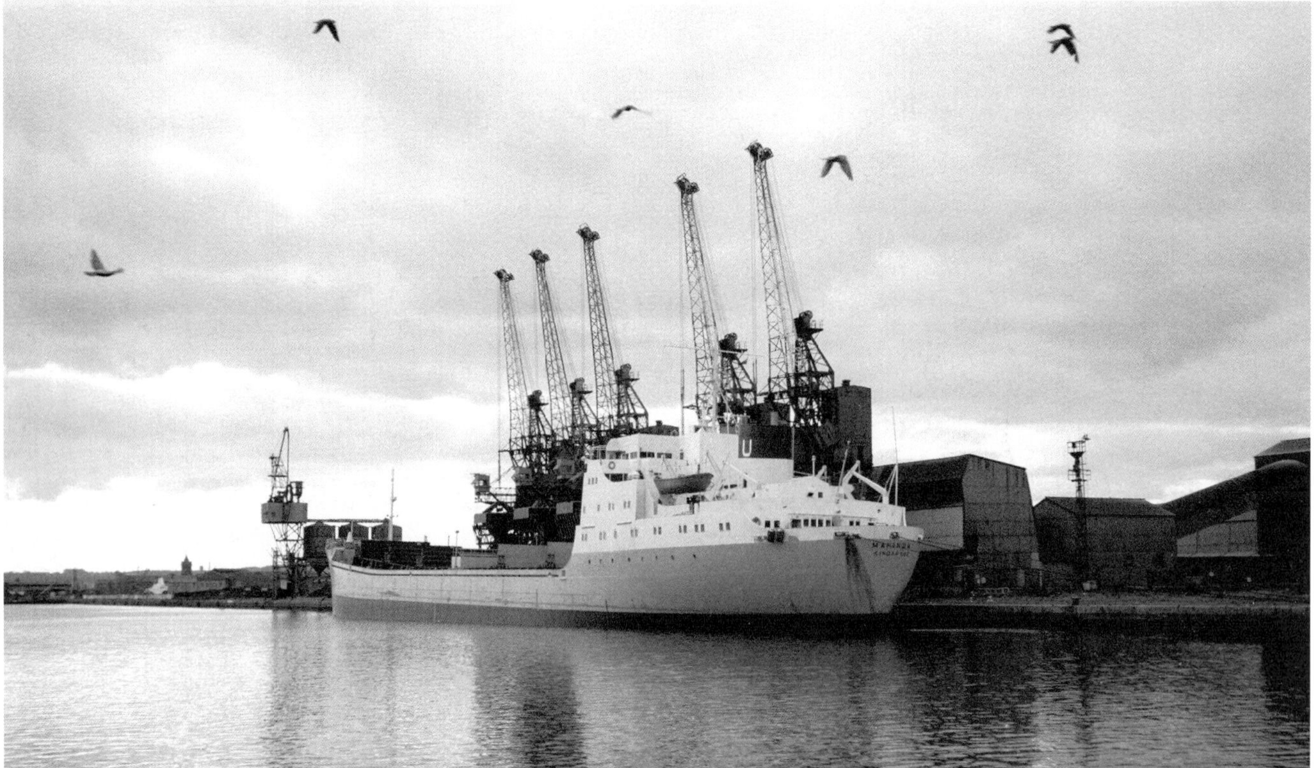

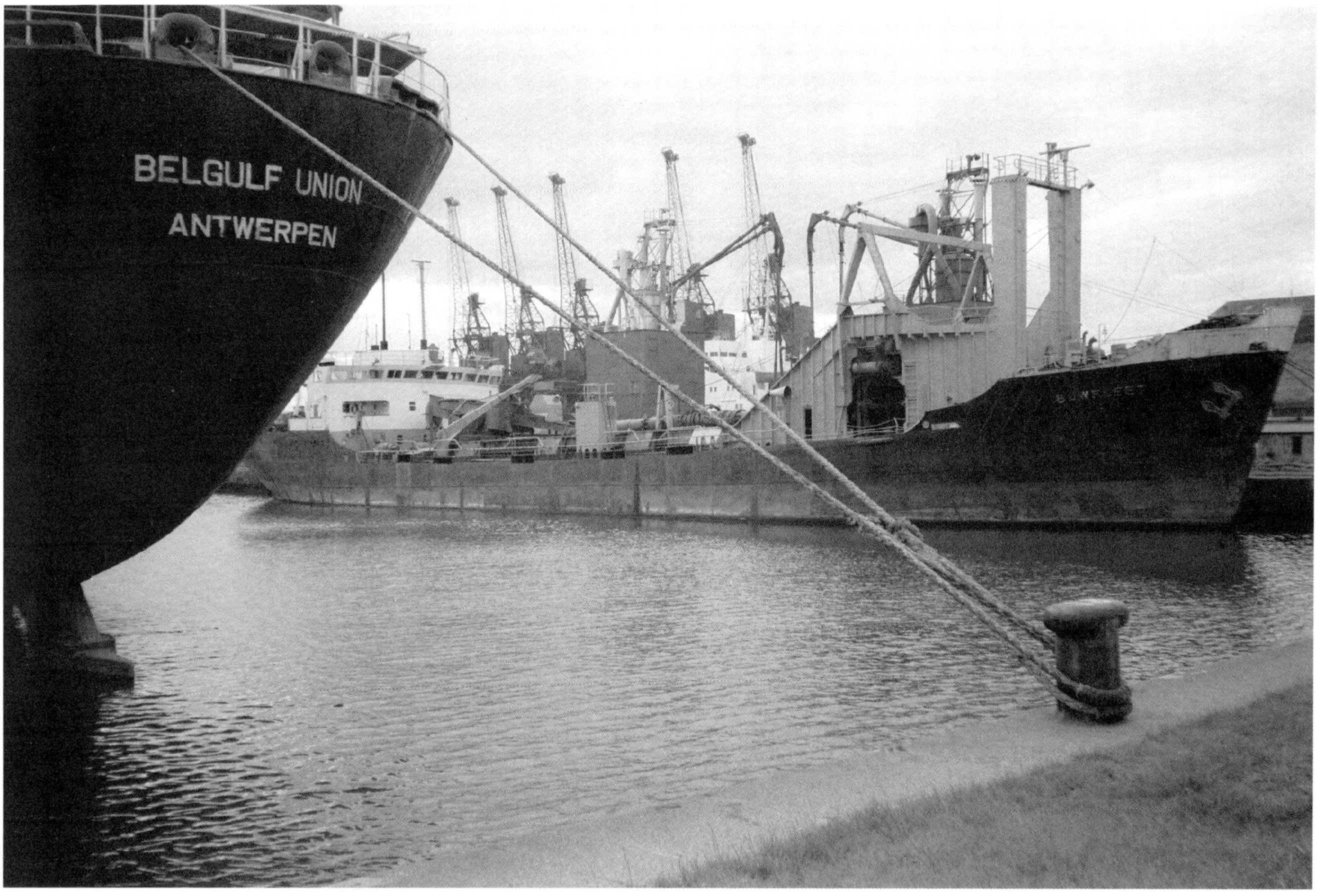
BELGULF UNION
ANTWERPEN

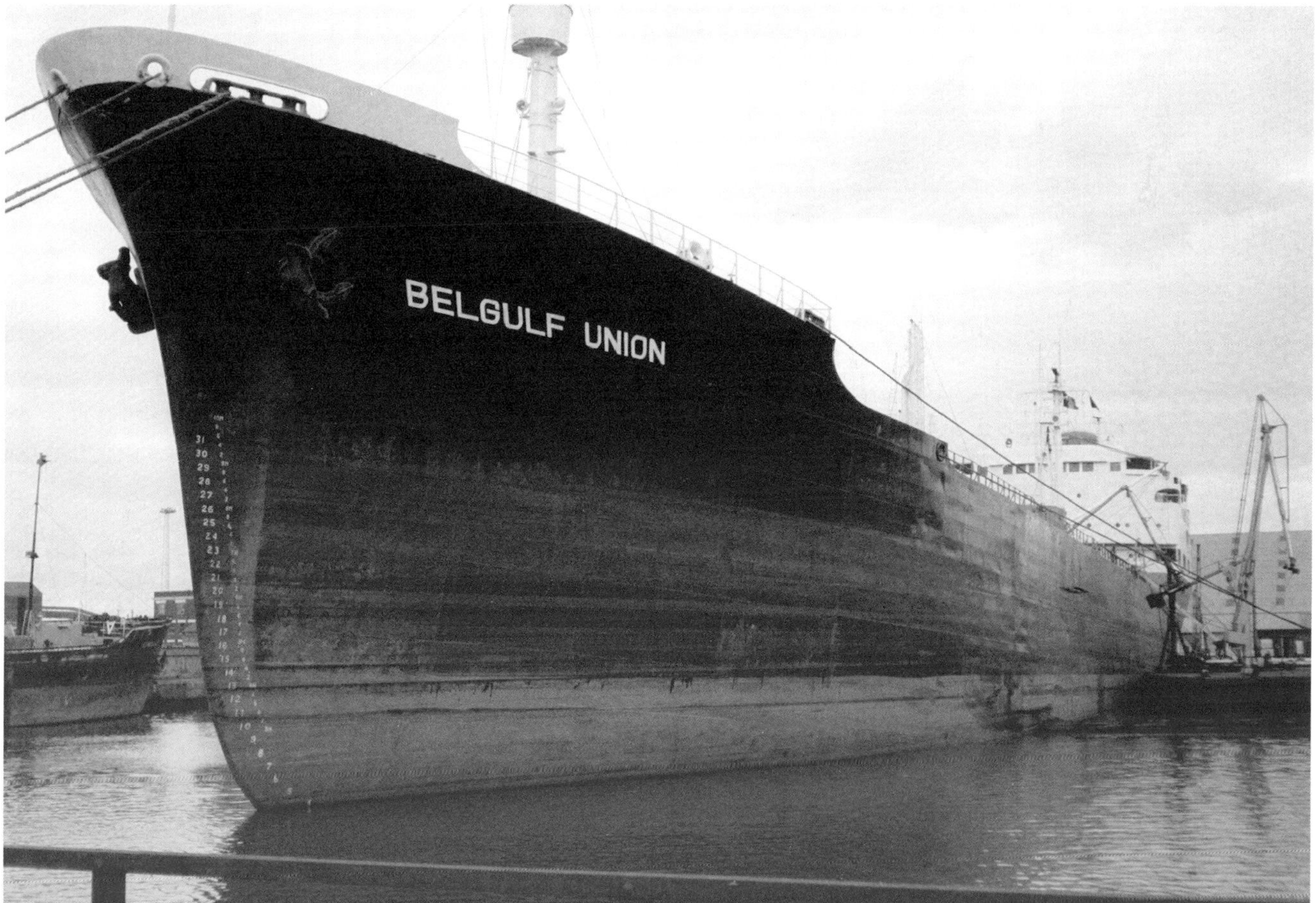
BELGULF UNION

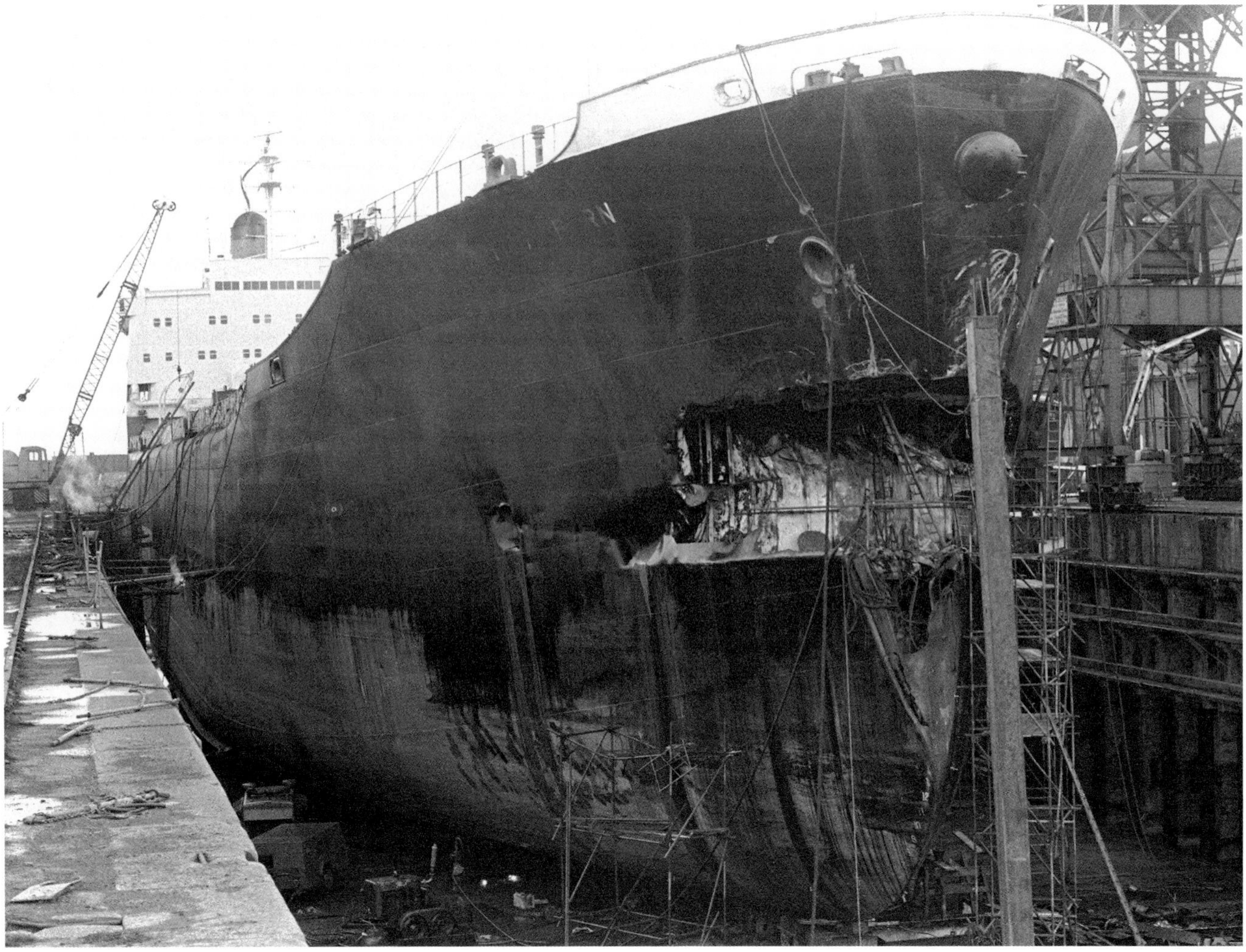

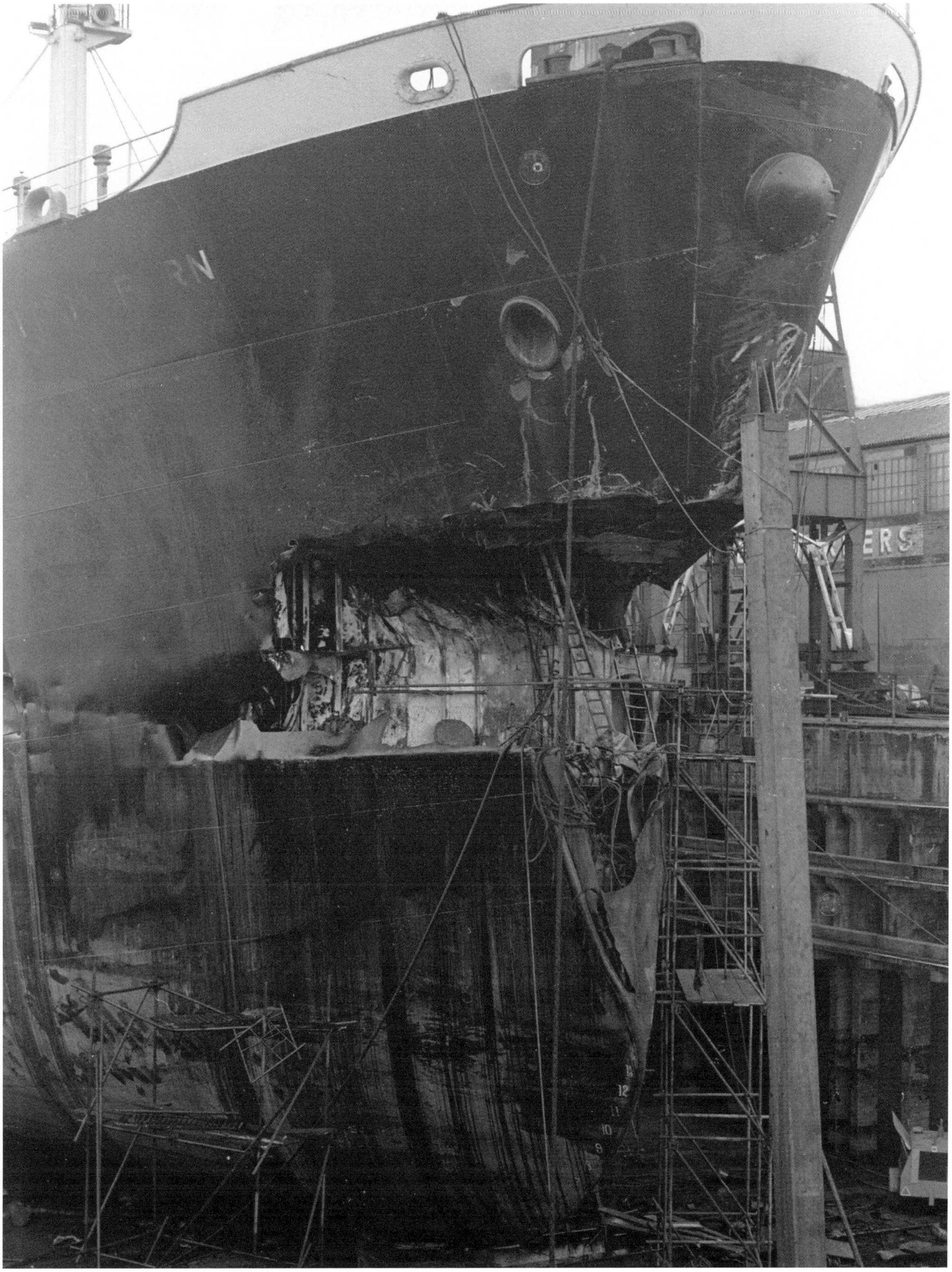

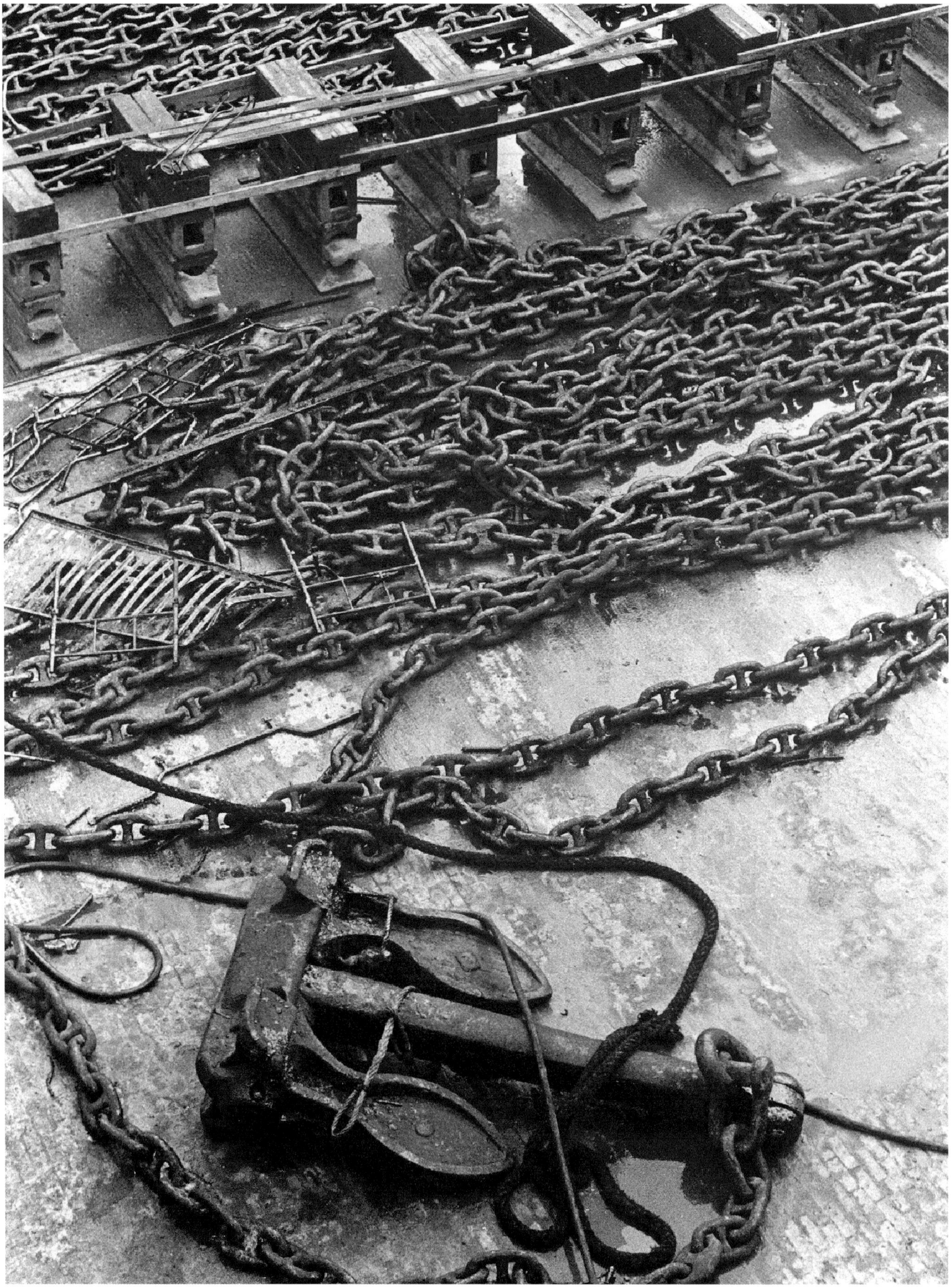

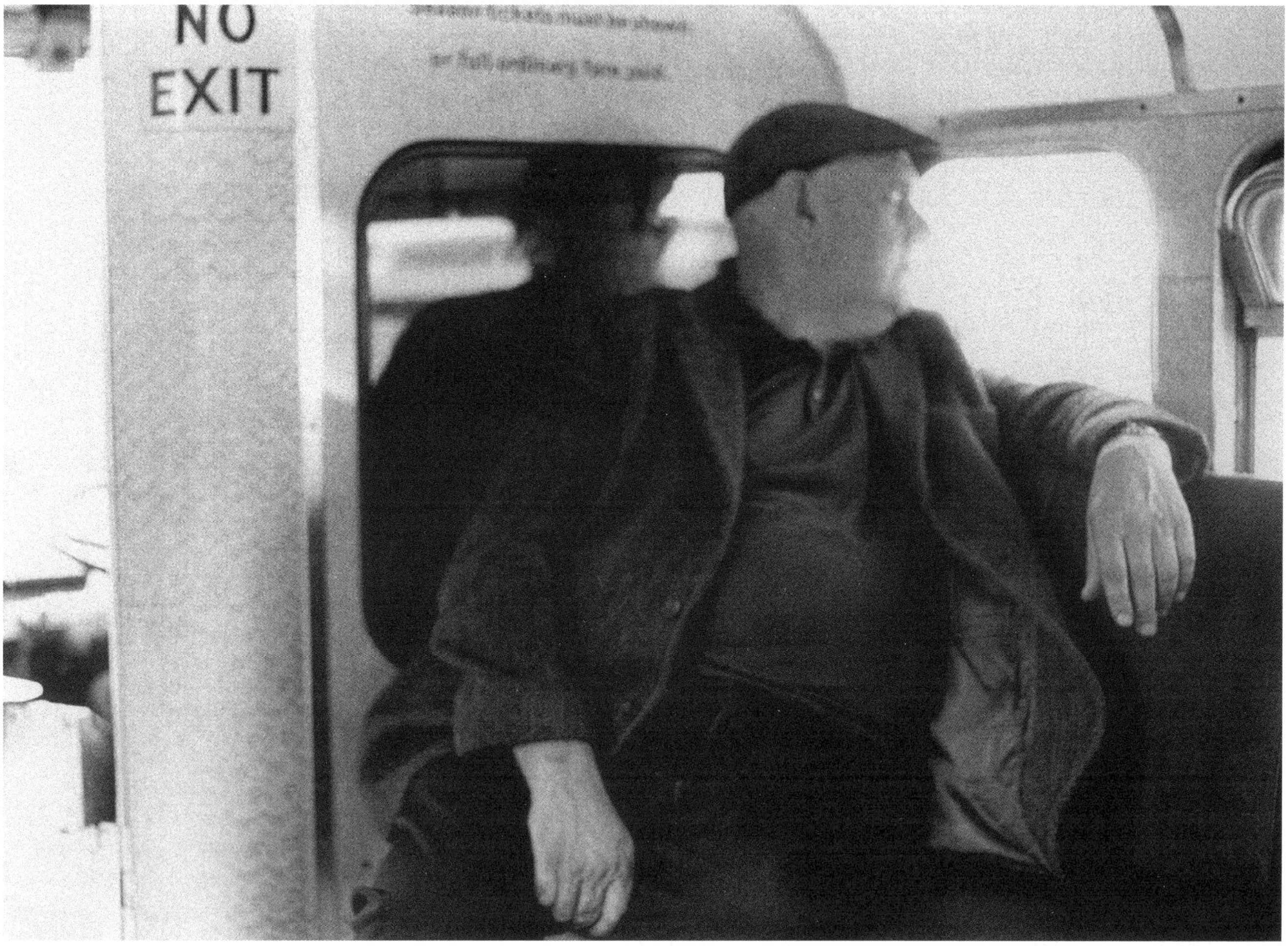
NO
EXIT

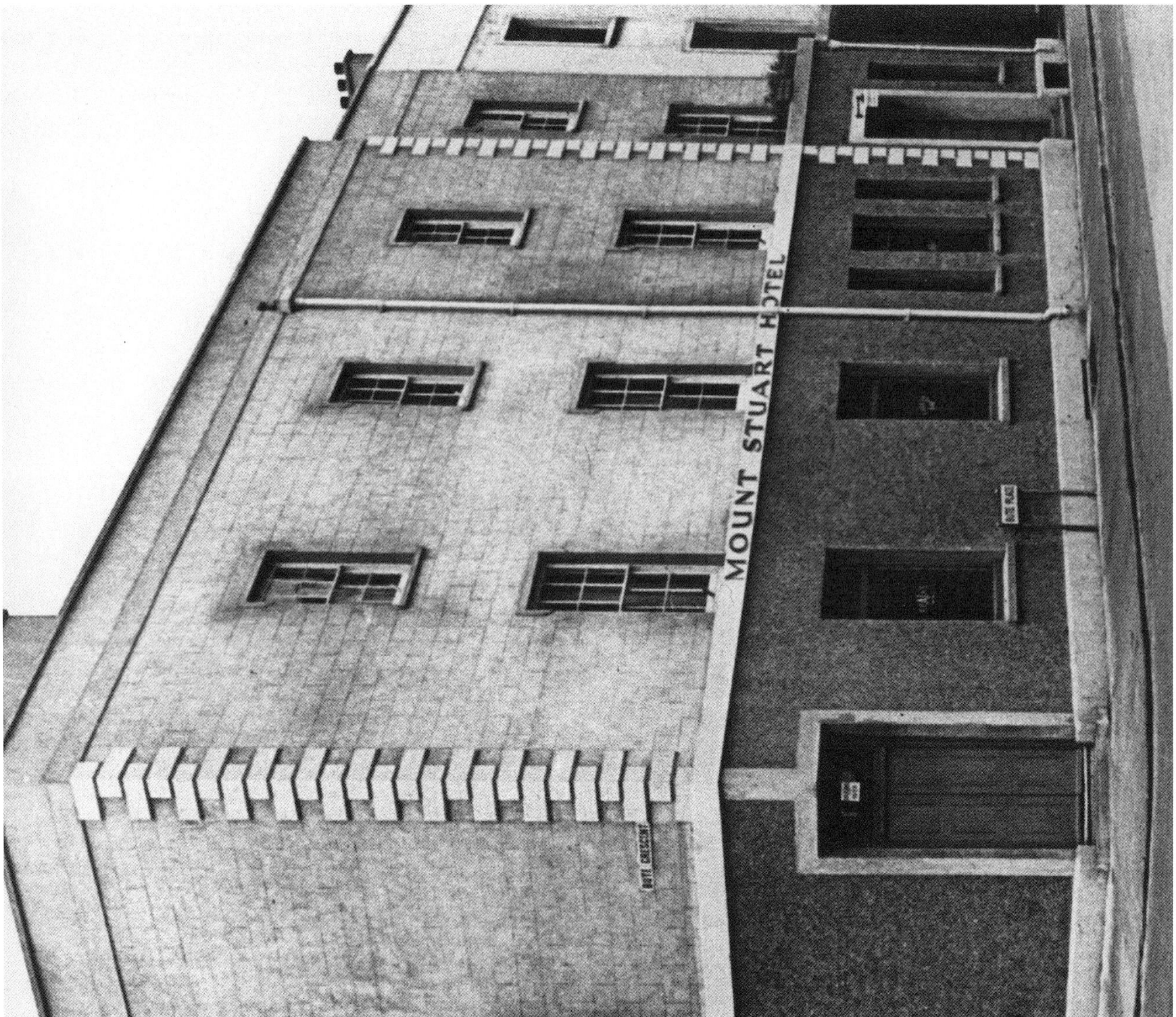
MOUNT STUART HOTEL
MOUNT STUART HOTEL

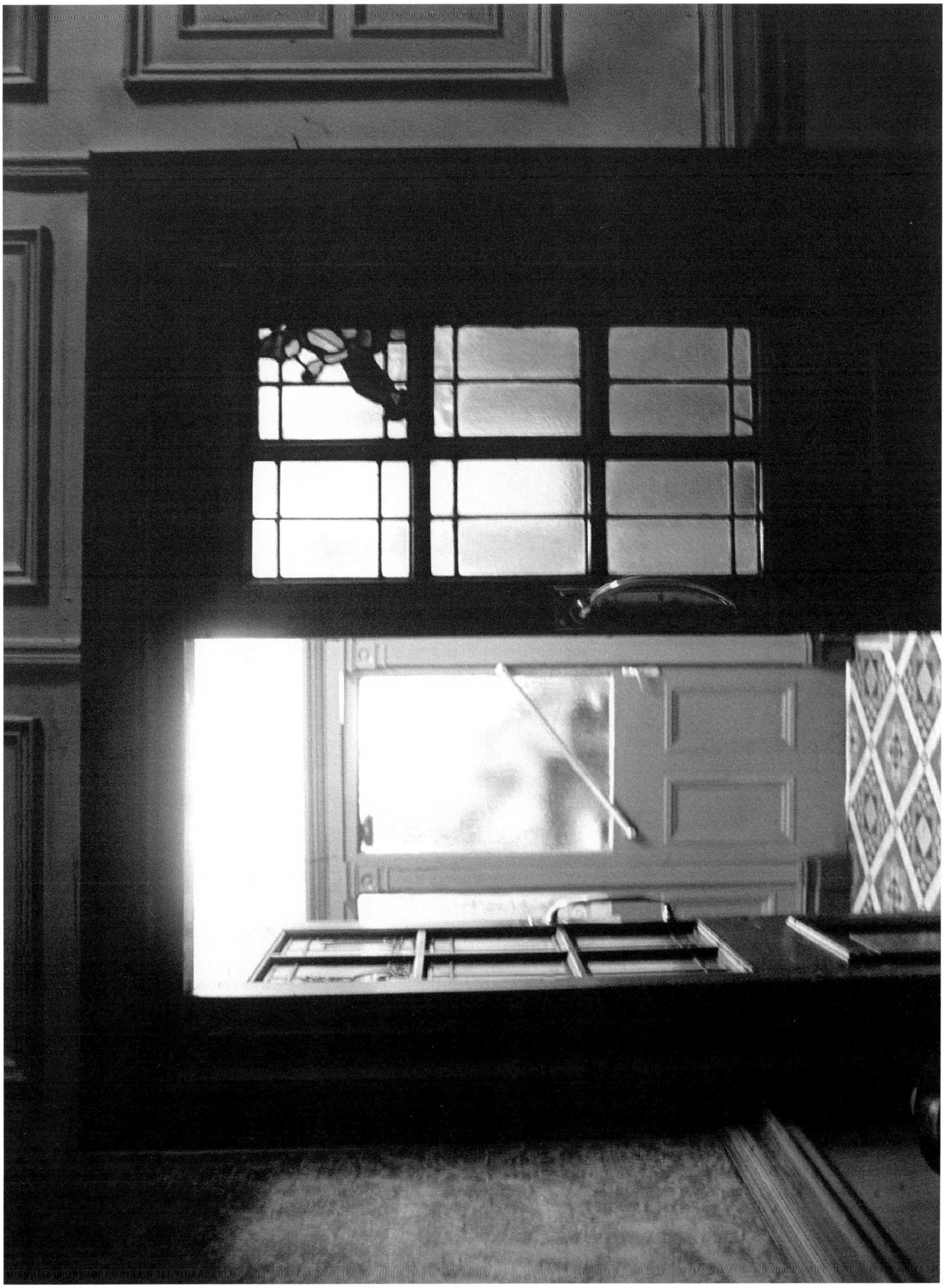

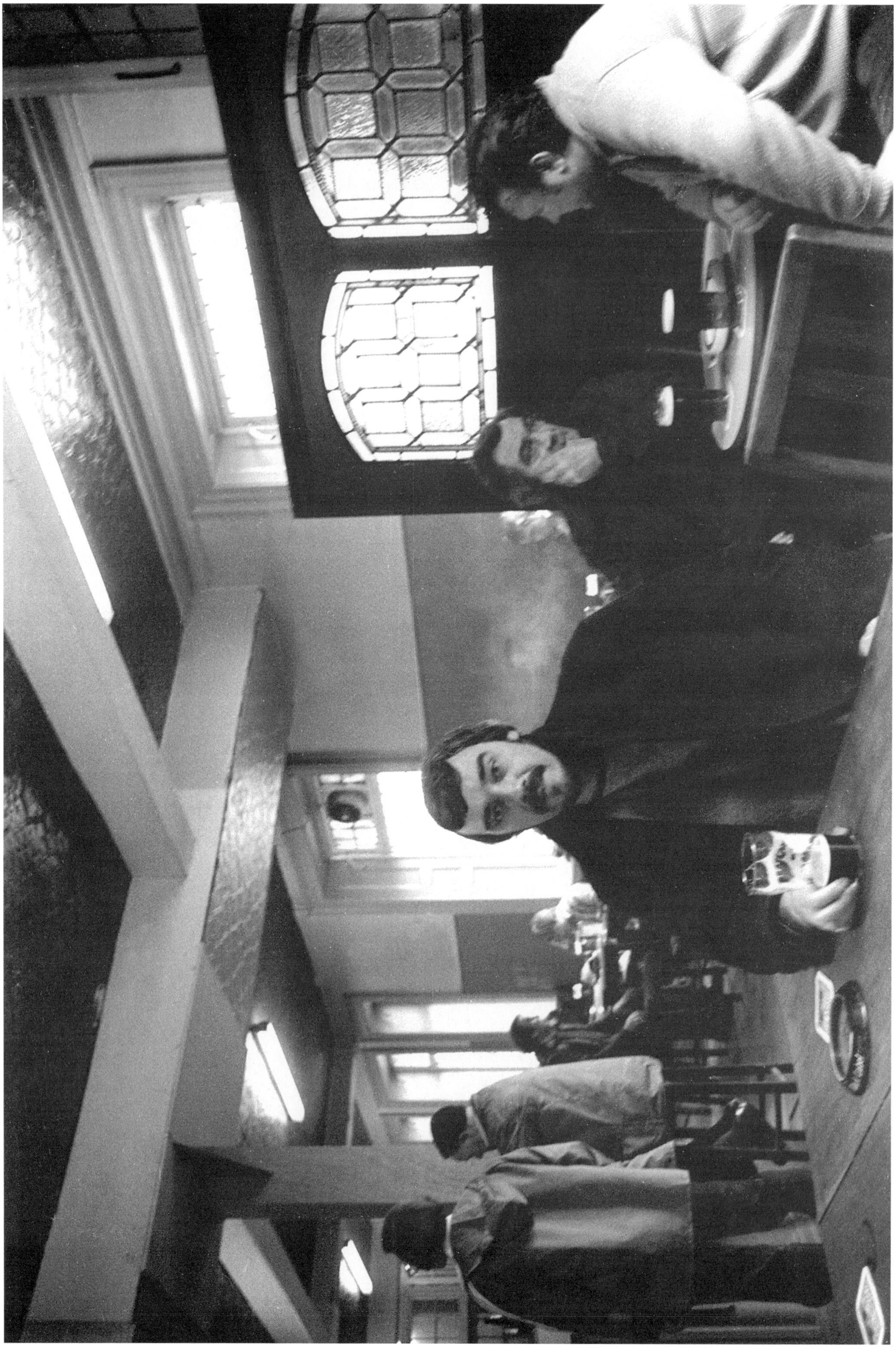

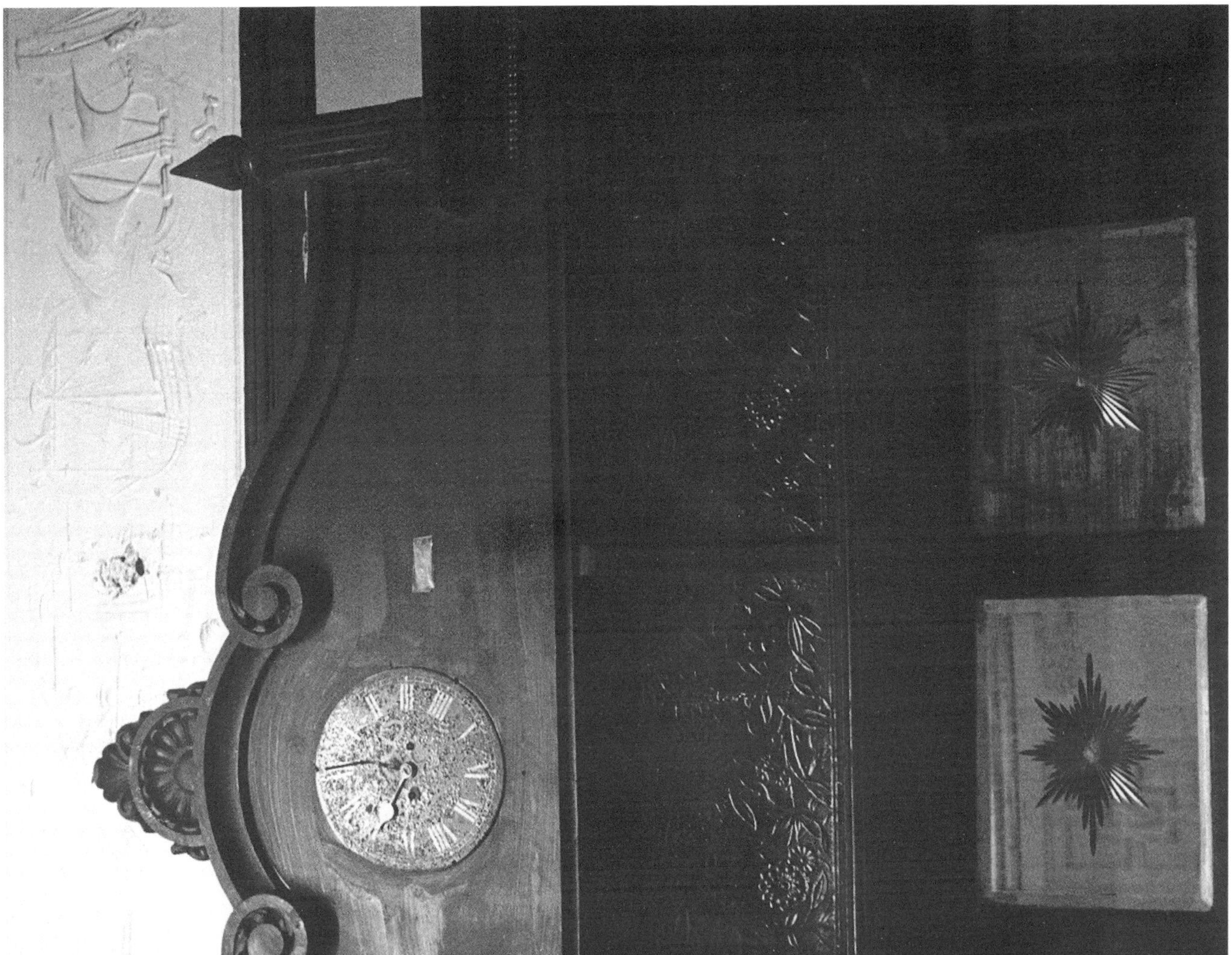

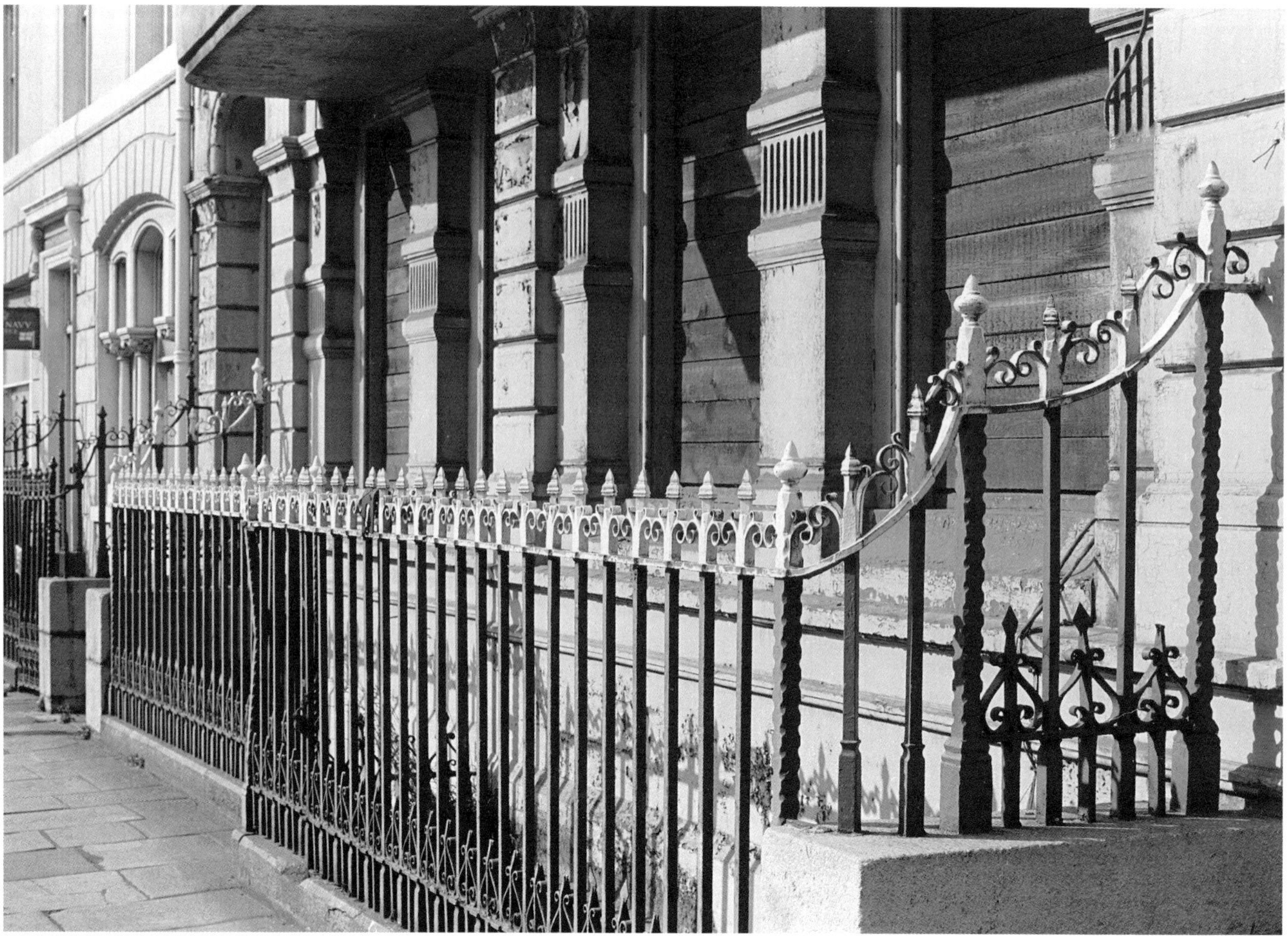

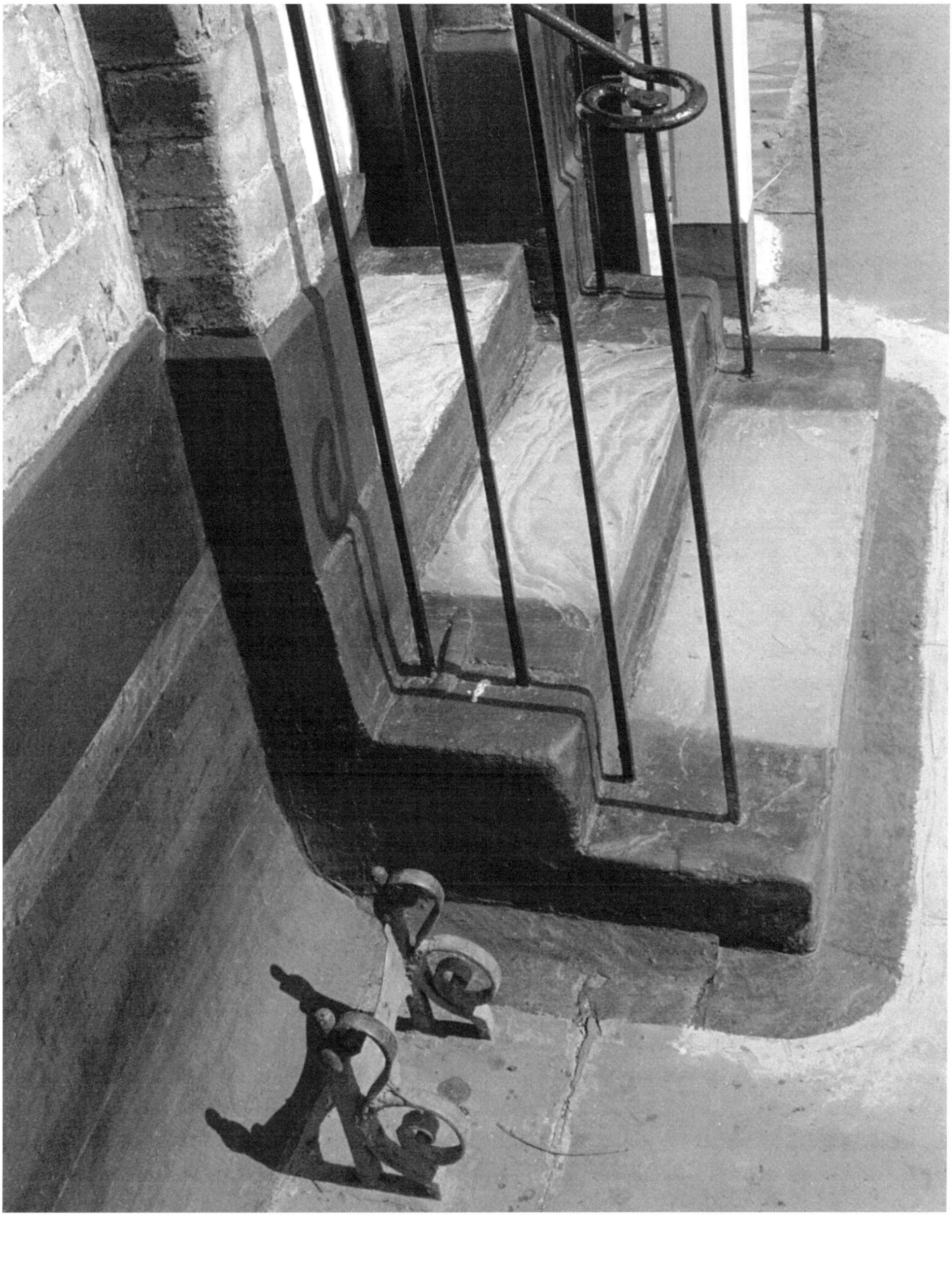

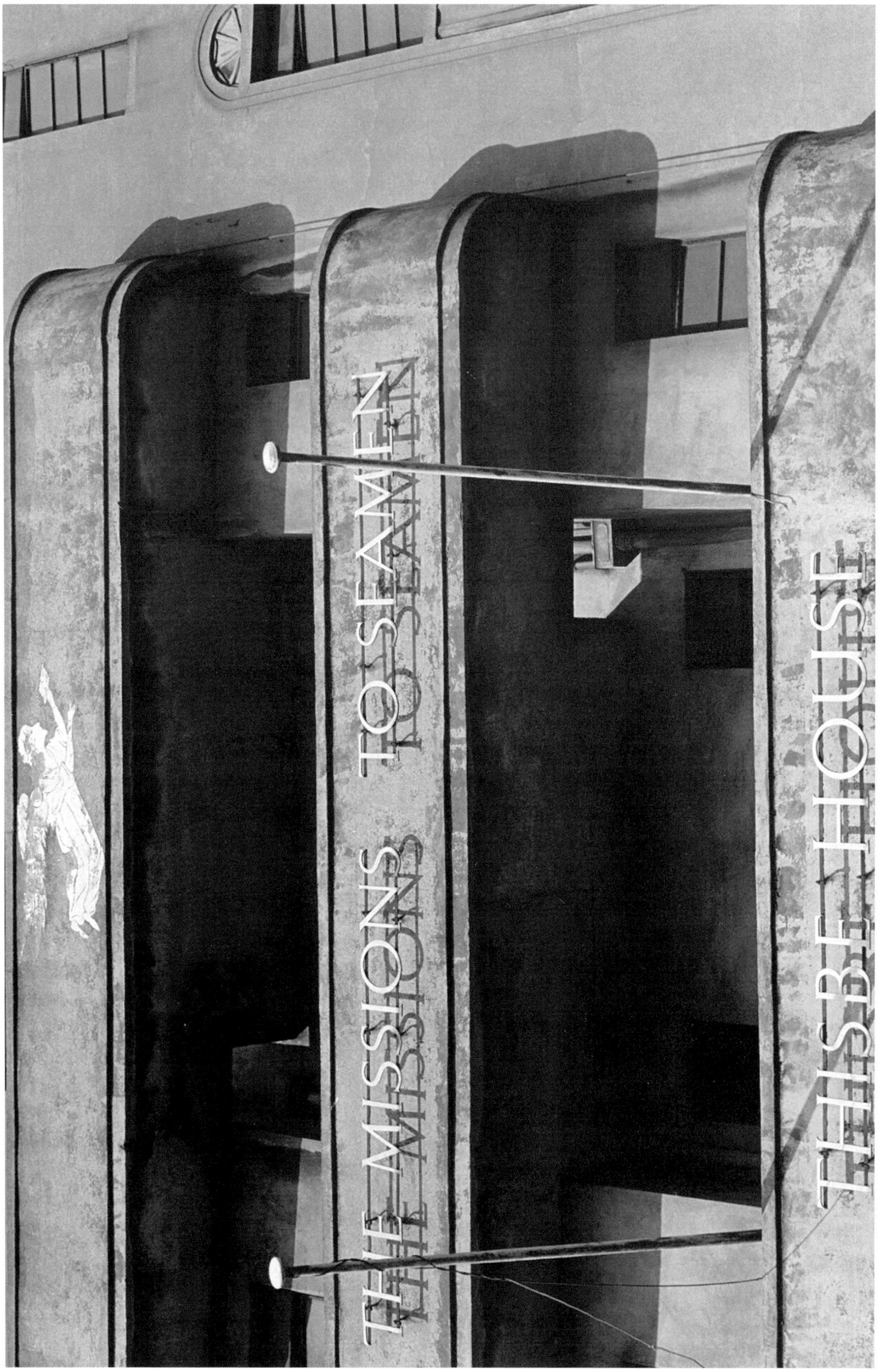
THE MISSIONS TO SEAMEN
THISBE HOUSE

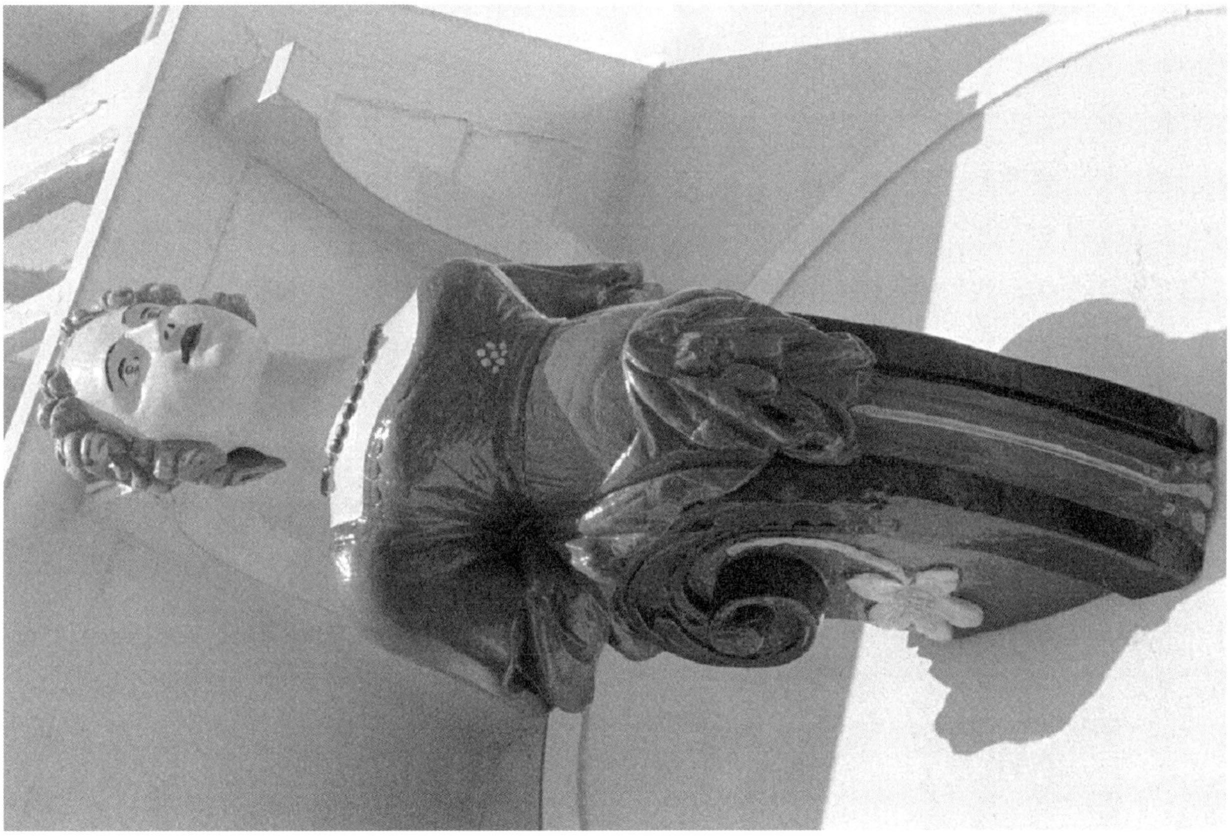

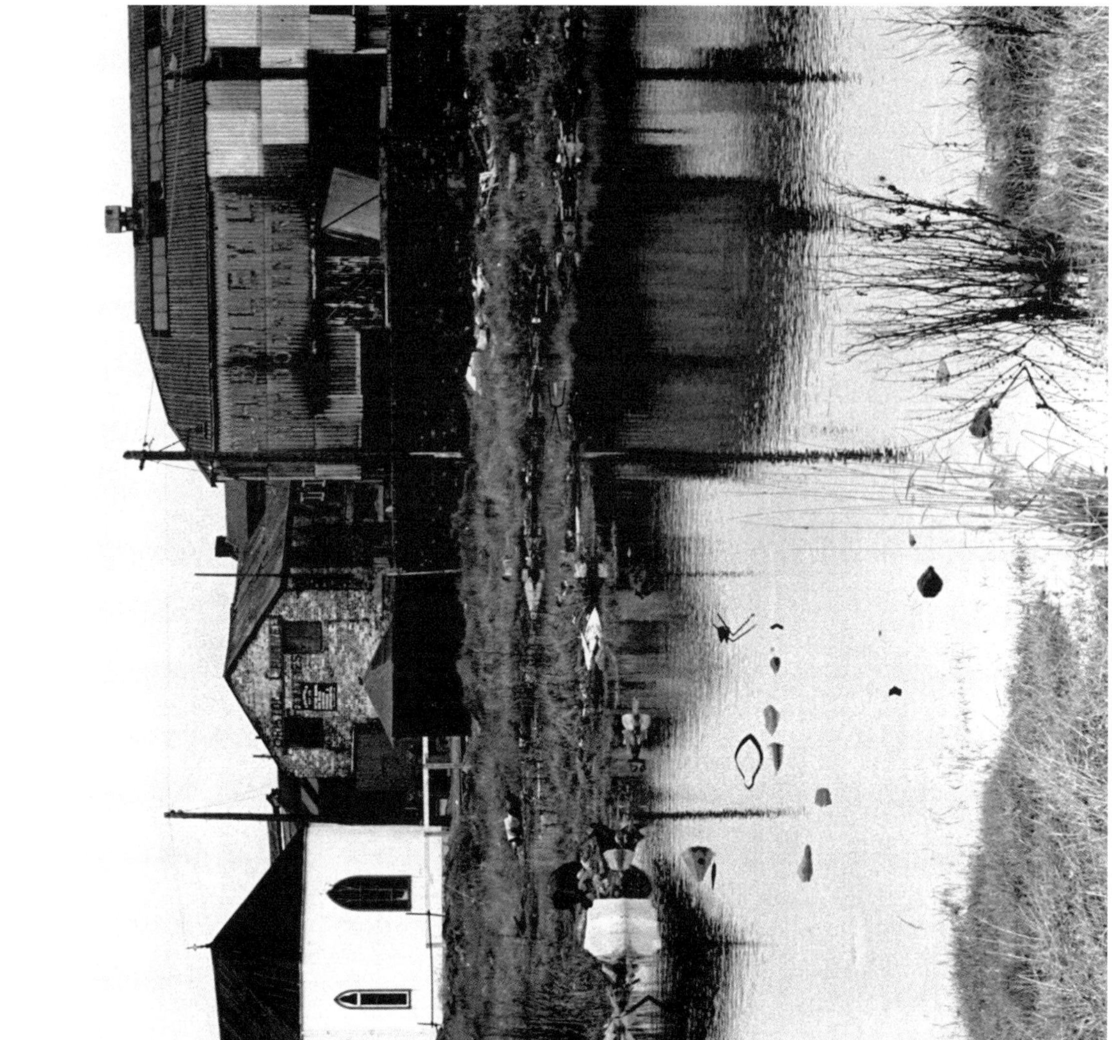
EXPRESS
STEELS

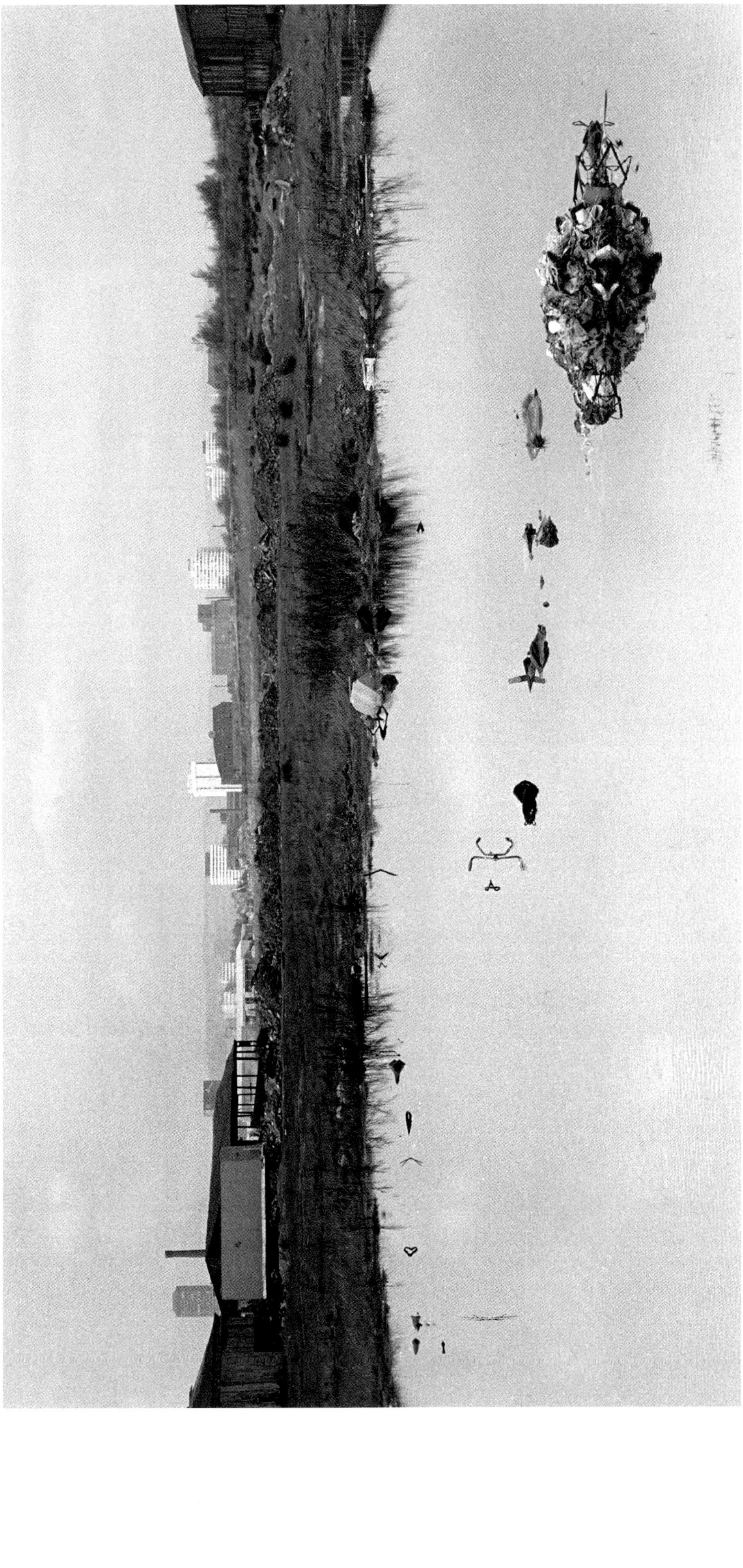

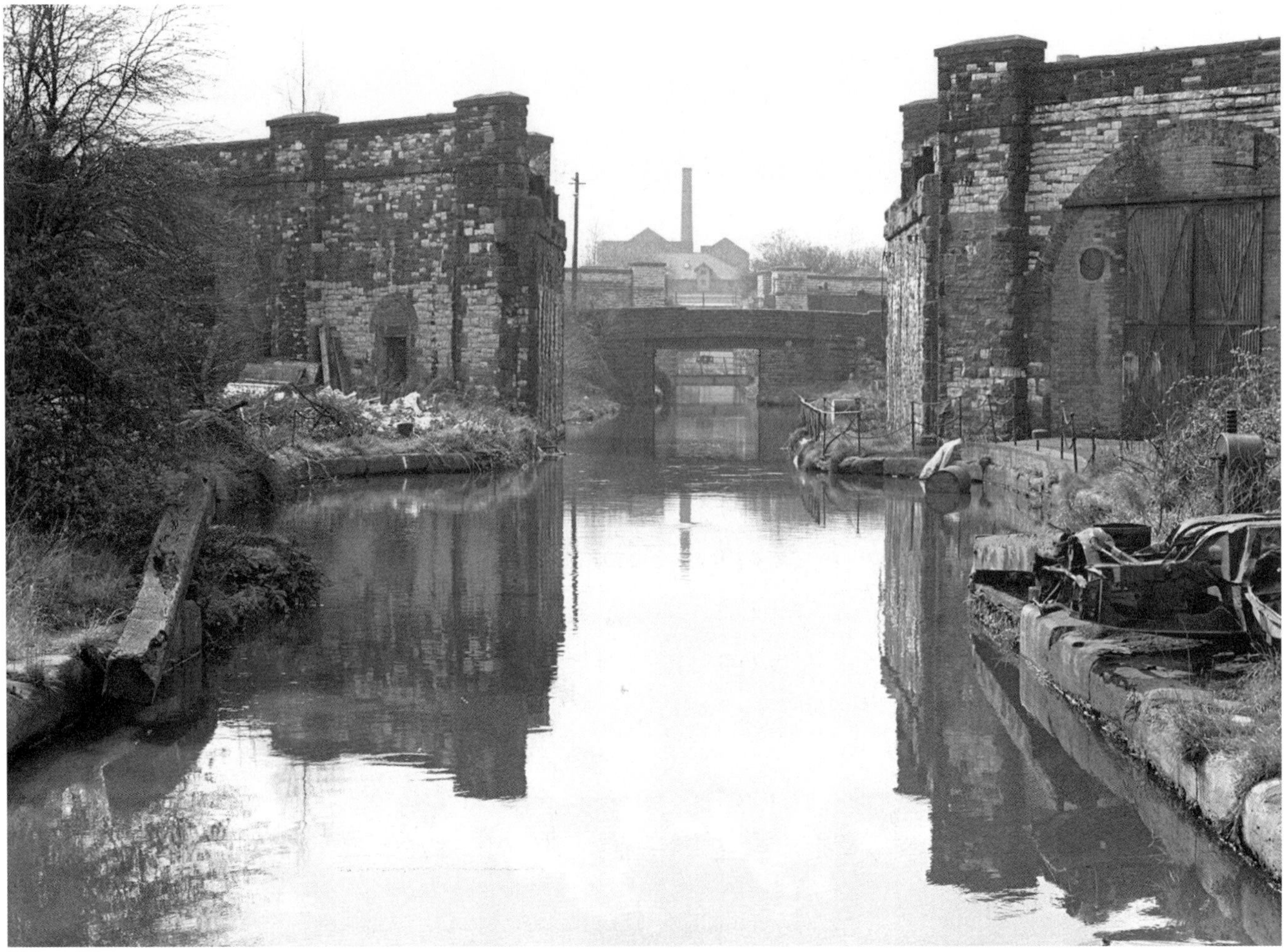

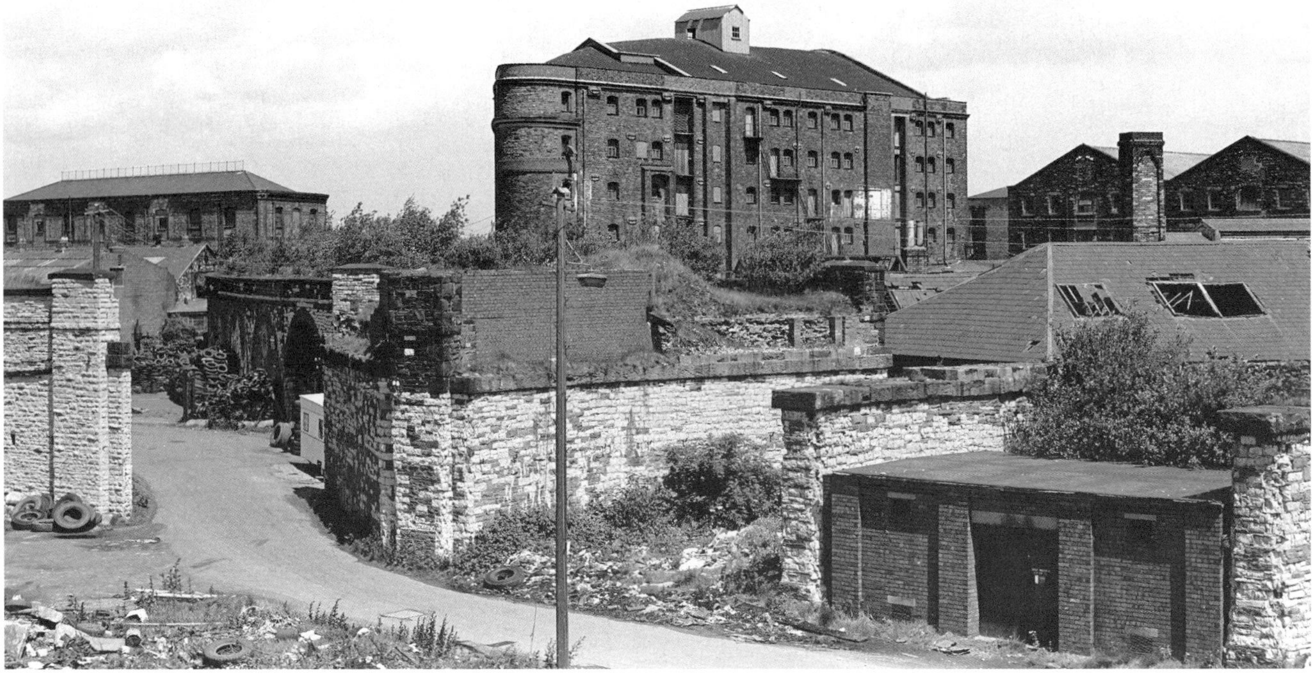

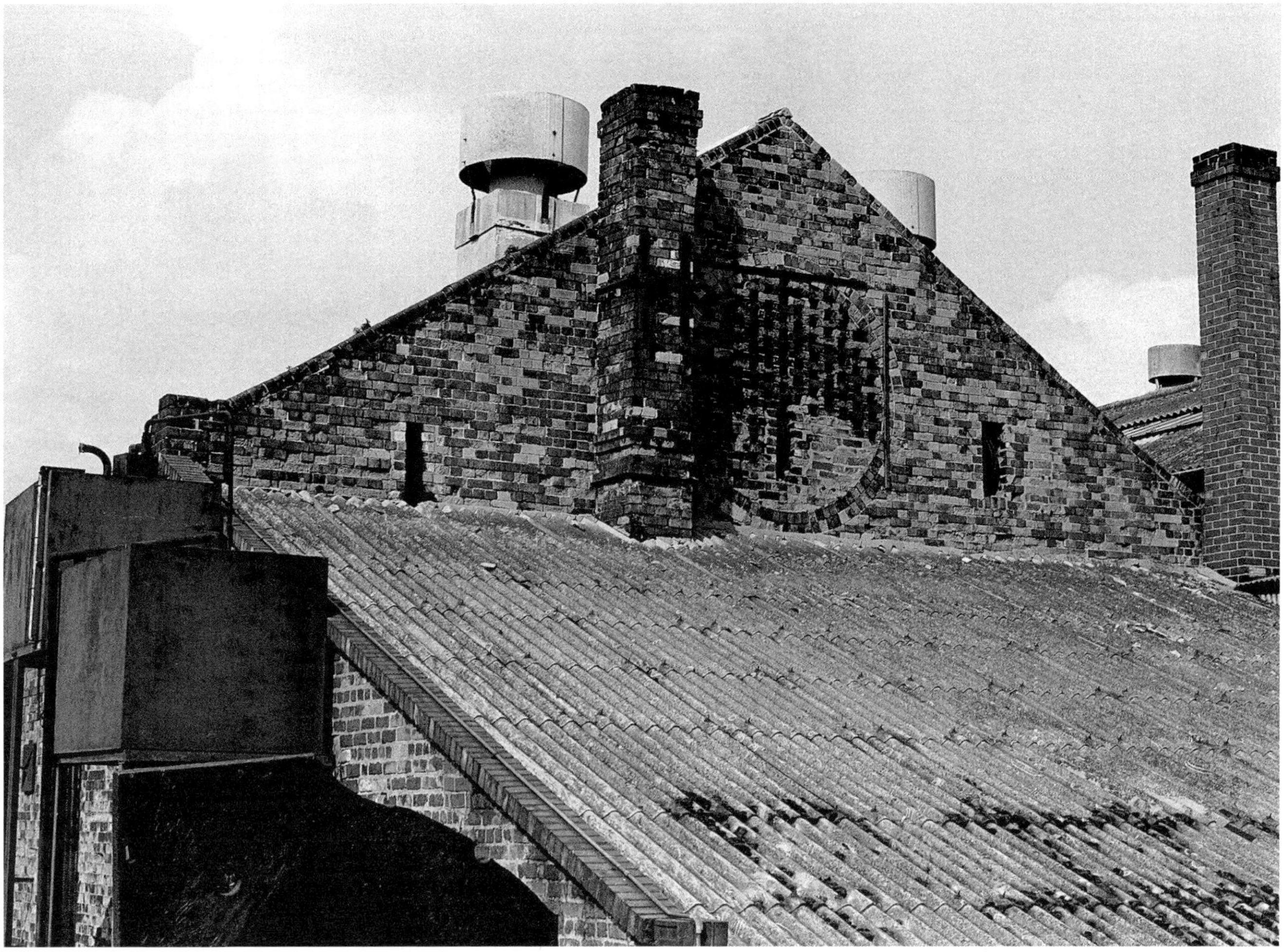

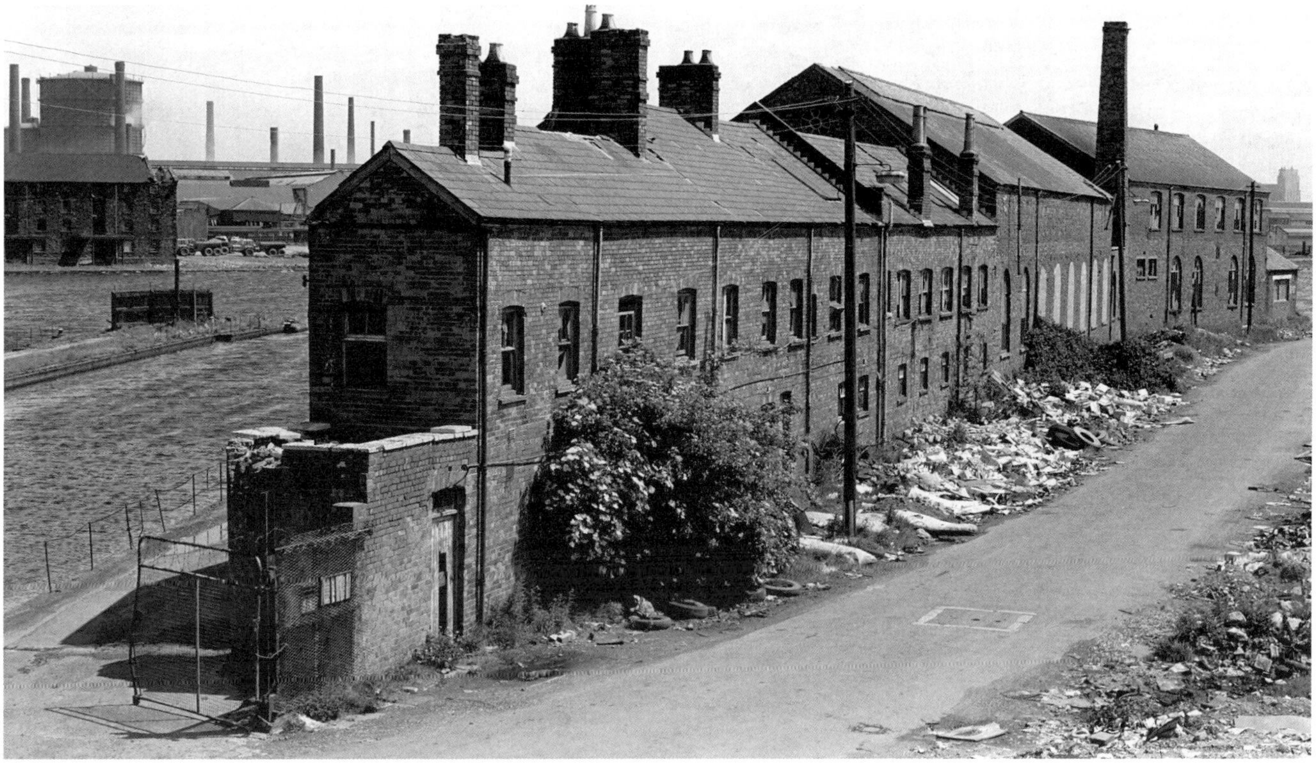

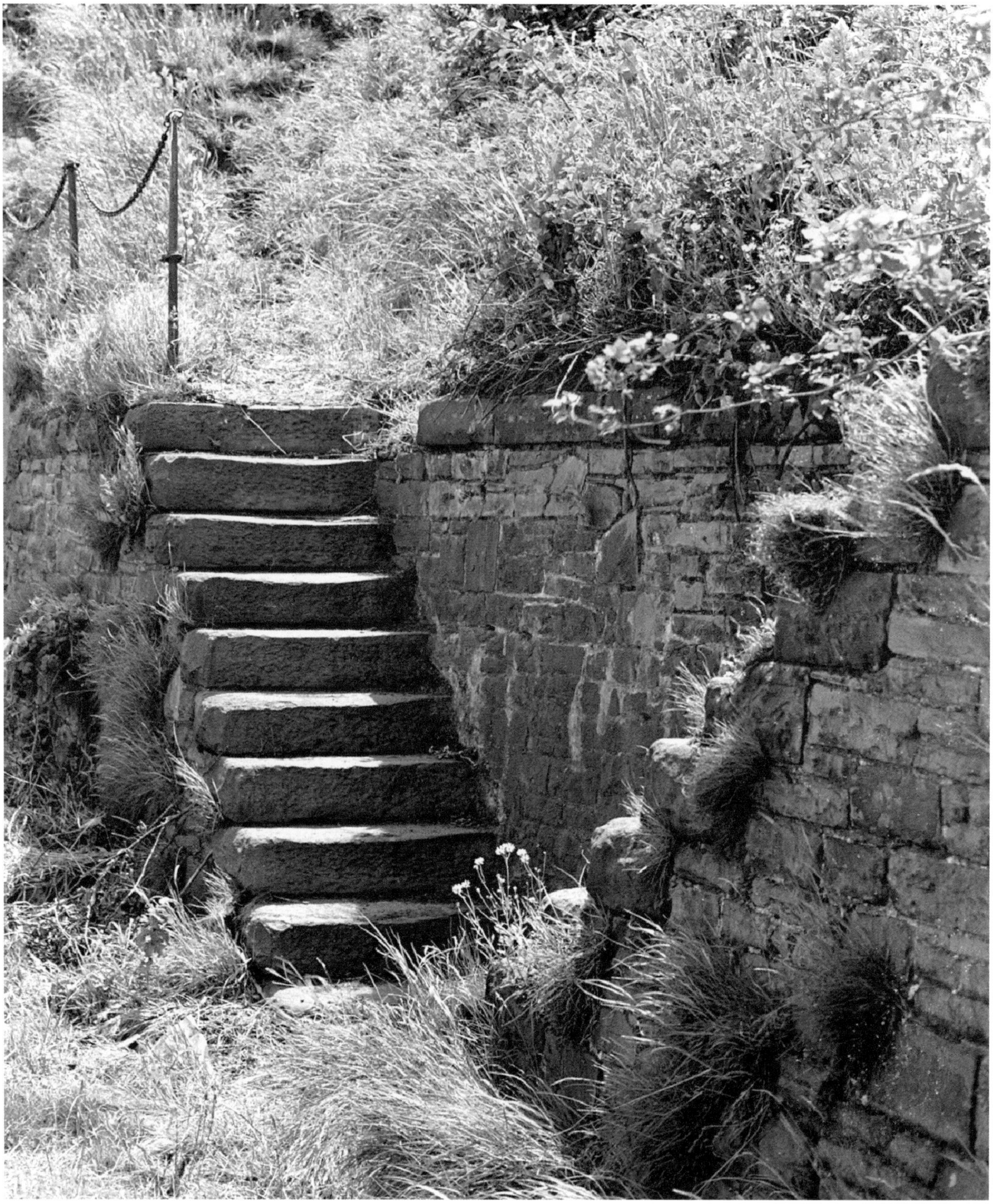

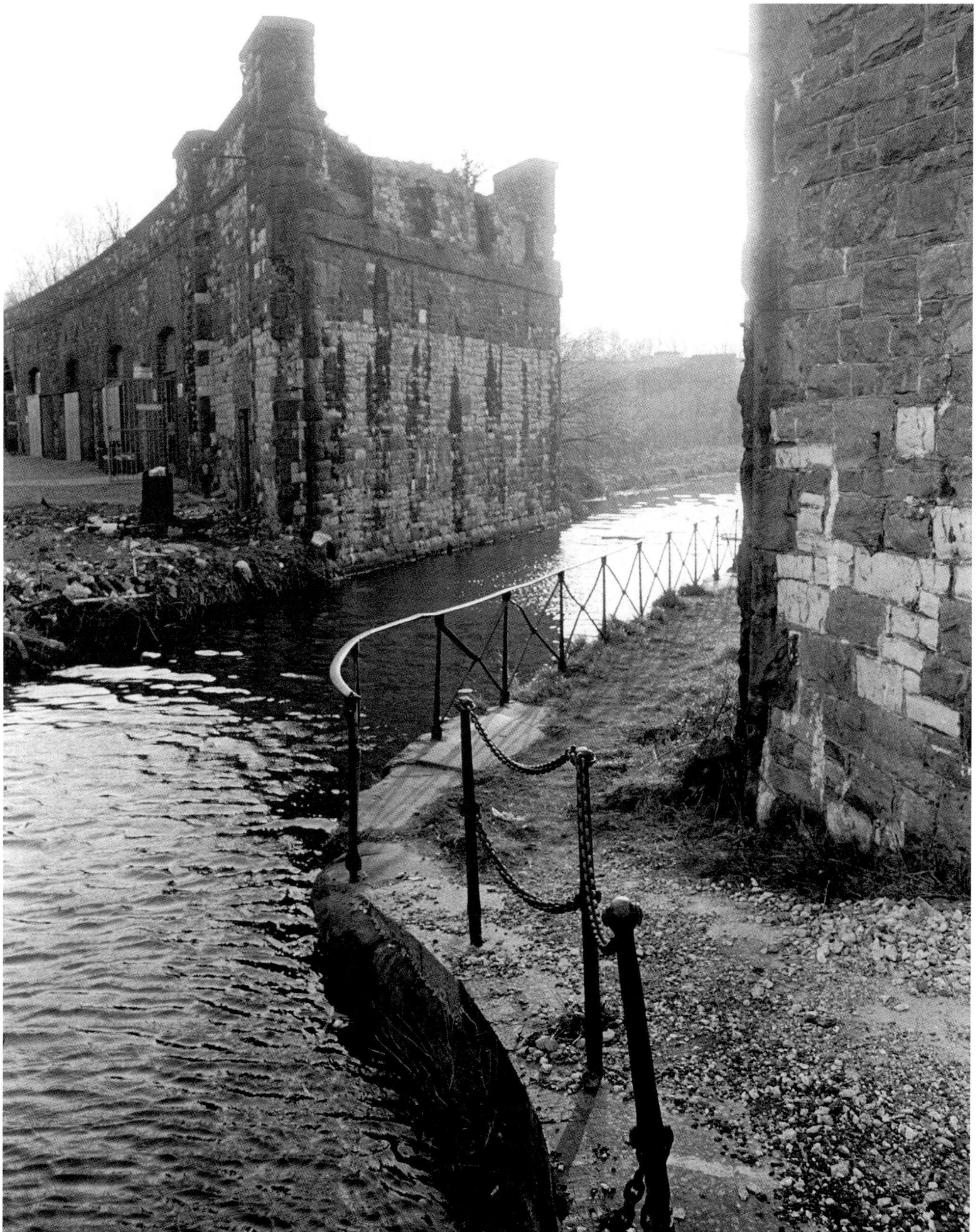

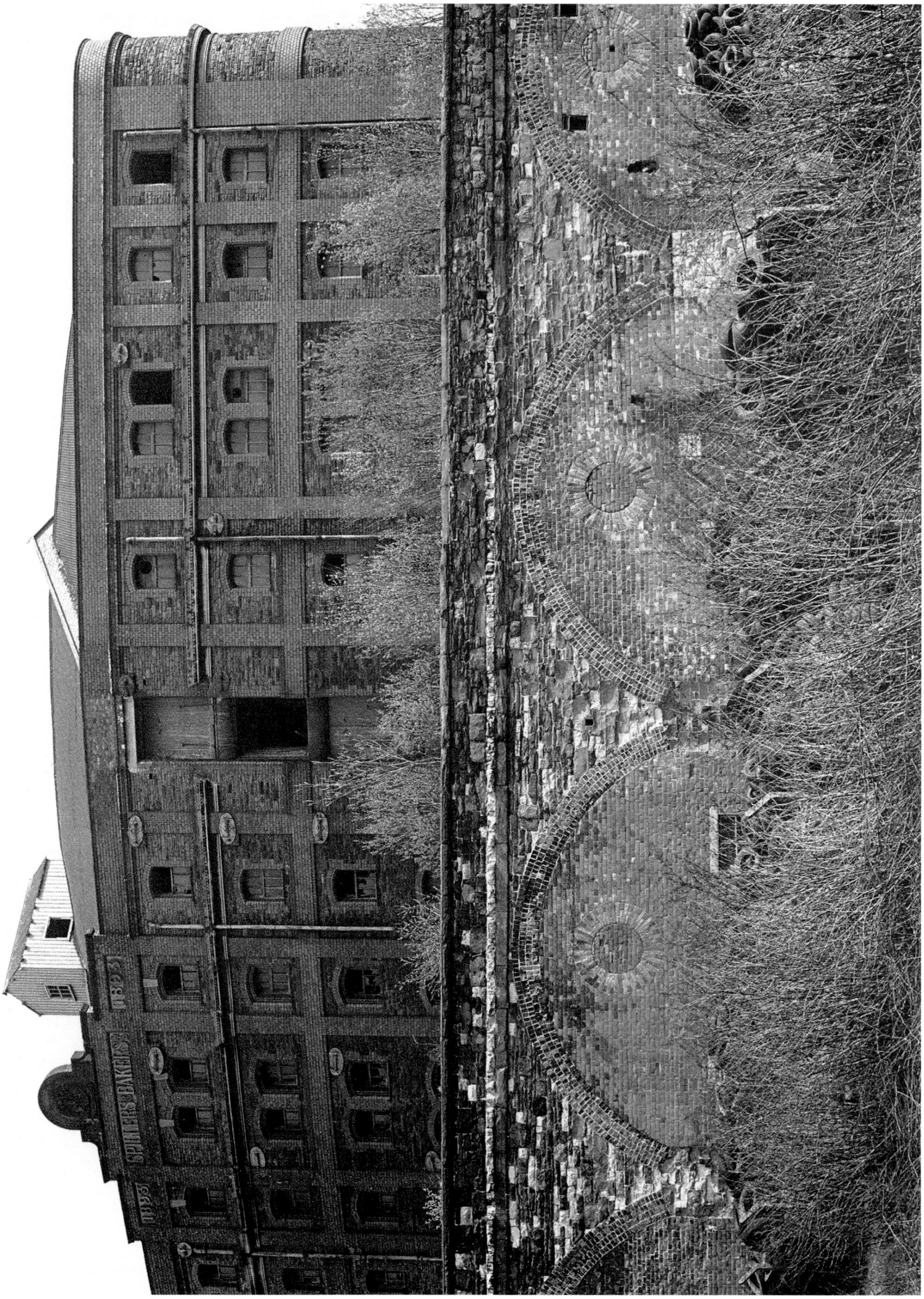

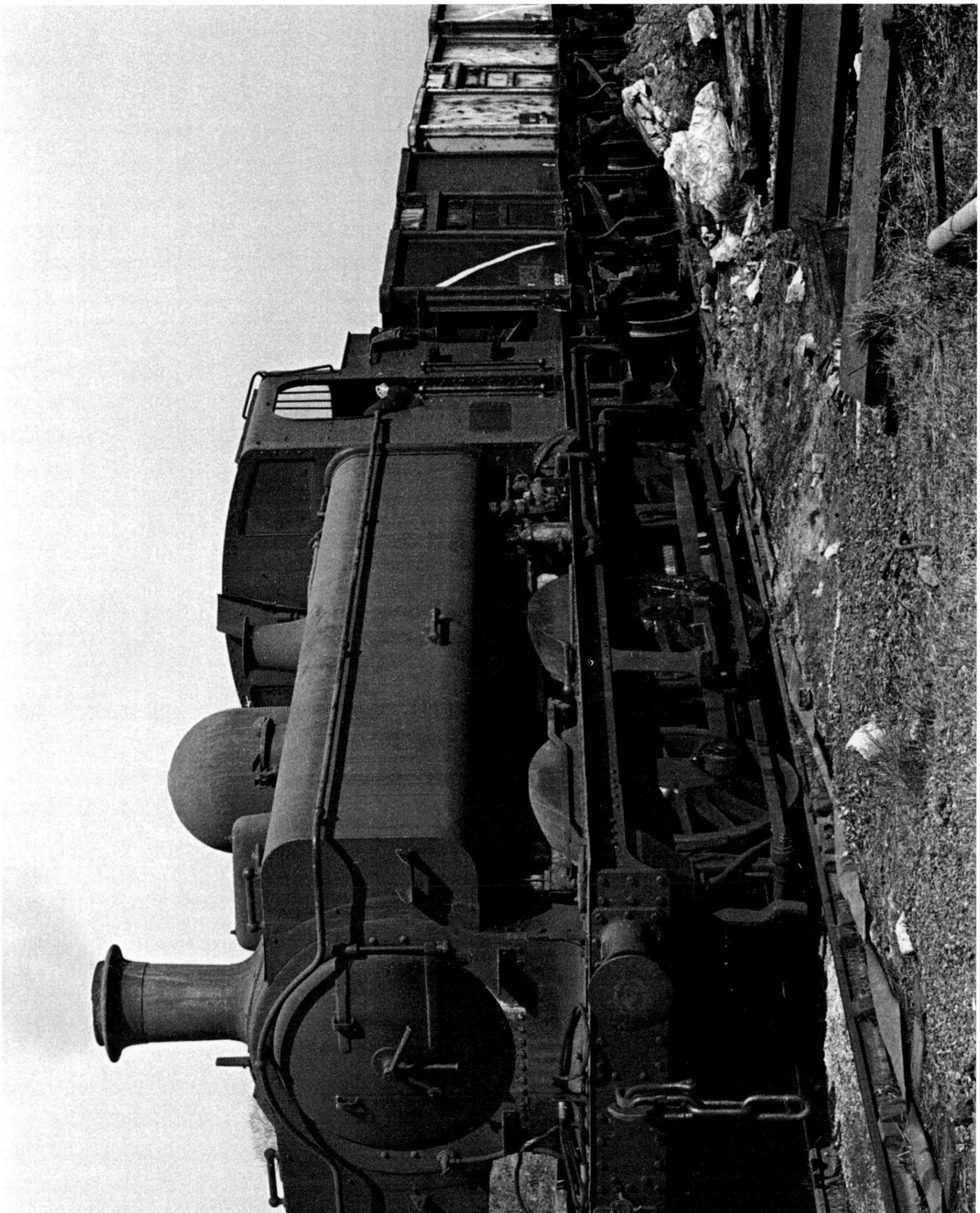

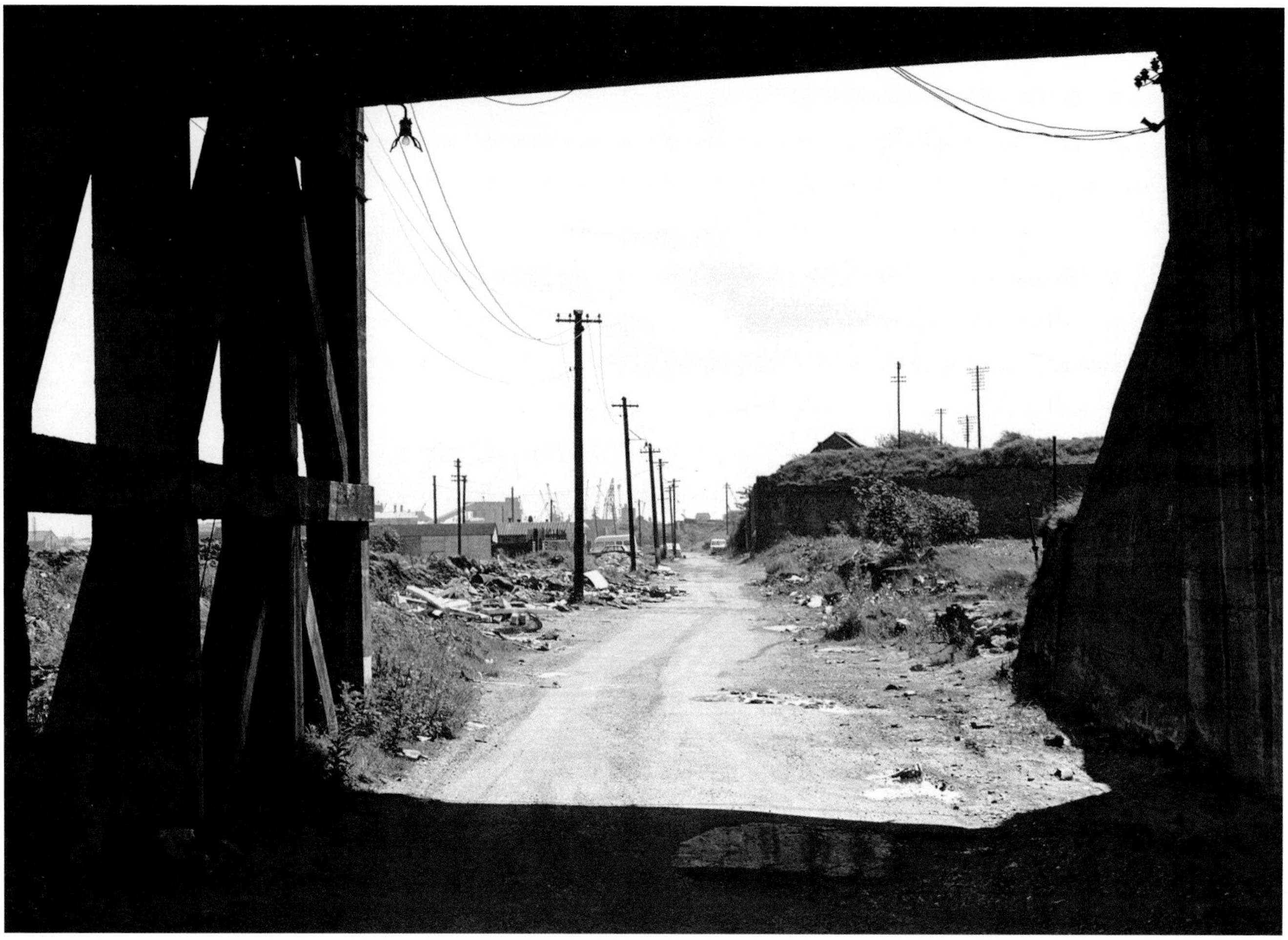

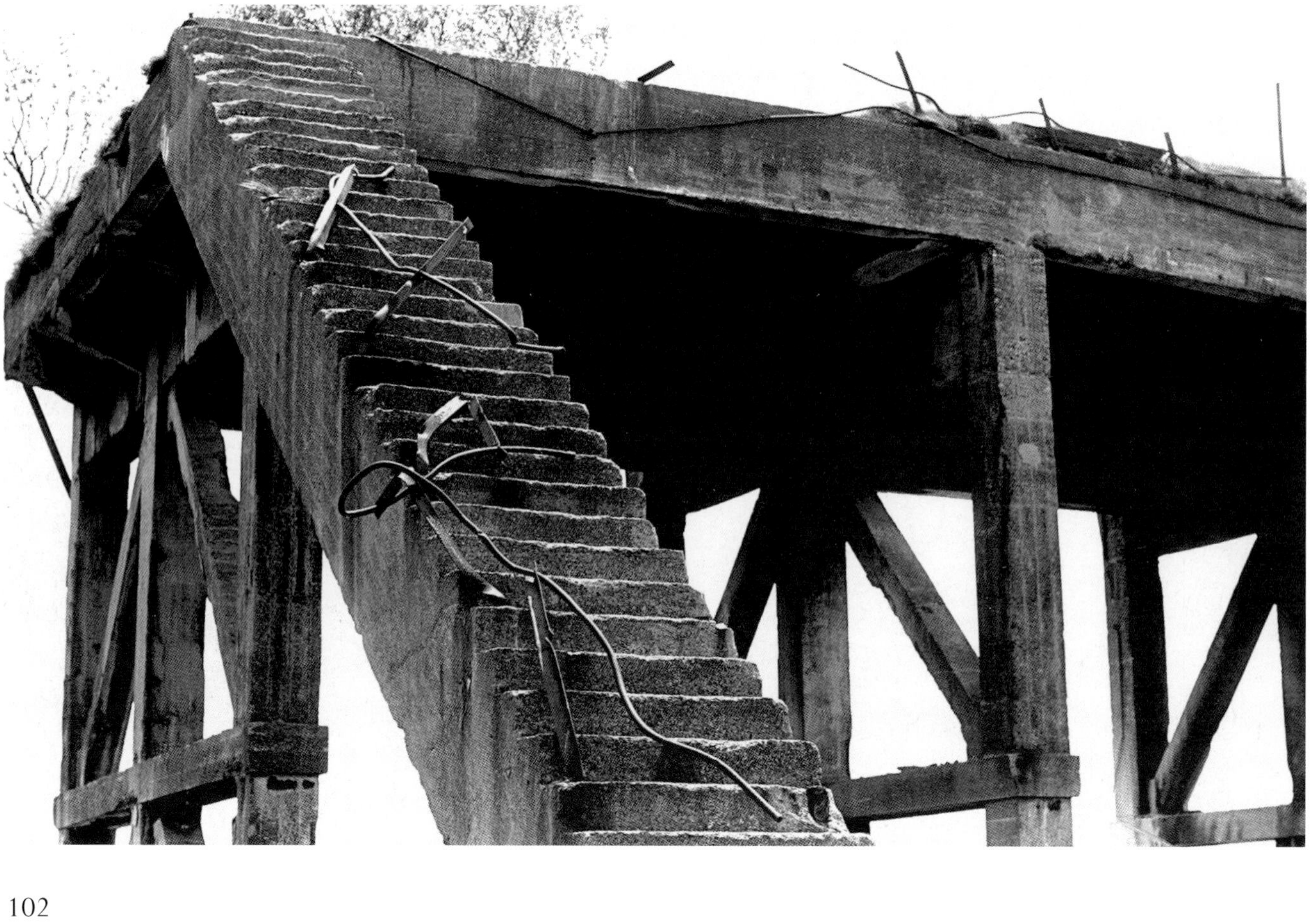

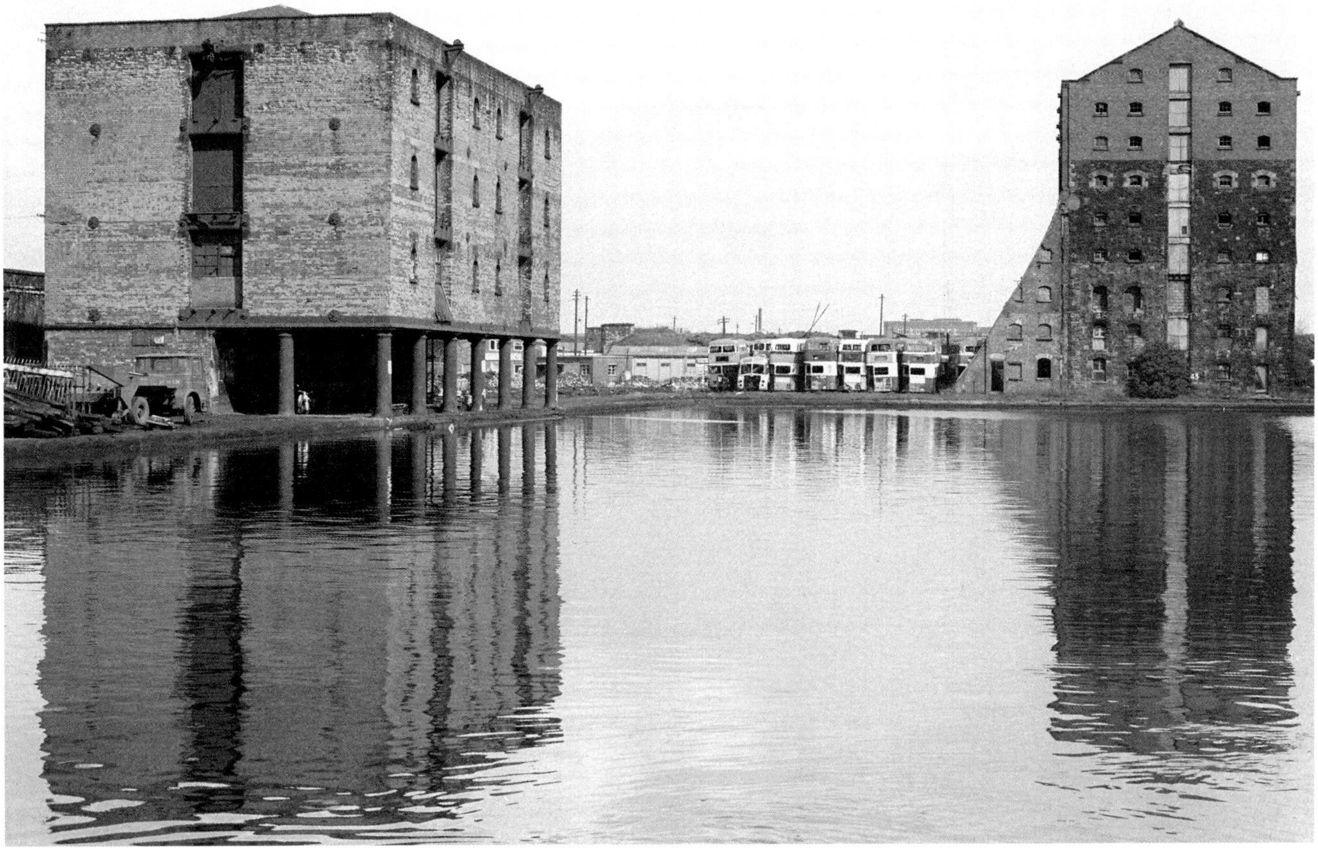

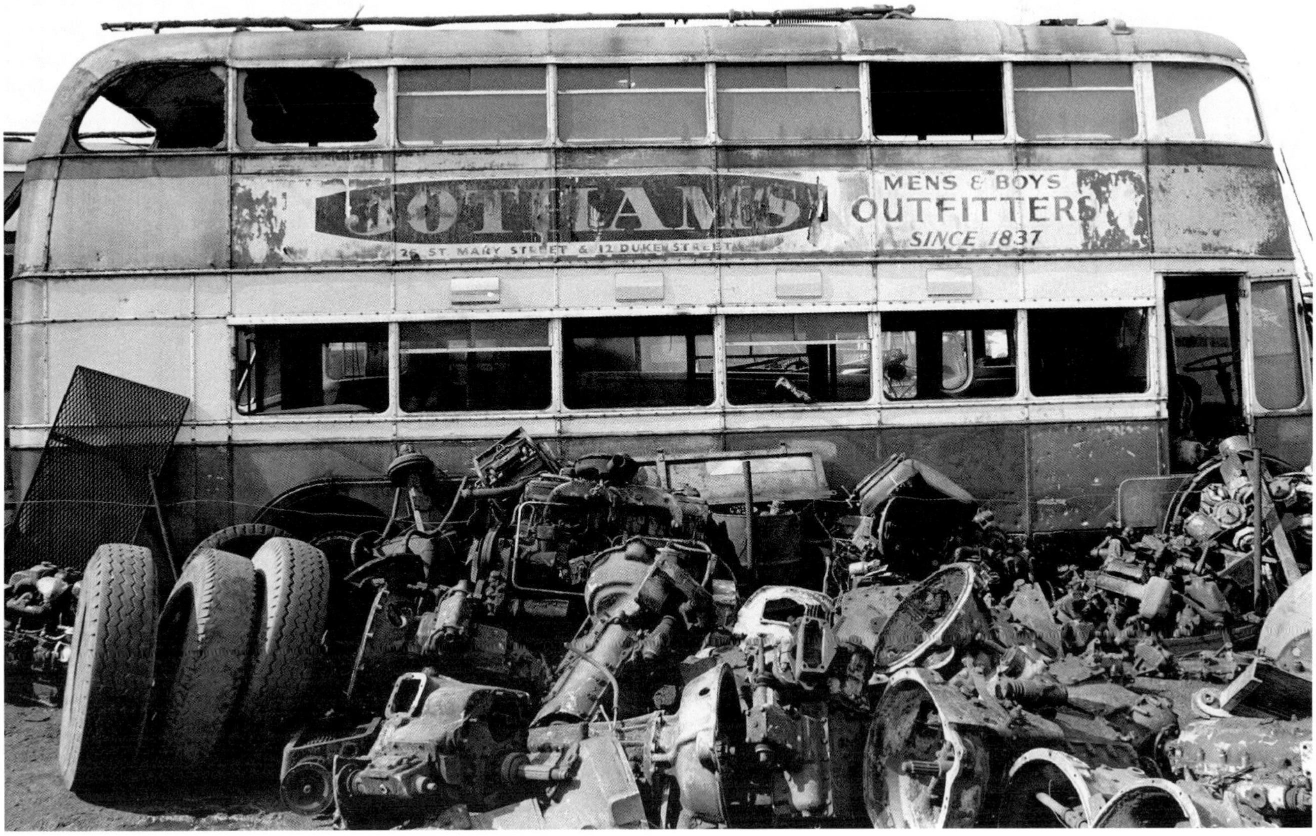
MENS & BOYS
OUTFITTERS
SINCE 1837

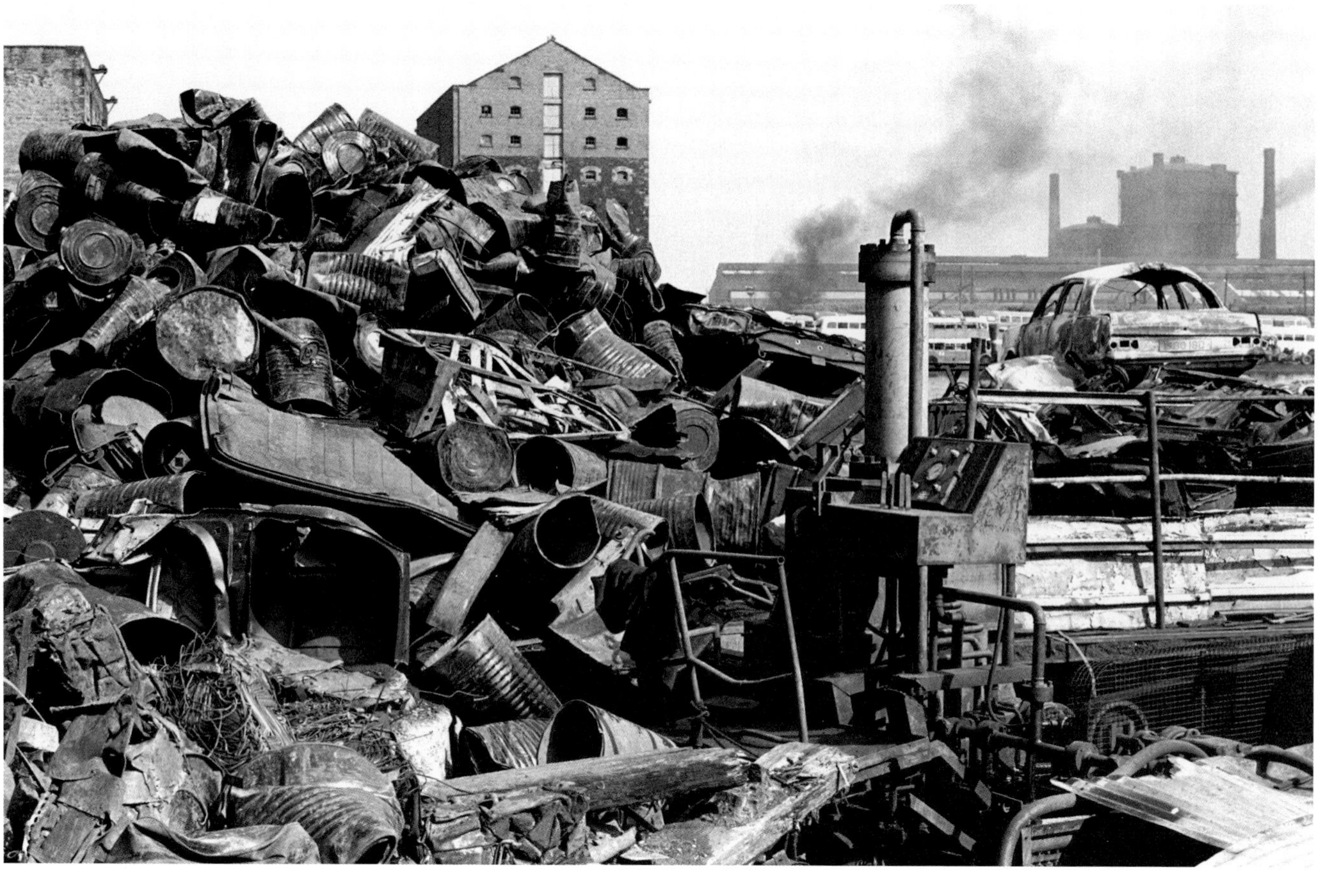

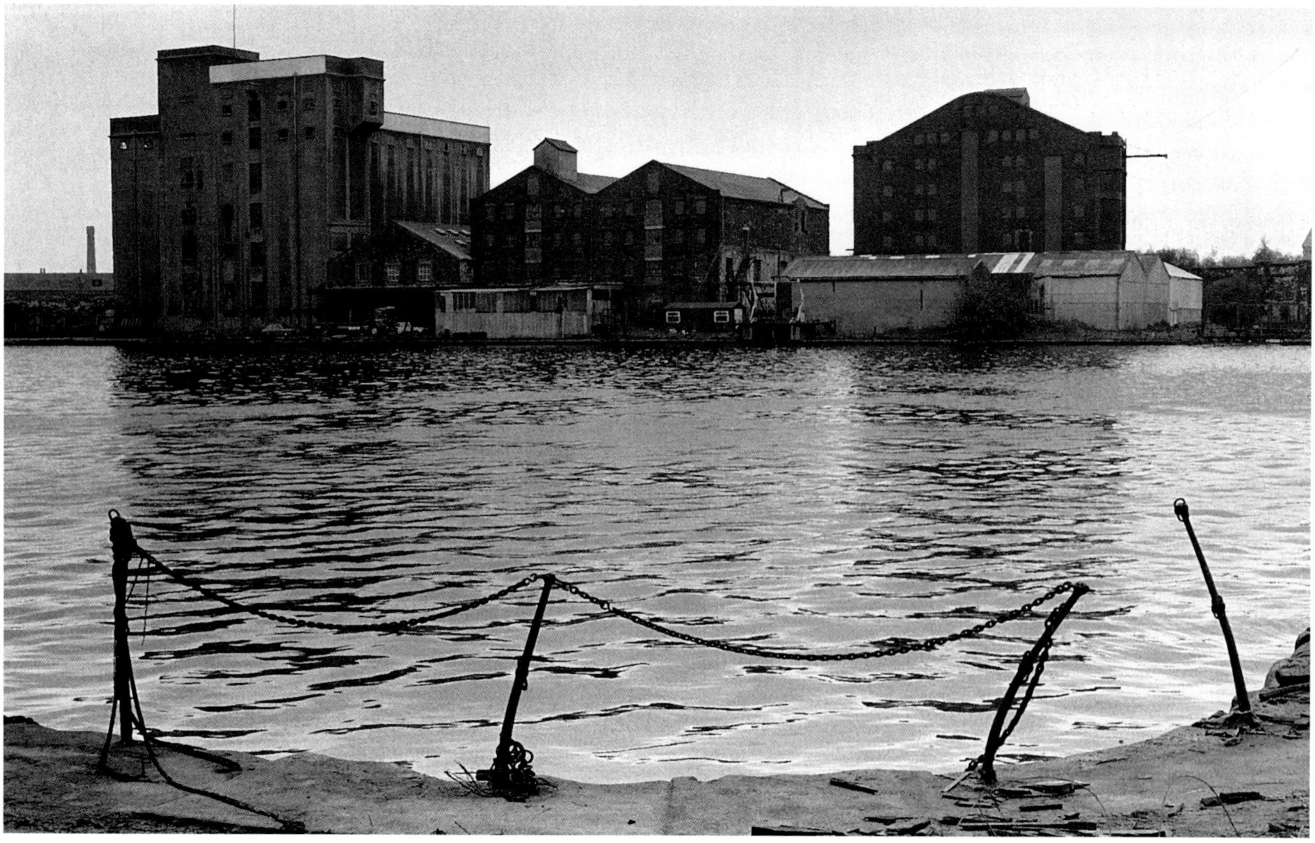

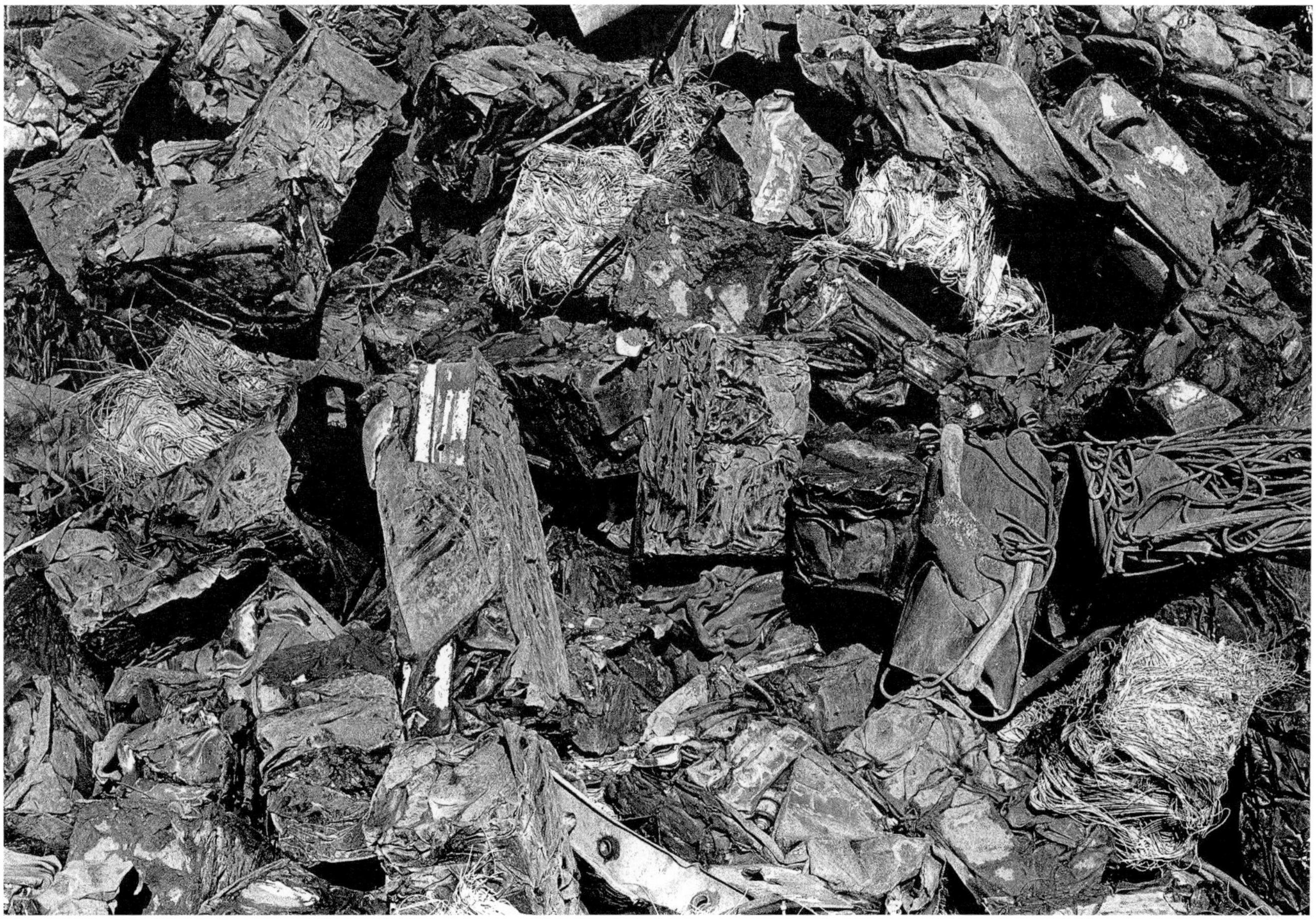

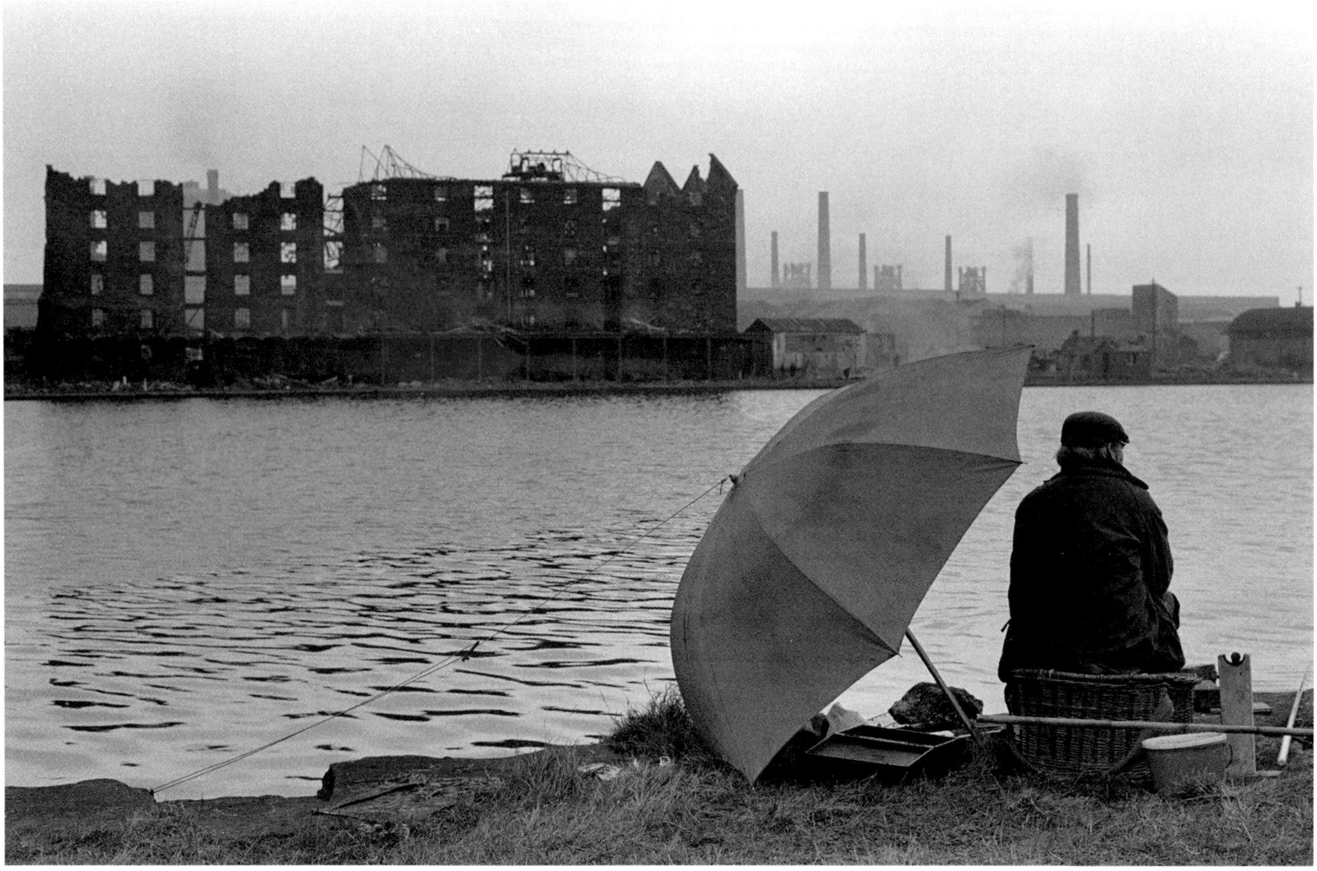

Three

Docklands and Steelworks Communities

Having covered Tiger Bay in Part 1, the photographs of this section provide visual insights into other docklands communities as well as that of Lower Splott built beside the steelworks of East Moors. Perhaps even the average Cardiffian is not aware of the number of communities whose livelihood and existence were intimately bound up with the rise and decline of Cardiff's port and its port-based industries. Two large and distinct sectors on the west side of the first Bute Dock – Tiger Bay and the Lower Docks – were complemented by another three – Newtown, Adamsdown and Splott, to create a network surrounding the docks on its land-based sides. Though they all served Cardiff's maritime and steel industries, each of these locales had its own distinct boundaries, history and character.

The residential area of the Lower Docks lay beyond Tiger Bay and, to be more precise, south of James Street. Its foundations having been laid in the 1830s by the Second Marquess of Bute, most of its streets, as those of Tiger Bay, bore the christian names of members of the Bute family – George, Louisa, (South) William, Adelaide, Dudley, Eleanor, Evelyn, and Margaret. The area was at one time characterised by great ethnic diversity – Spanish, Italian, Maltese, Arab and Irish families lived there amongst other nationalities. By 1974 they had almost all vanished and the streets were empty.

For me, the impression created by the sight of an entire section of the city lying abandoned as though the local population had fled in fear of an invader was nothing short of breathtaking. Once beyond the initial visual shock I began to look at and photograph the individual elements of this distinctive urban environment, one which had played host for over a century to great interaction between the docks and the local community. The old infrastructure was still basically in place: streets, houses, pubs, shops, schools, churches, hotels, the bookie's, the chippie, the hairdresser's, the library... Fascinating in its scope, in its detail and in its unfamiliarity it was bound to attract my photographic attention.

At the opposite (north) end of the docks, Newtown or 'Little Ireland' was purpose-built in the 1840s to house the families of Irish labourers whose manpower was vital to the construction of the Bute East Dock. Irish women also worked on the Edward England potato wharves of the Bute West Dock, just across Herbert Street from the houses of Newtown. This was a tightly self-contained community of just six streets in the area covered today by the units of the Tyndall Street Industrial Estate. Part of one street remains – Ellen Street at its westernmost fringe – and in place of its old terraced houses are now the Driscoll Workshops, dedicated to the memory of Newtown's most famous son, the boxer 'Peerless' Jim Driscoll. The houses were all knocked down in in 1966, not long after the closure of the Bute West Dock in 1964 and not long after the replacement of old houses by new ones in nearby Tiger Bay. In contrast to Tiger Bay, however, there were no plans to rehouse the people of Newtown and, their homes gone, they were dispersed to other parts of Cardiff. By the time I first gazed over Newtown from the Pellett Street footbridge in 1974, its former site was waste ground covered by the caravans and scrap metal heaps of a gypsy site. Before my incredulous eyes, two gypsy children amid the squalor posed for my camera. The only original remains of Newtown were those of the former All Saints' Church, long-since converted to a transport depot, and the abandoned relic of Jim's Transport Café in Tyndall Street. Not much, but starkly compelling to explore, photograph and make one wonder about what had gone before.

The third community I visited a number of times was Adamsdown, especially the part of it along the railway line north-east of Newtown. The 'Black Bridge', a nineteenth-century footbridge reaching over the railway to link Adamsdown with the Bute Docks and the East Moors steelworks, provided a particular point of photographic interest. There were also the nearby side-streets whose names were an immediate source of fascination: metals – Zinc Street, Lead Street, Copper Street, etc. – and celestial bodies – Constellation Street, Planet Street, System Street, etc. Given the state of some of the houses in these streets, it appeared to me that if the old terraces were still inhabited, they might not be for much longer. I was right, as it turned out. The new Anderson Place estate (what's wrong with metals and heavenly bodies?) has been built on the site of many former houses and some streets such as Cycle Street have been buried. Nonetheless the people of Adamsdown were able to stay put. They perhaps considered themselves more fairly done by than the inhabitants of other long-standing communities who had no choice but to move when compulsory purchase orders were put on their houses and the bulldozers moved in.

The area known as Lower Splott also and inevitably became part of my photographic itinerary. Just as the terraces of Butetown had been laid out and built to house the docks workforce so too was Lower Splott built to accommodate the workers of the Dowlais Steelworks, newly erected next to Cardiff Docks in 1880. As the area grew many of the streets were given names of Welsh towns: Aberystwyth, Pontypridd, Neath and Swansea amongst others. Standing as they did next to the steelworks and bordering on the road and rail routes to and from the docks, they housed generations of steel, railway and docks workers until 1978. With the docks trade in almost terminal decline during the 1960s and 1970s, inevitably the steelworks shut and another Cardiff working-class community was doomed. Yet whilst the streets of the Lower Docks were those of a ghost-town, and those of Newtown were non-existent, some of those in Lower Splott still carried on as normal. It was in this mid-70s scenario that I arrived and was able to record some scenes of everyday activity. The steelworks loomed dramatically over the houses in a still-classic south Wales scene where, as in so many other mining and steeltowns, human life and habitation existed literally in the shadow of gigantic industrial structures. Where the houses had already gone and weed-covered remains of streets lead only to oblivion, the scene was rendered just as spell-binding by the presence of skeletal, disused playground equipment, a sign on the steelworks wall warning parents of 'death lurking inside' for their children, and finally, those same children having nothing but empty wasteland as a place to play. The campaign to save the remaining streets of the community was long and impassioned. It is said that in the end, those who shouted the loudest were the ones who kept their homes. Indeed, a few of the streets – Aberystwyth, Aberdovey, Neath and Swansea – survive today in truncated form. I wish I'd been to some of the community meetings with my camera at the time.

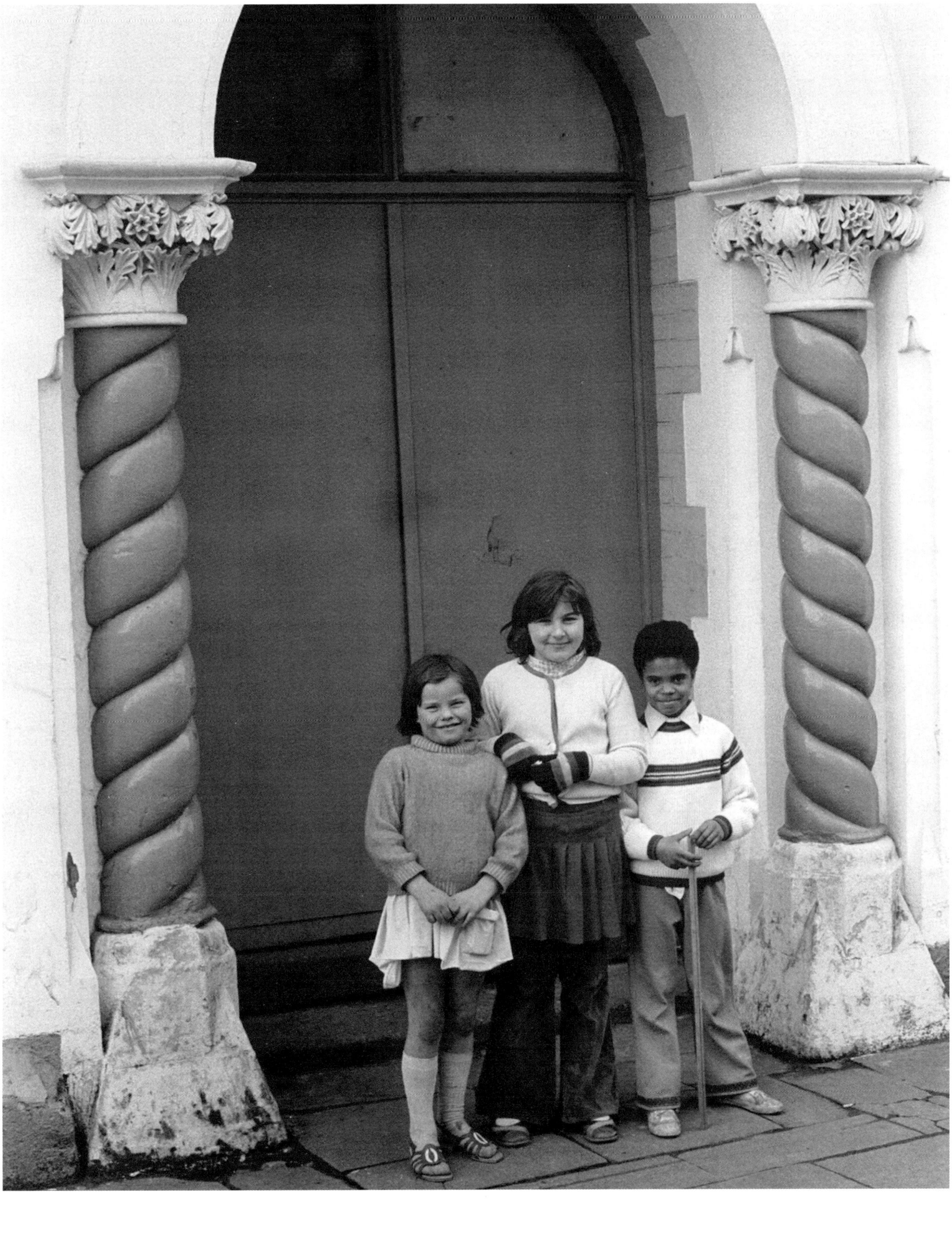

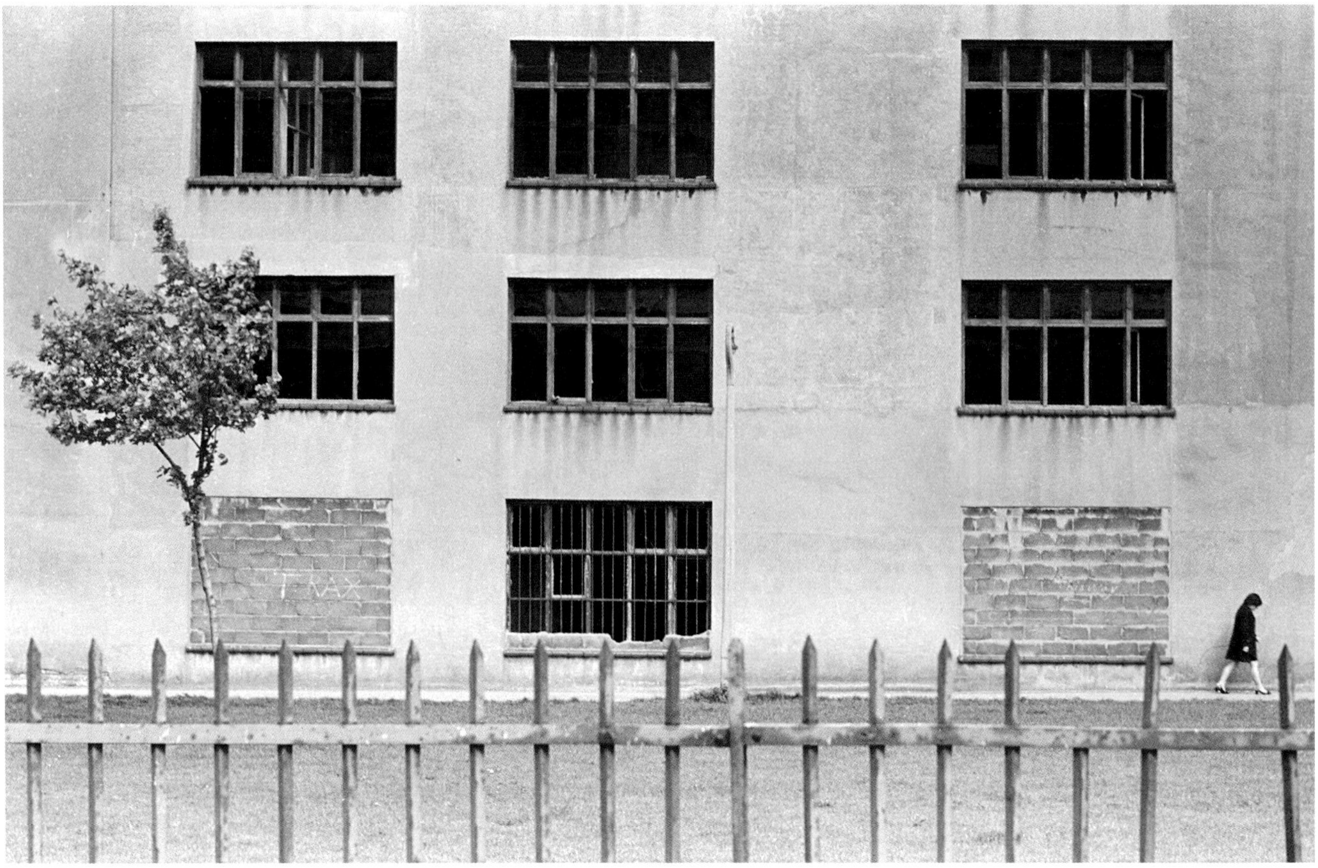

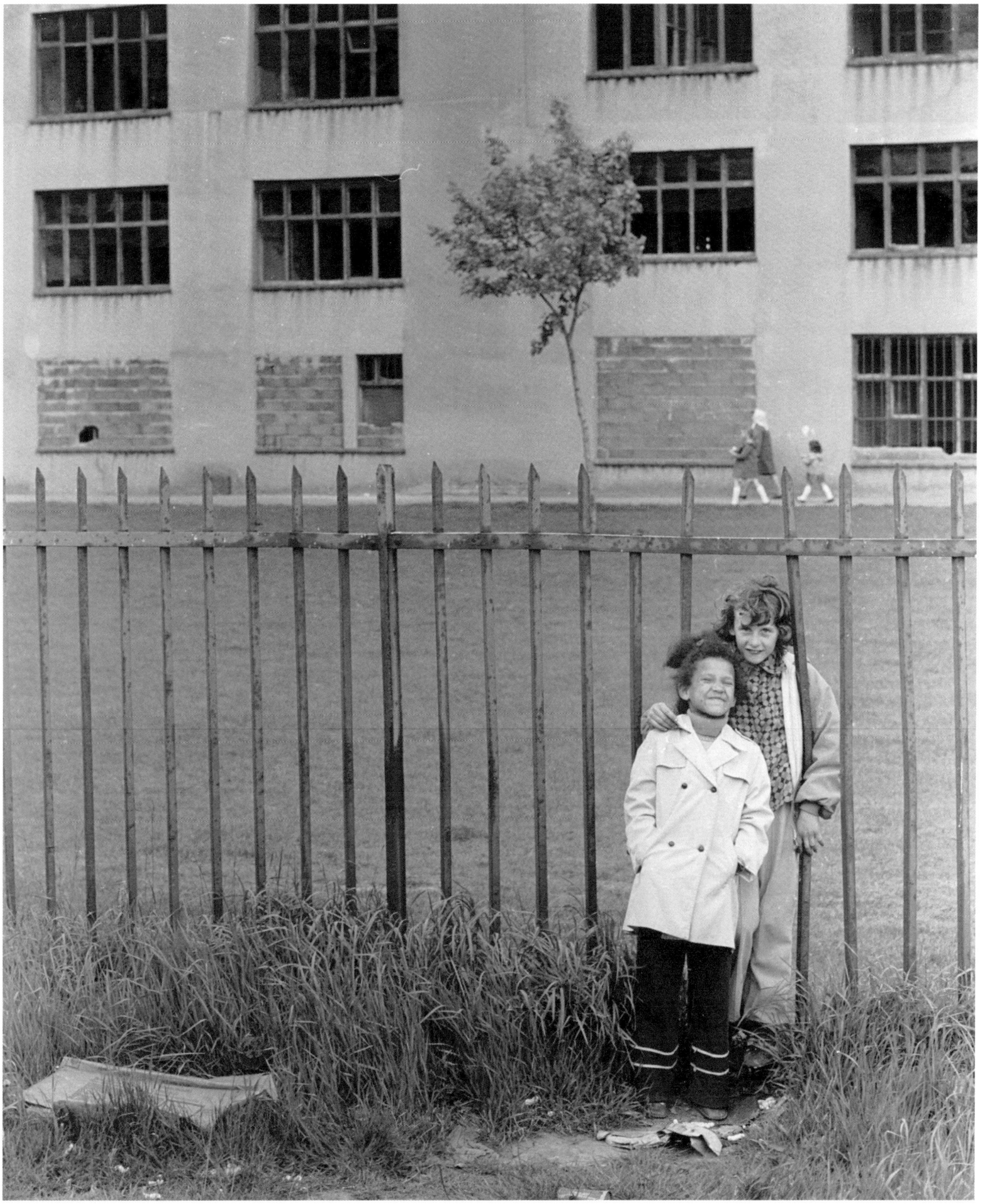

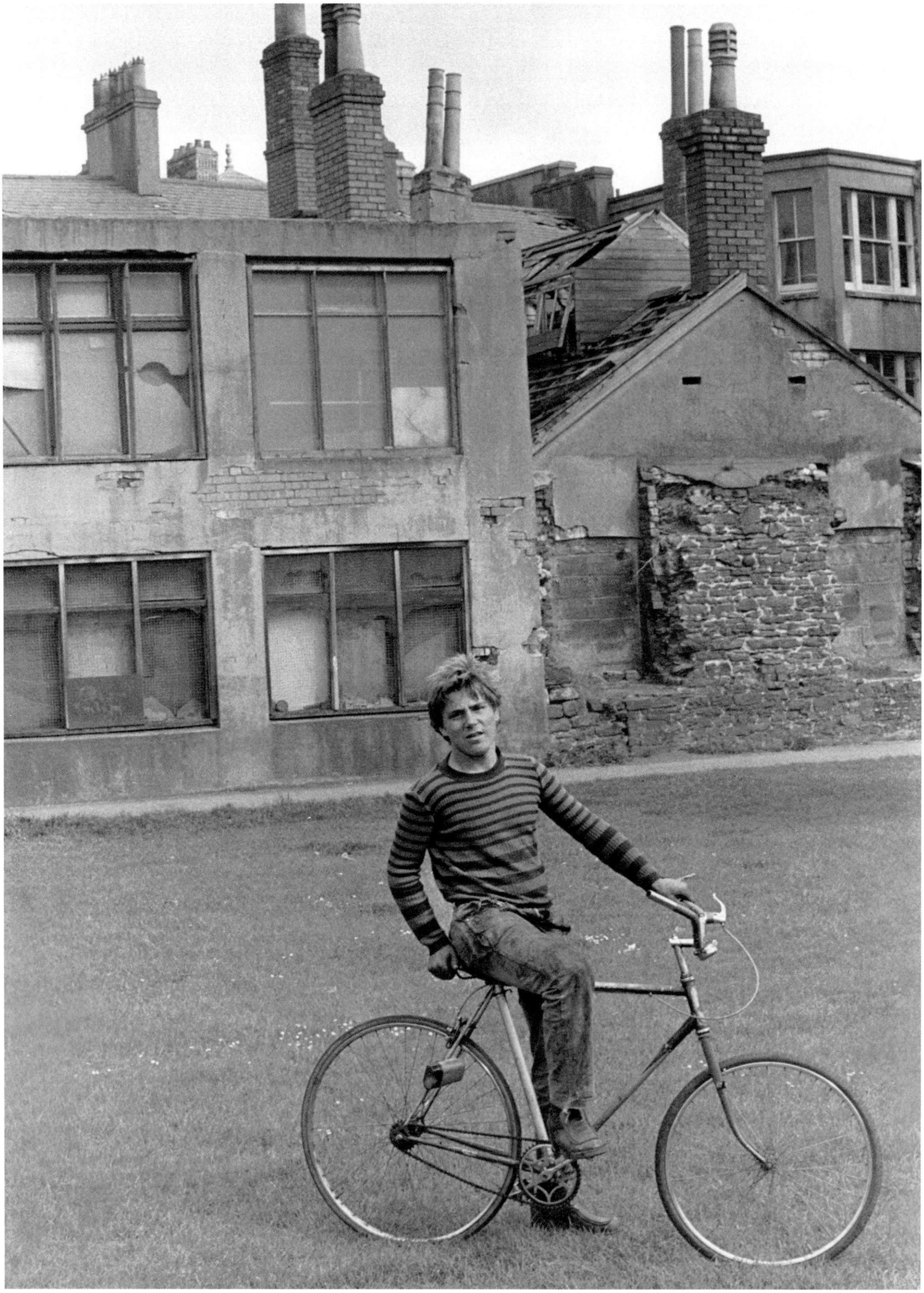

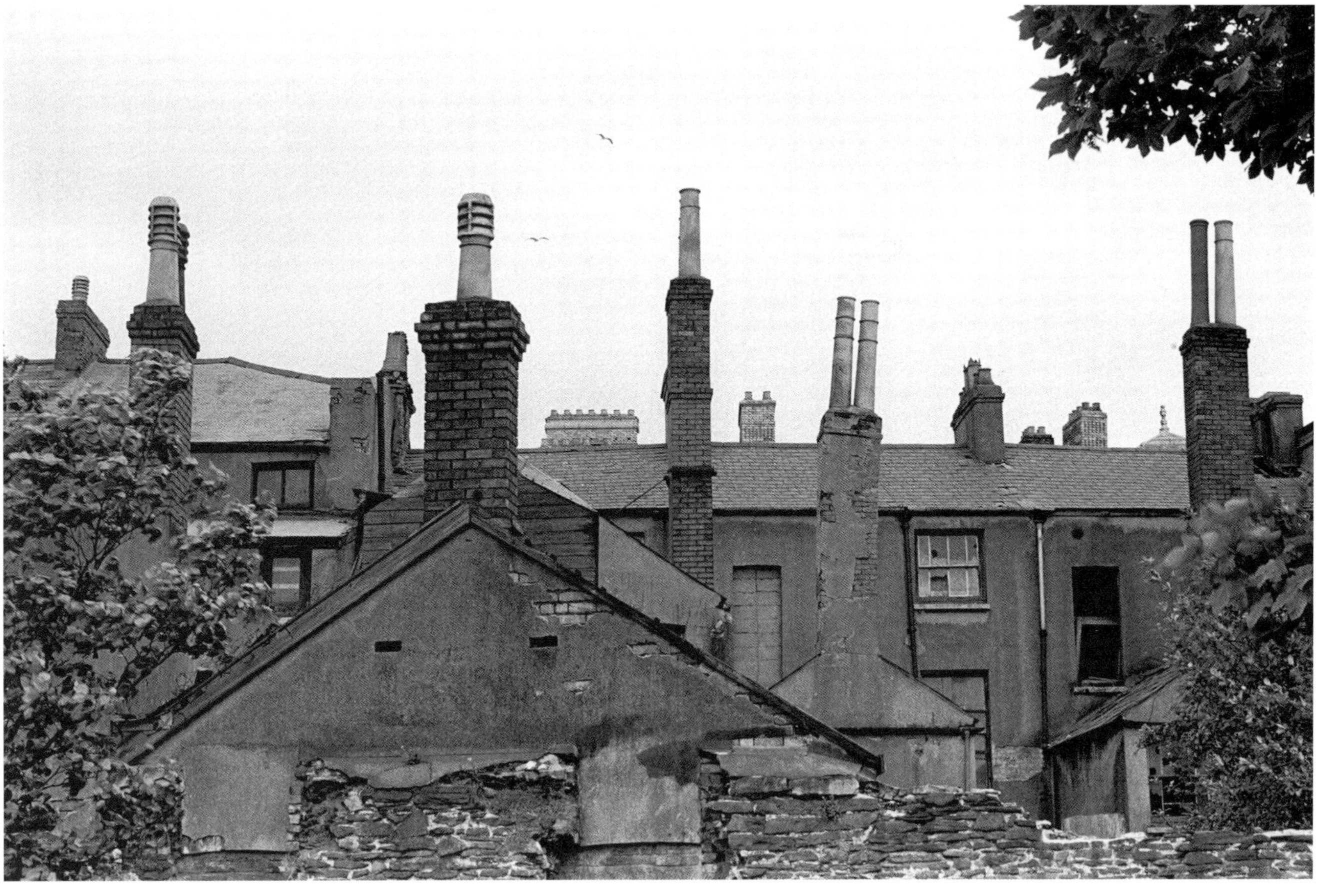

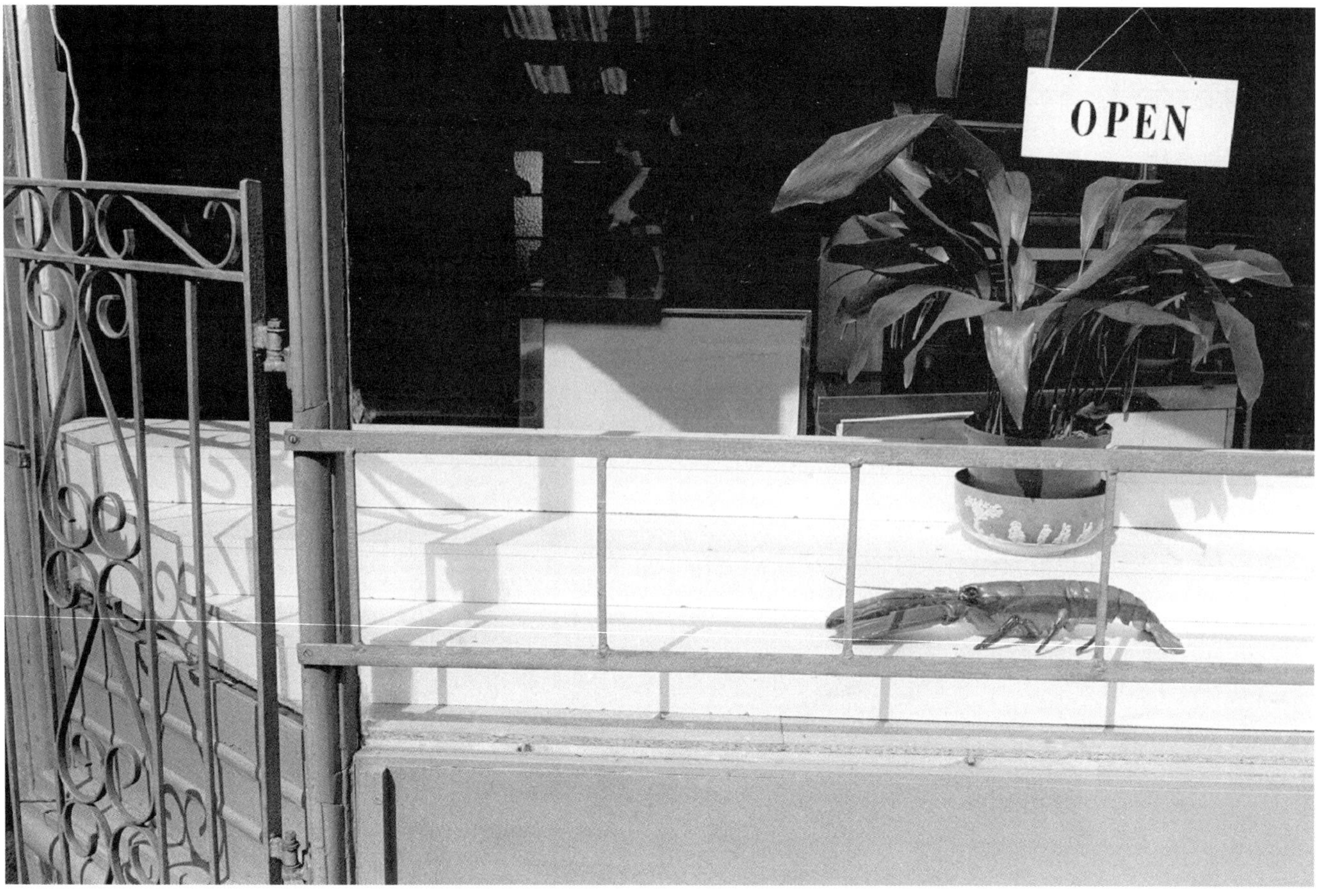
OPEN

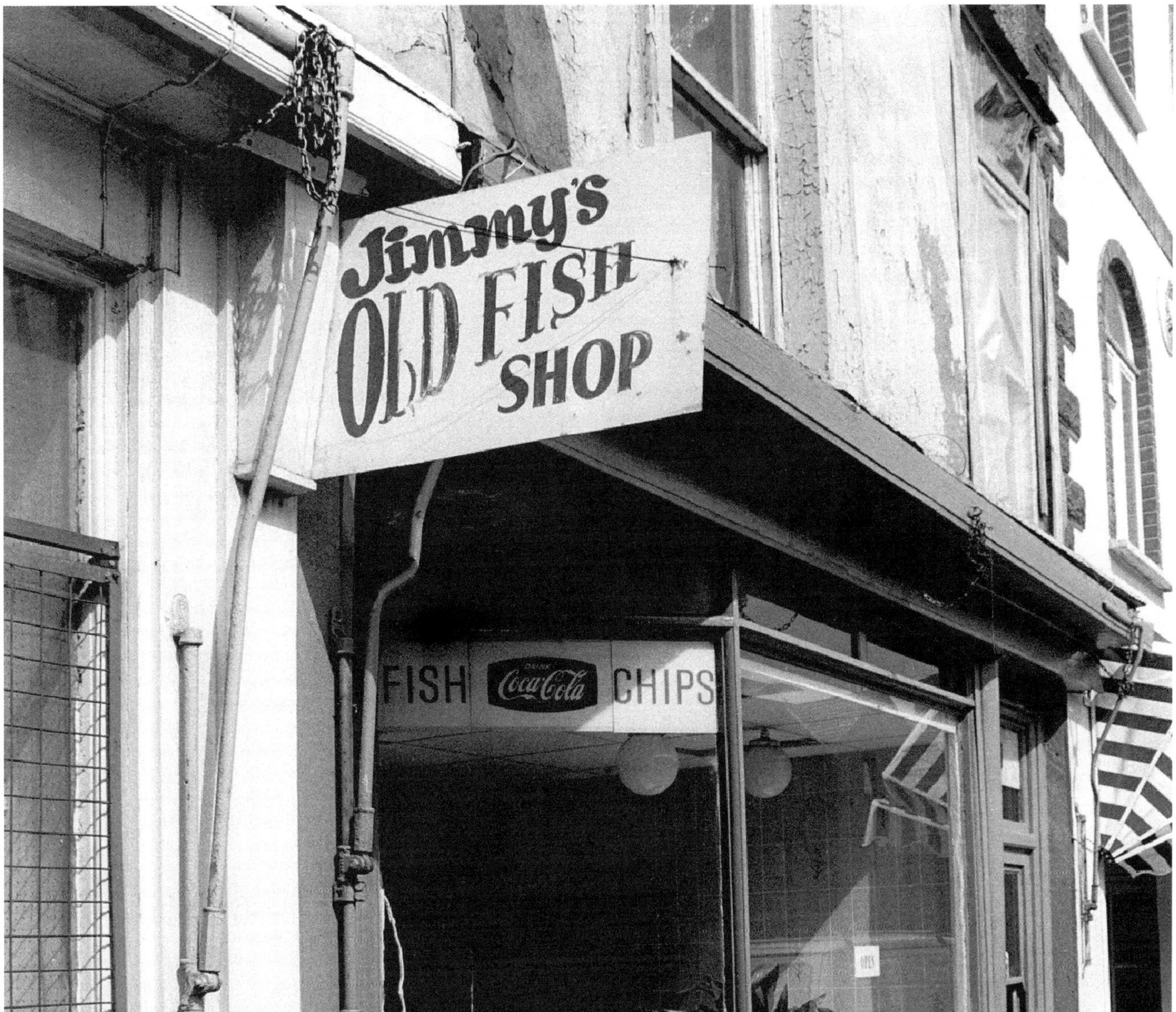
Jimmy's
OLD FISH
SHOP
FISH
Coca-Cola
CHIPS

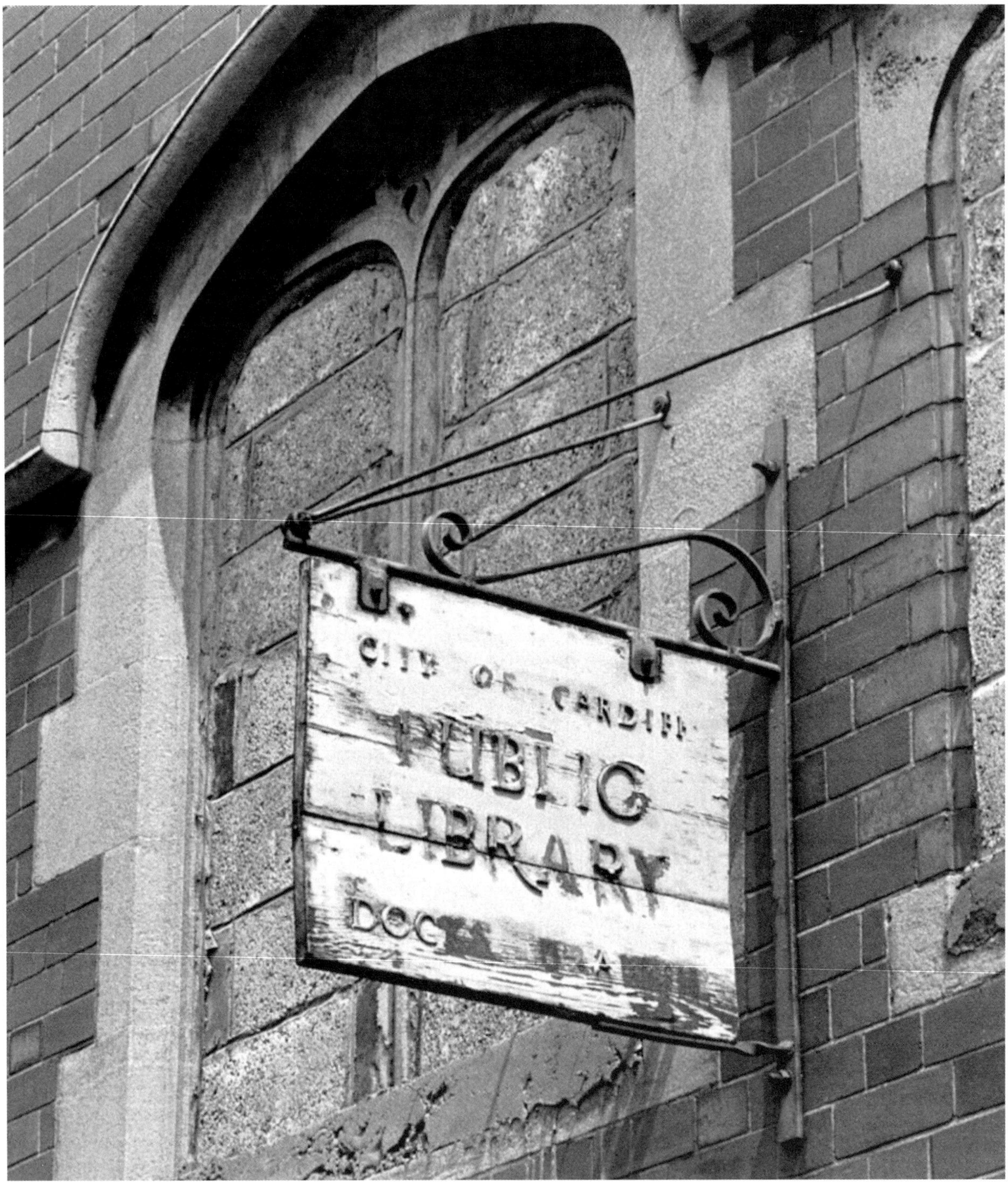
CITY OF CARDIFF
PUBLIC
LIBRARY

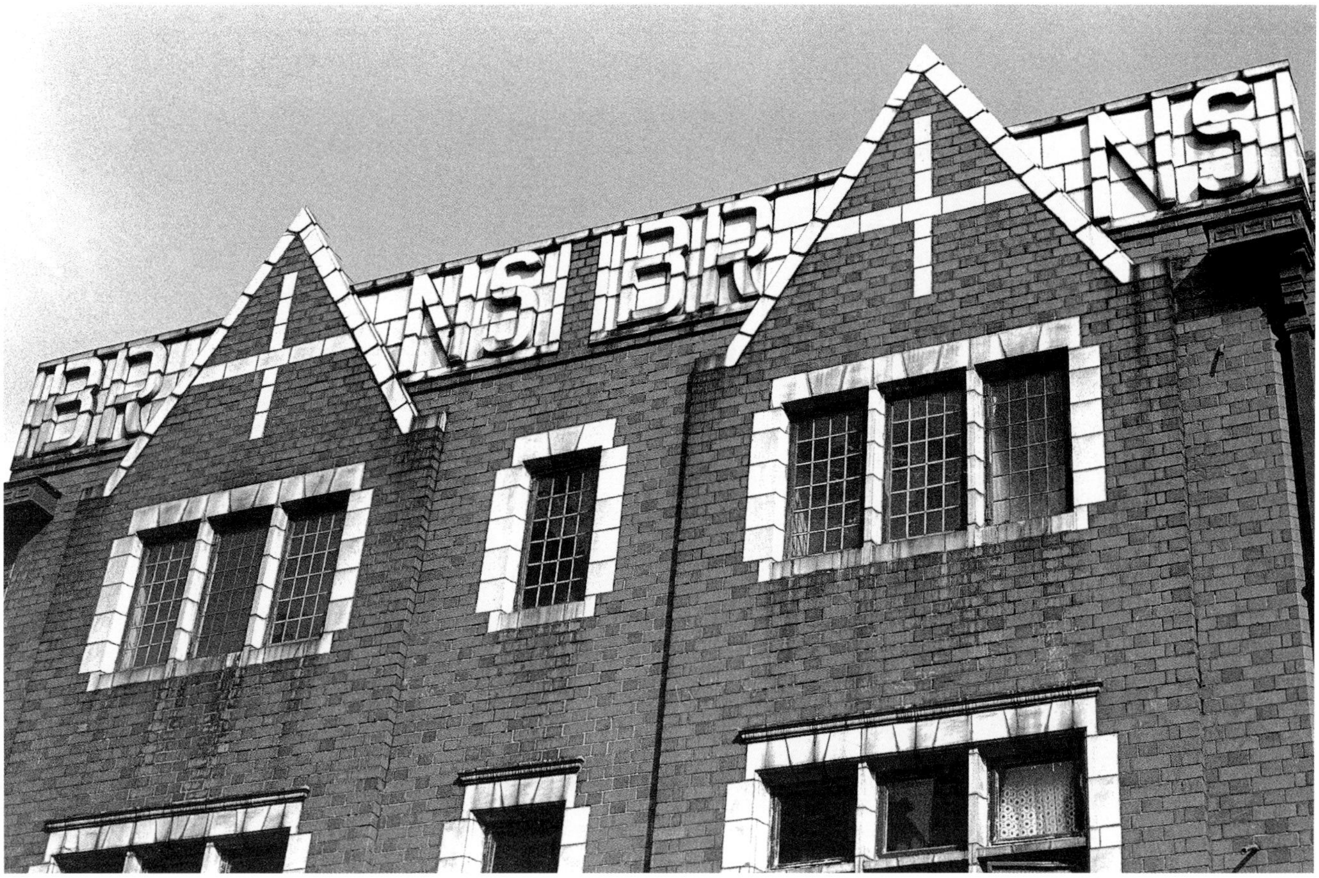
BR
NS BR
NS

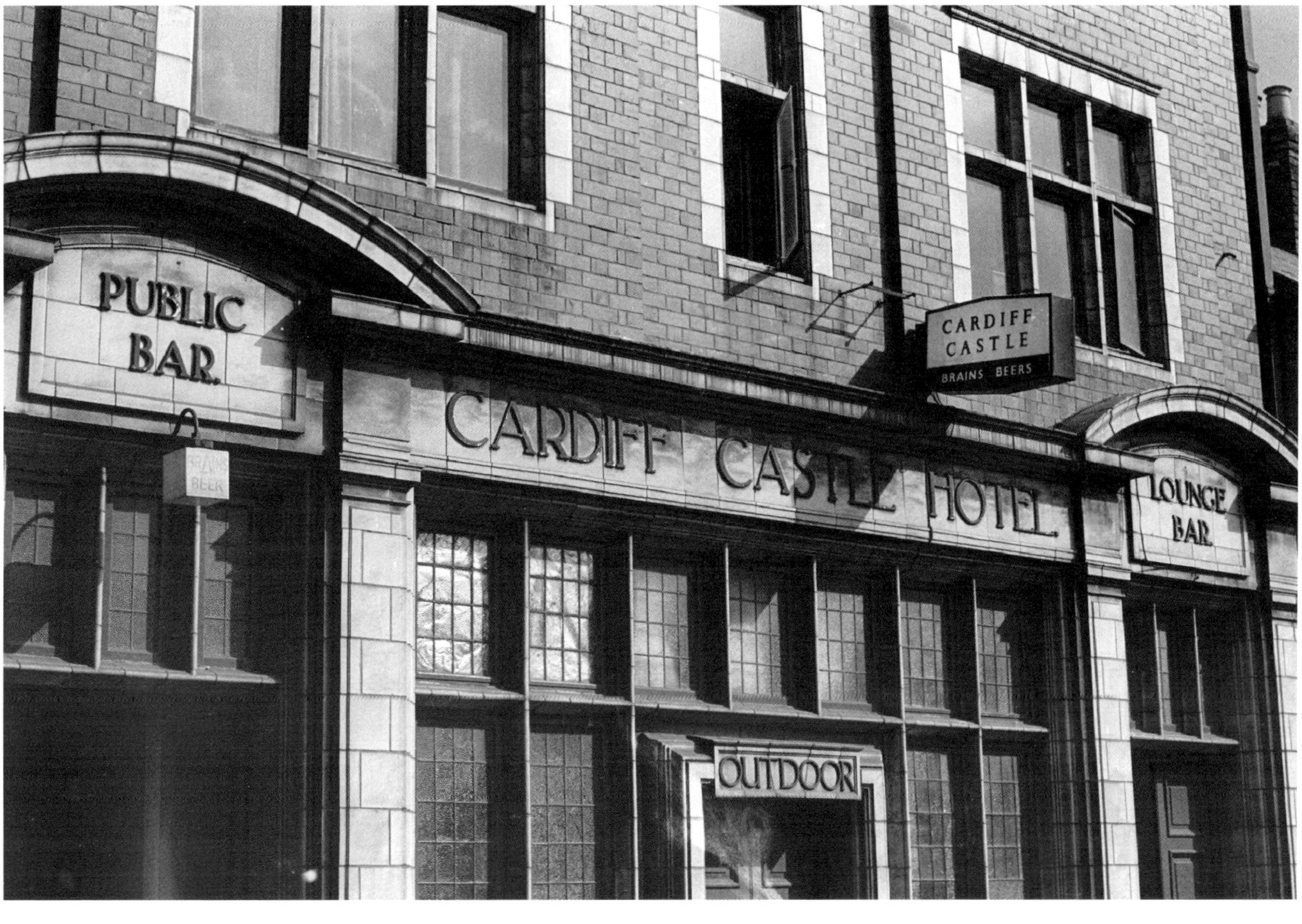
PUBLIC
BAR.
CARDIFF
CASTLE
BRAINS BEERS
CARDIFF CASTLE HOTEL.
LOUNGE
BAR.
OUTDOOR

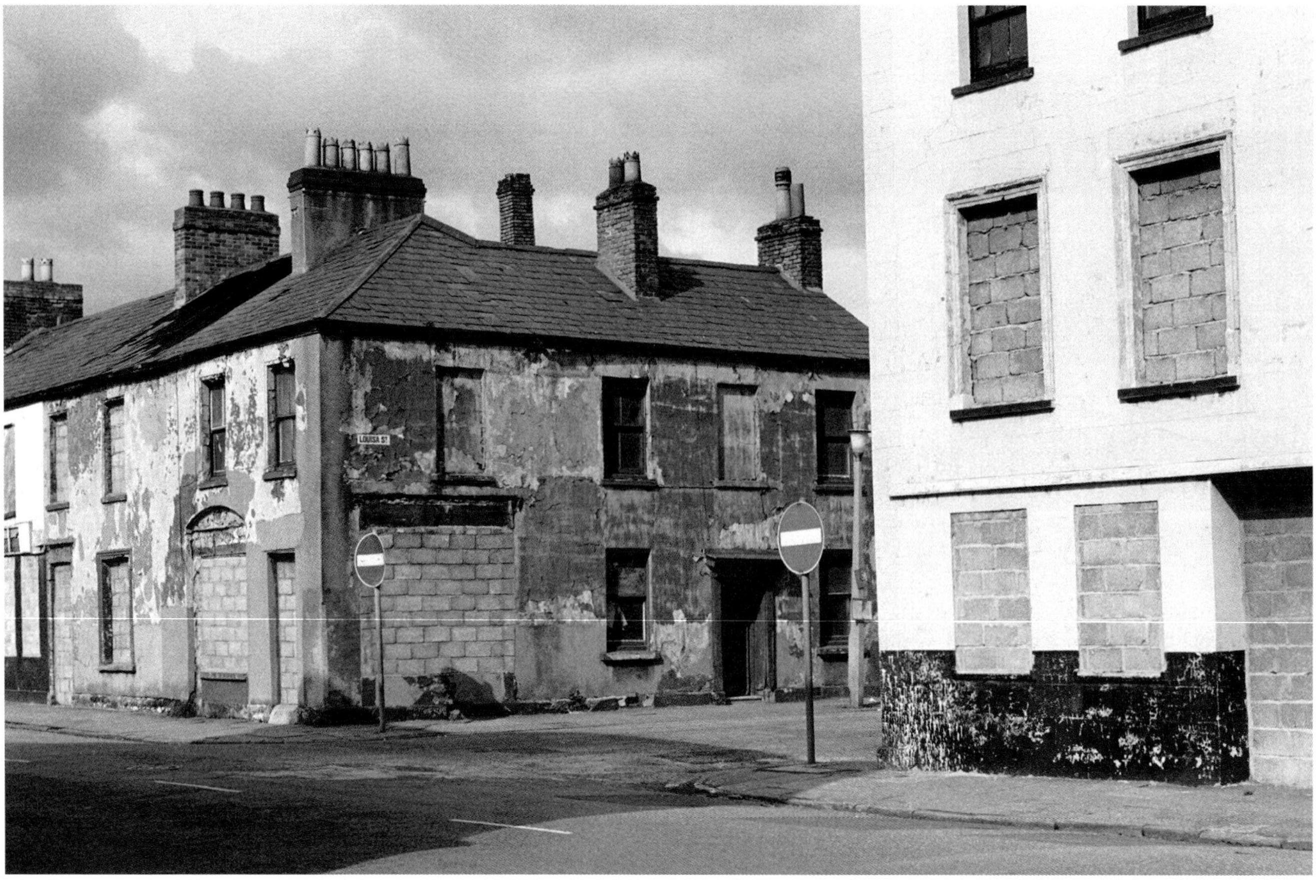
LOUISA ST

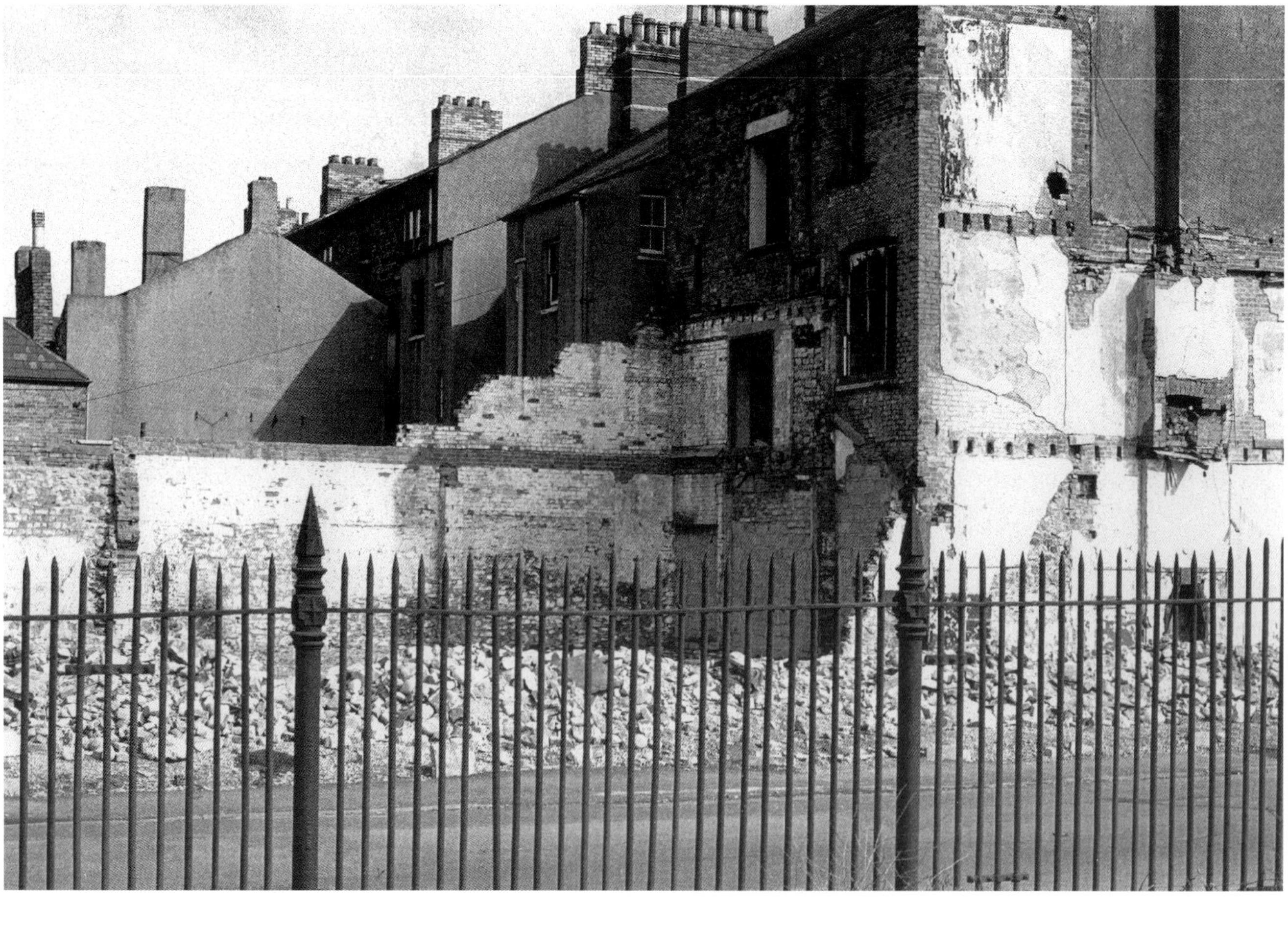

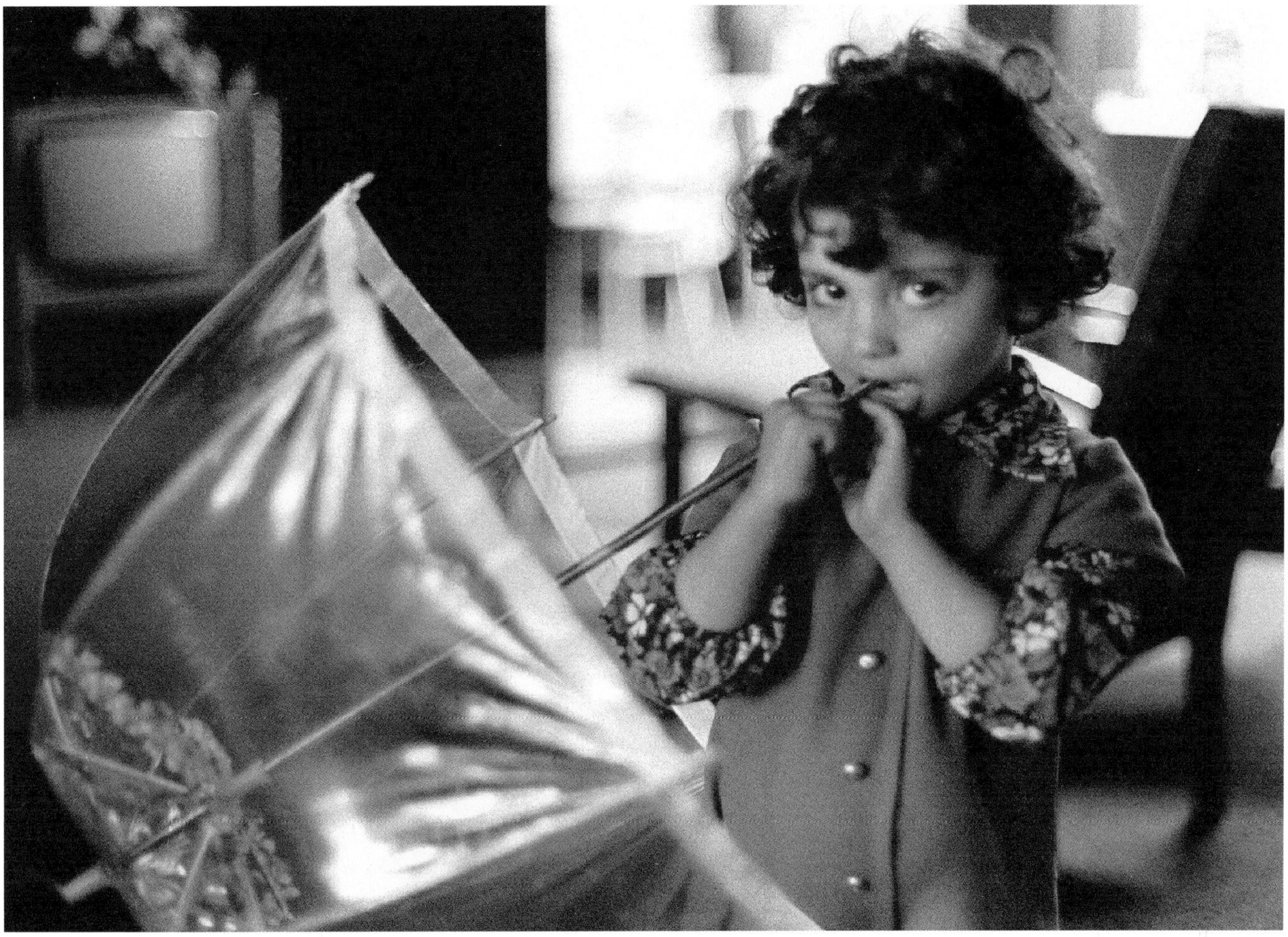

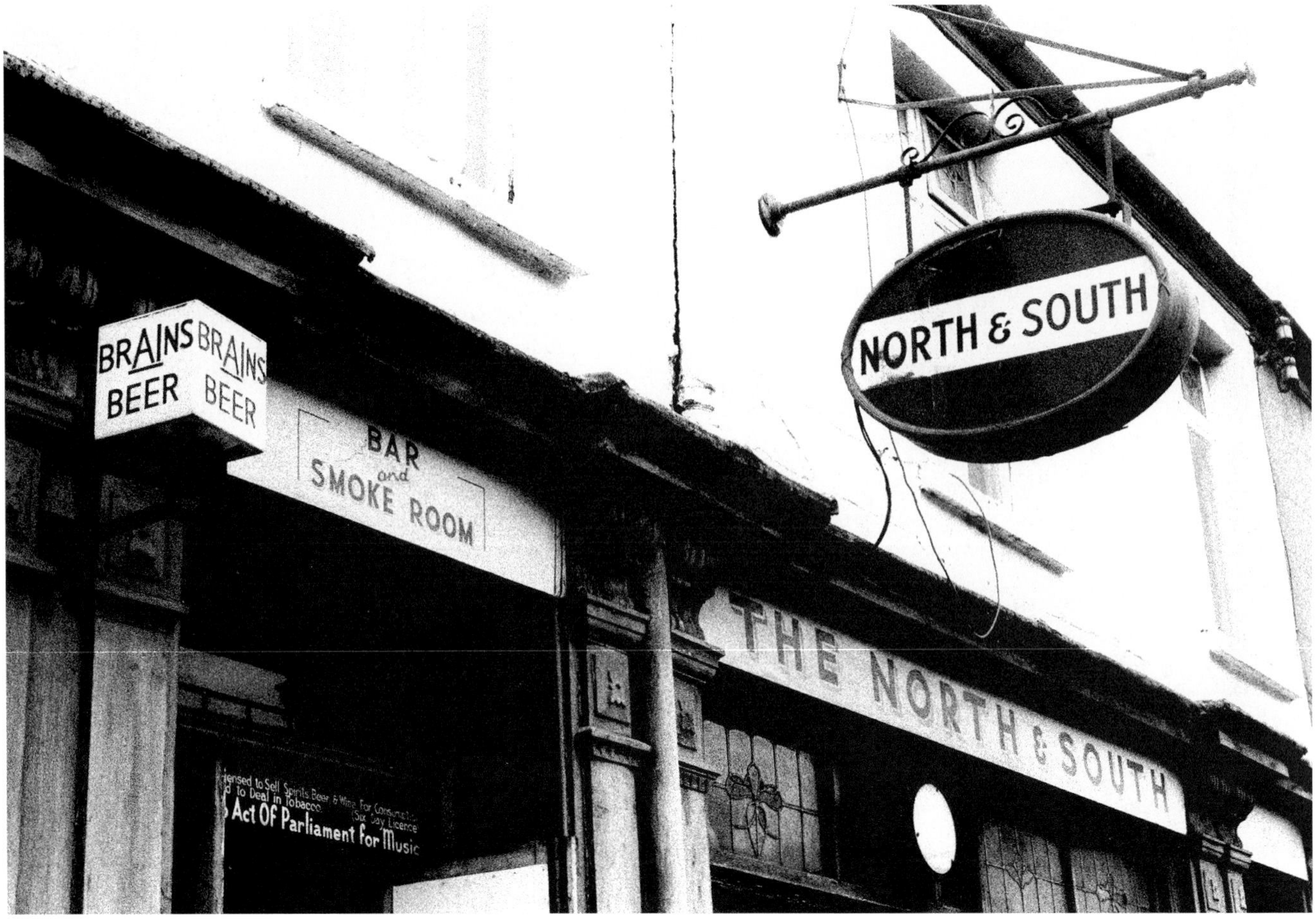
BRAINS BEER
BRAINS BEER
BAR and SMOKE ROOM
NORTH & SOUTH
Act Of Parliament for Music
THE NORTH & SOUTH

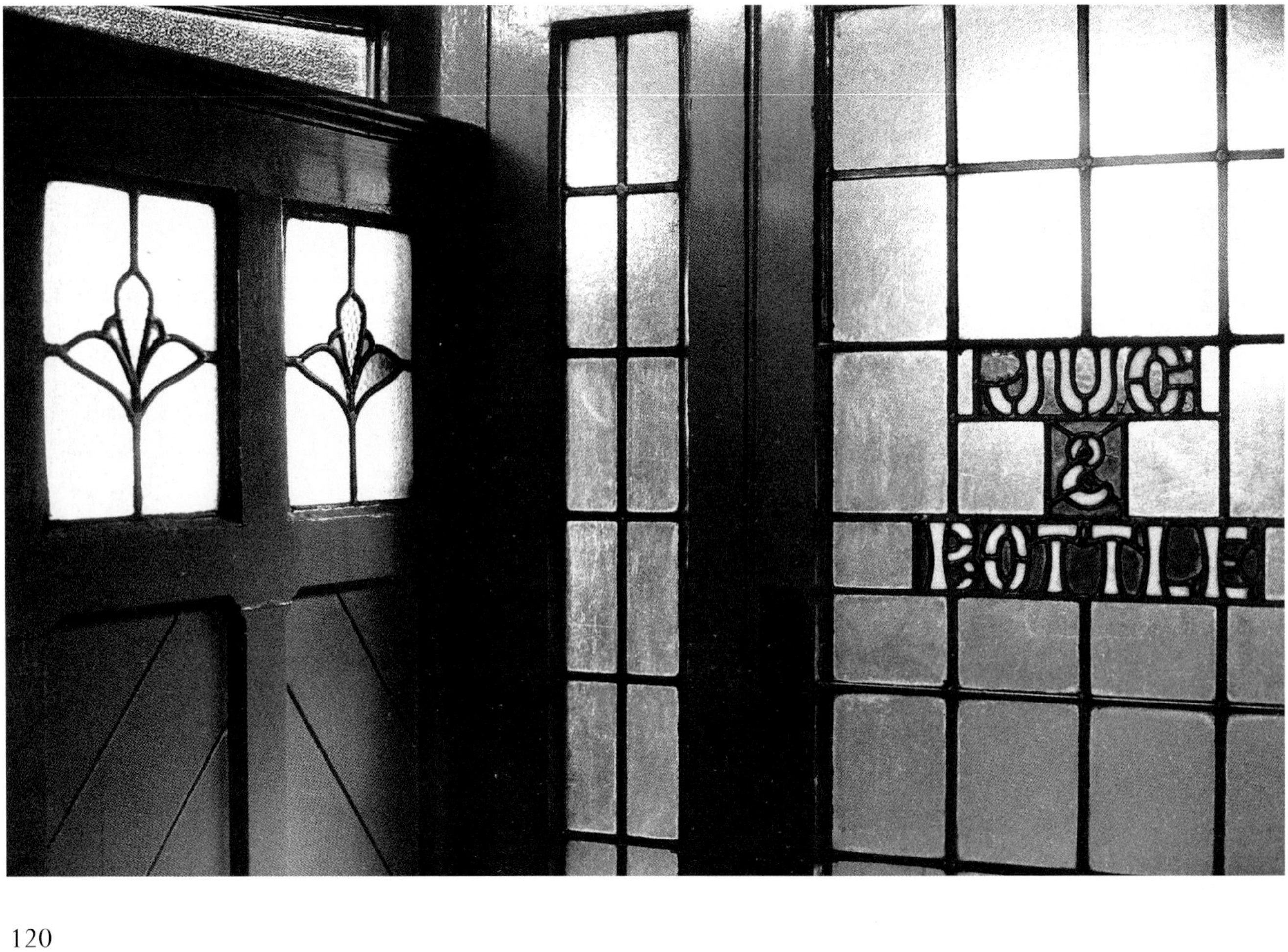
JUG & BOTTLE

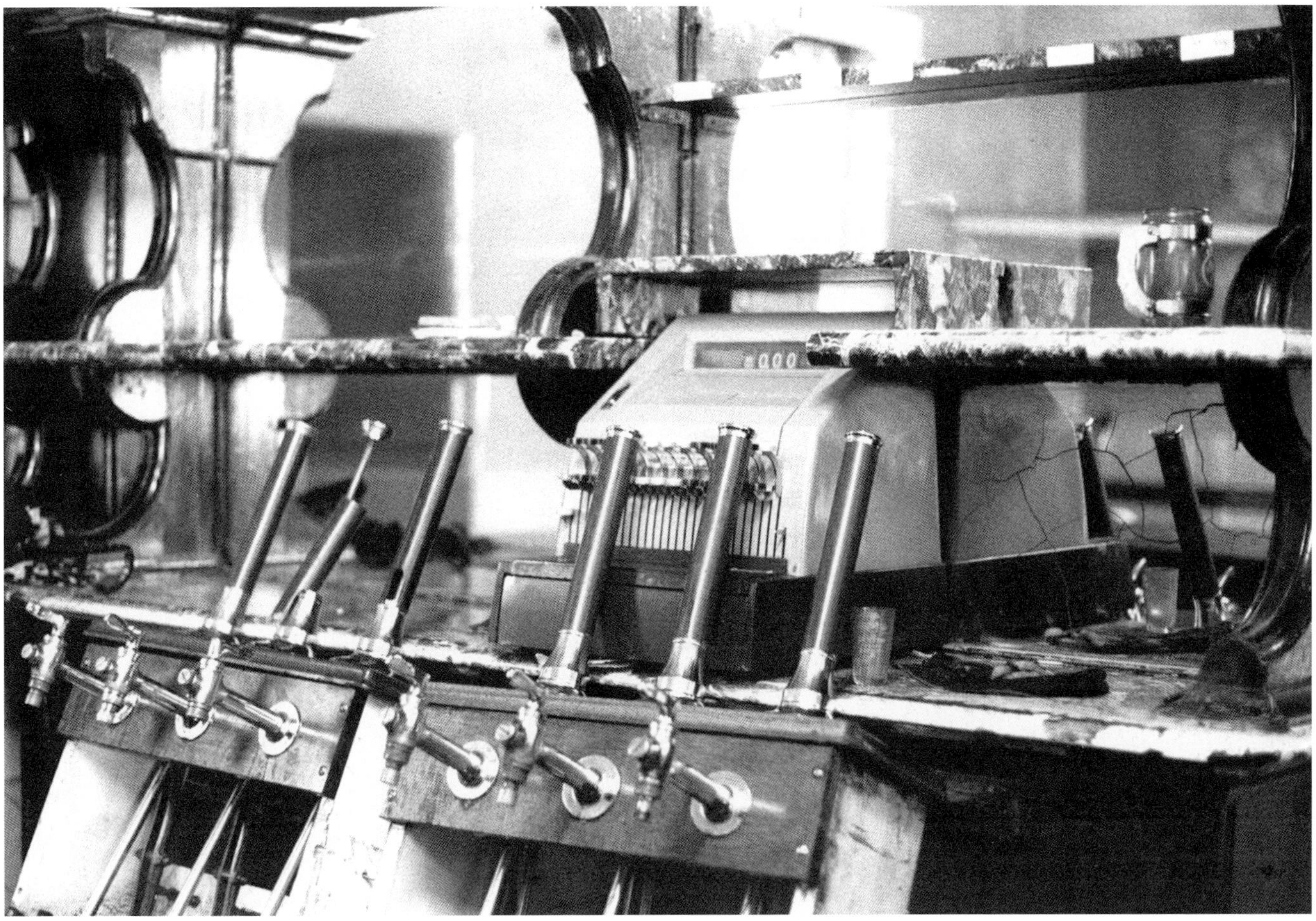

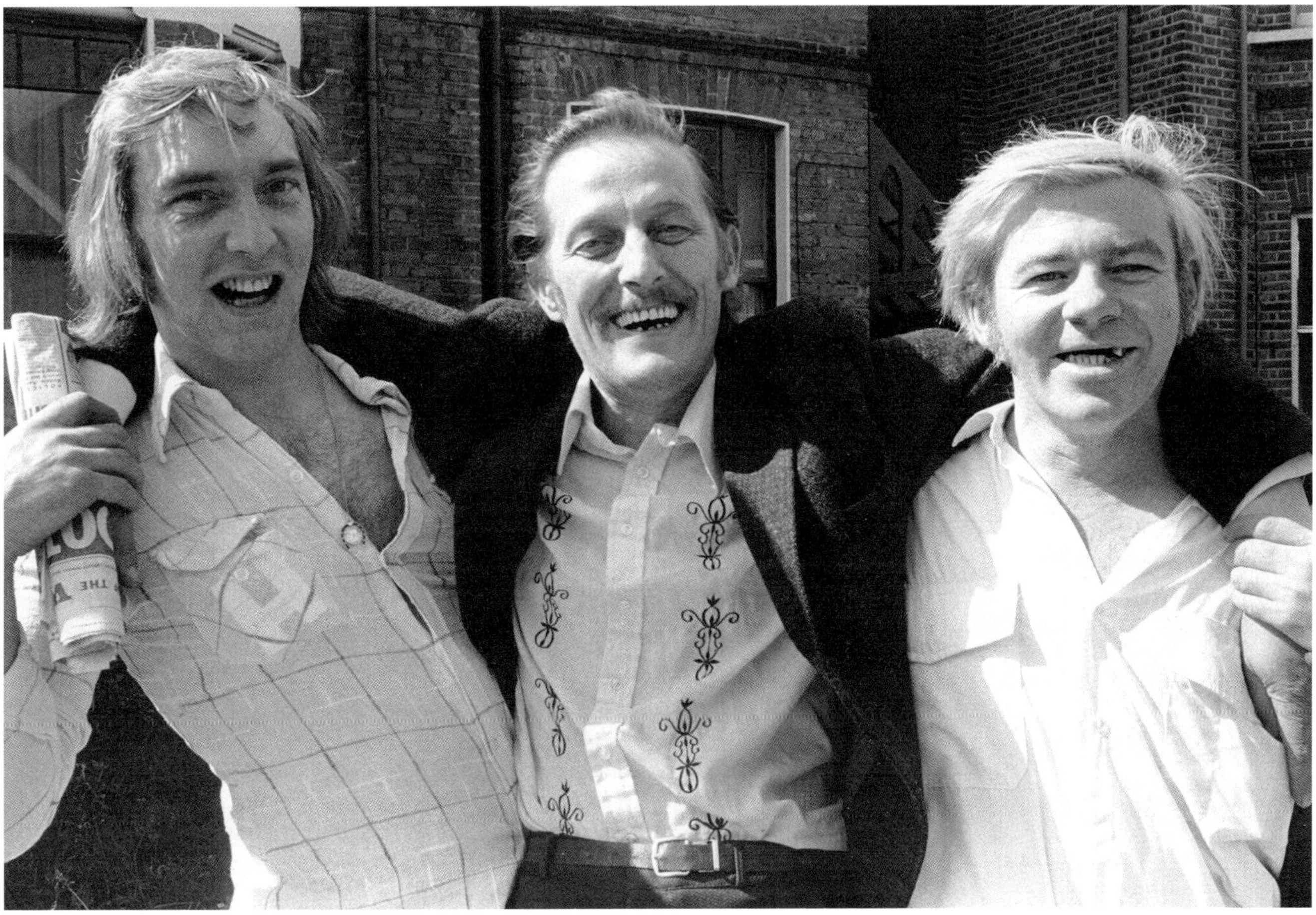

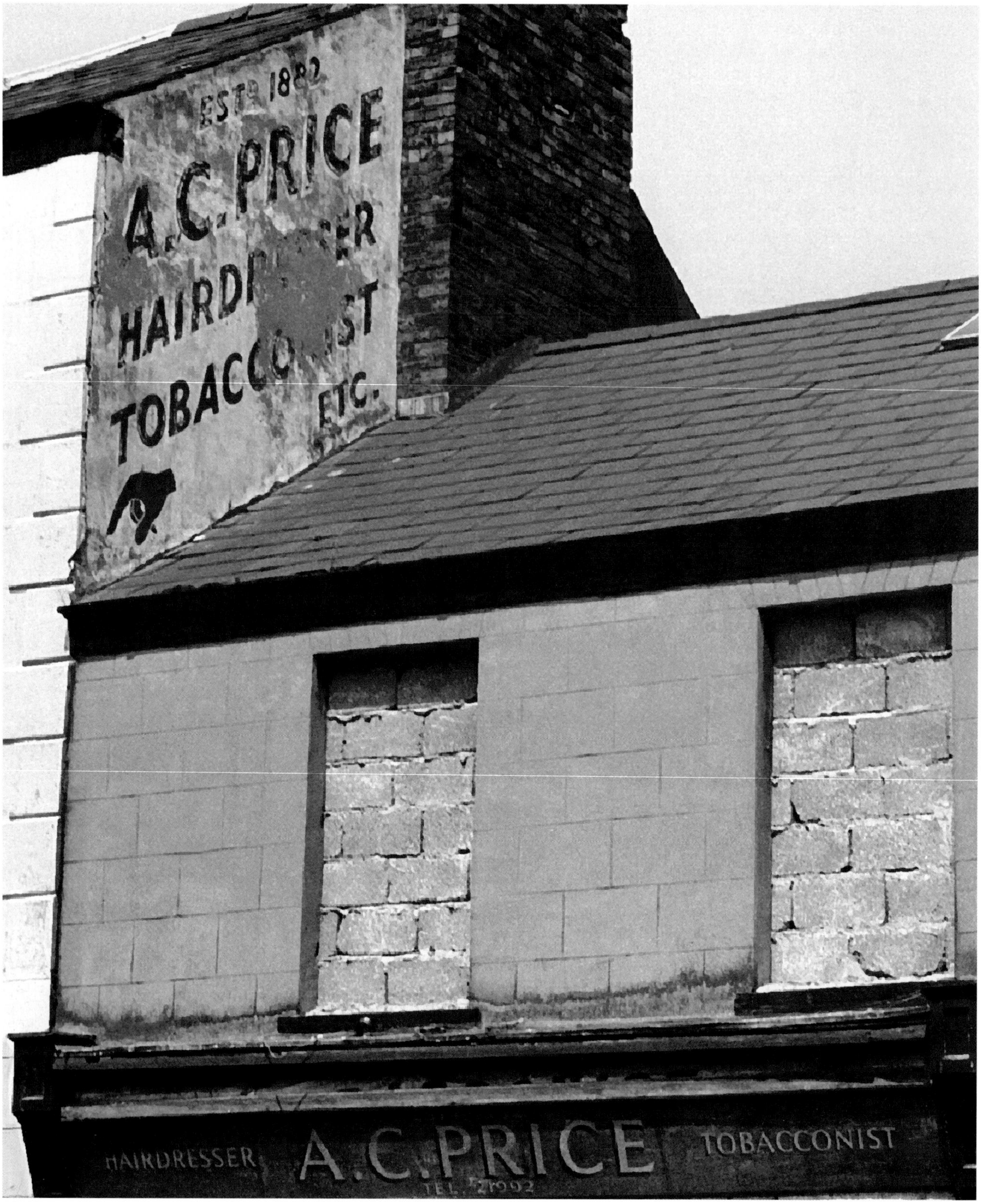
ESTD 1882
A.C. PRICE
HAIRDR ER
TOBACCO ST
ETC.
HAIRDRESSER
A.C. PRICE
TOBACCONIST
TEL. 21992

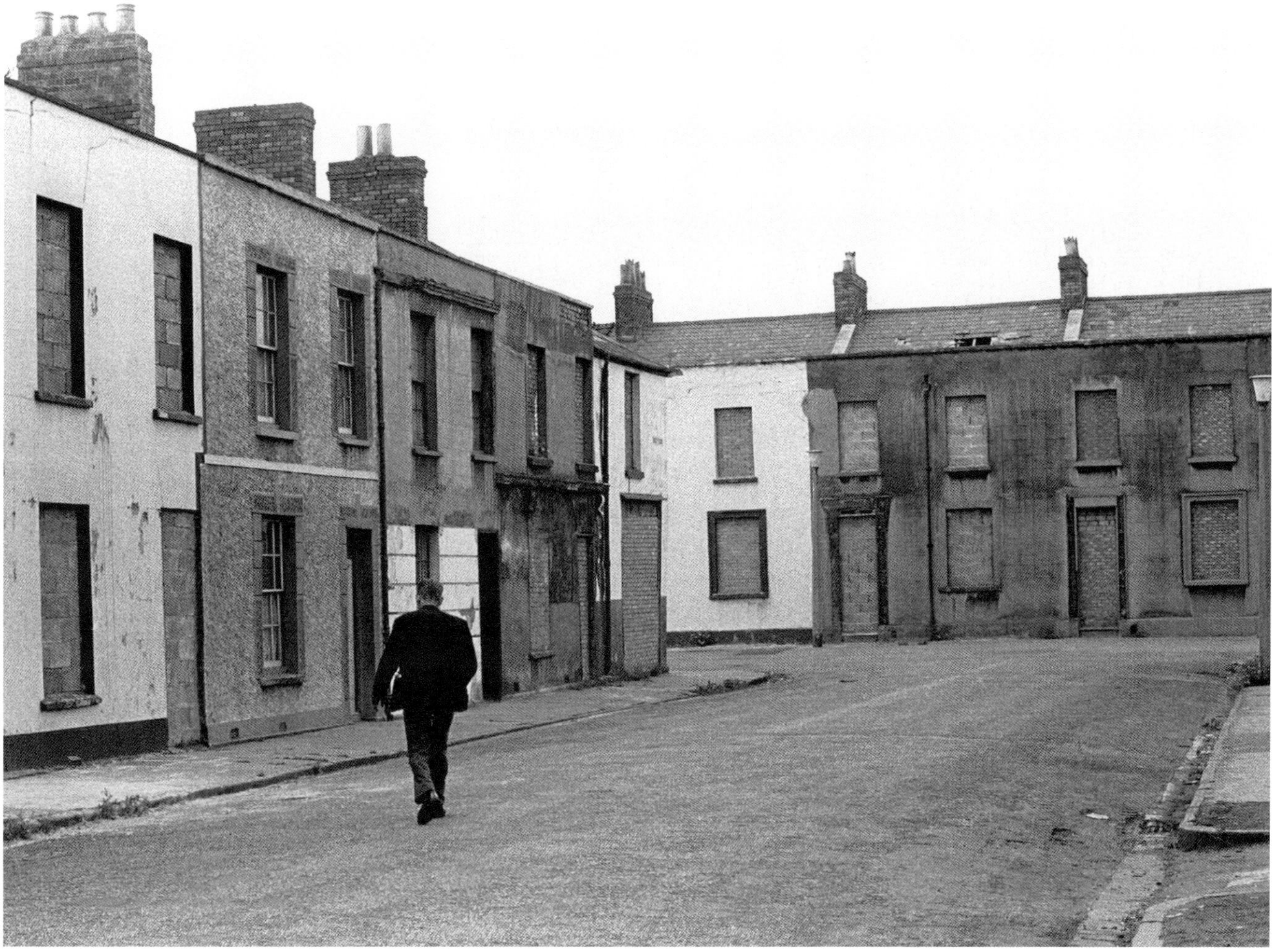

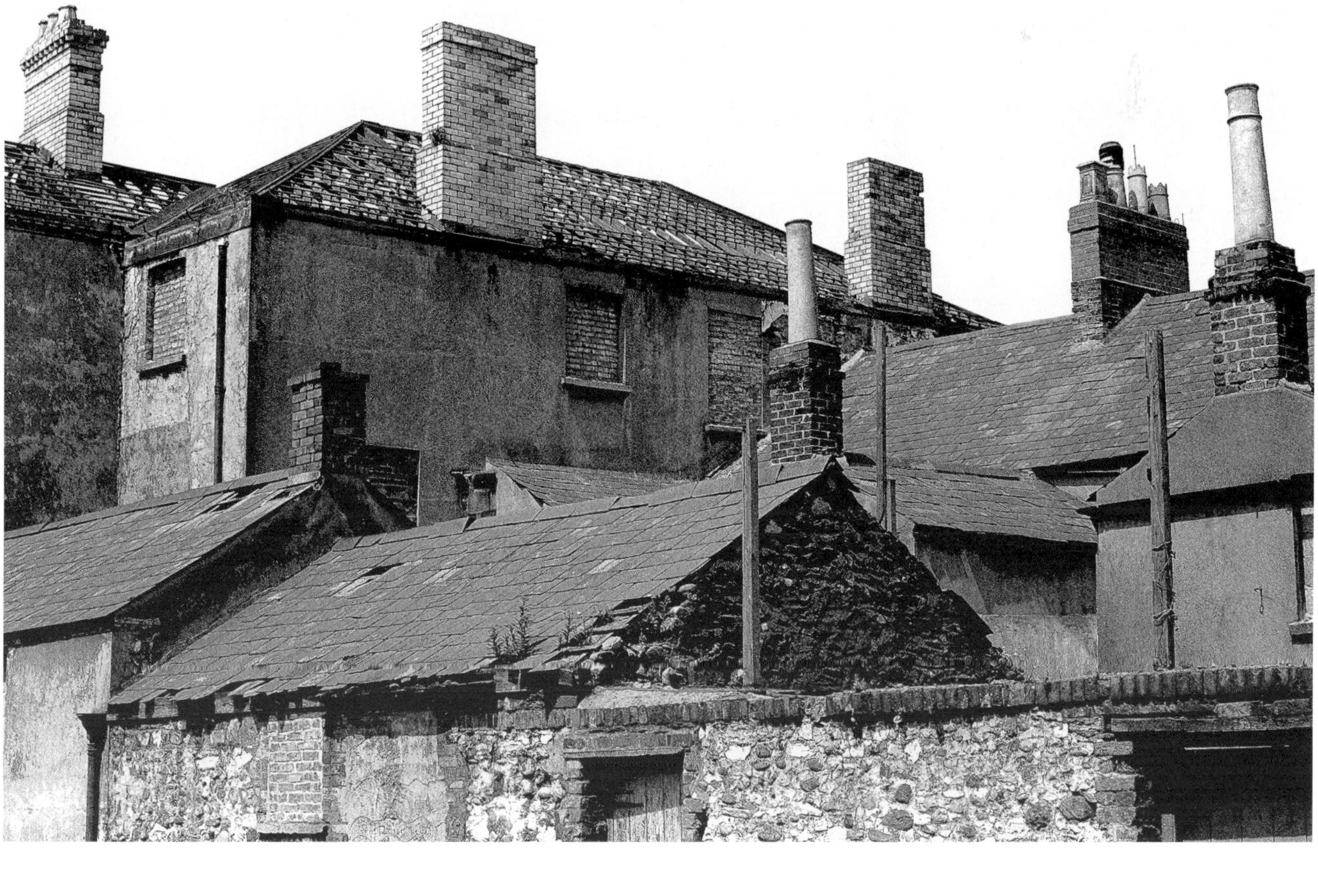

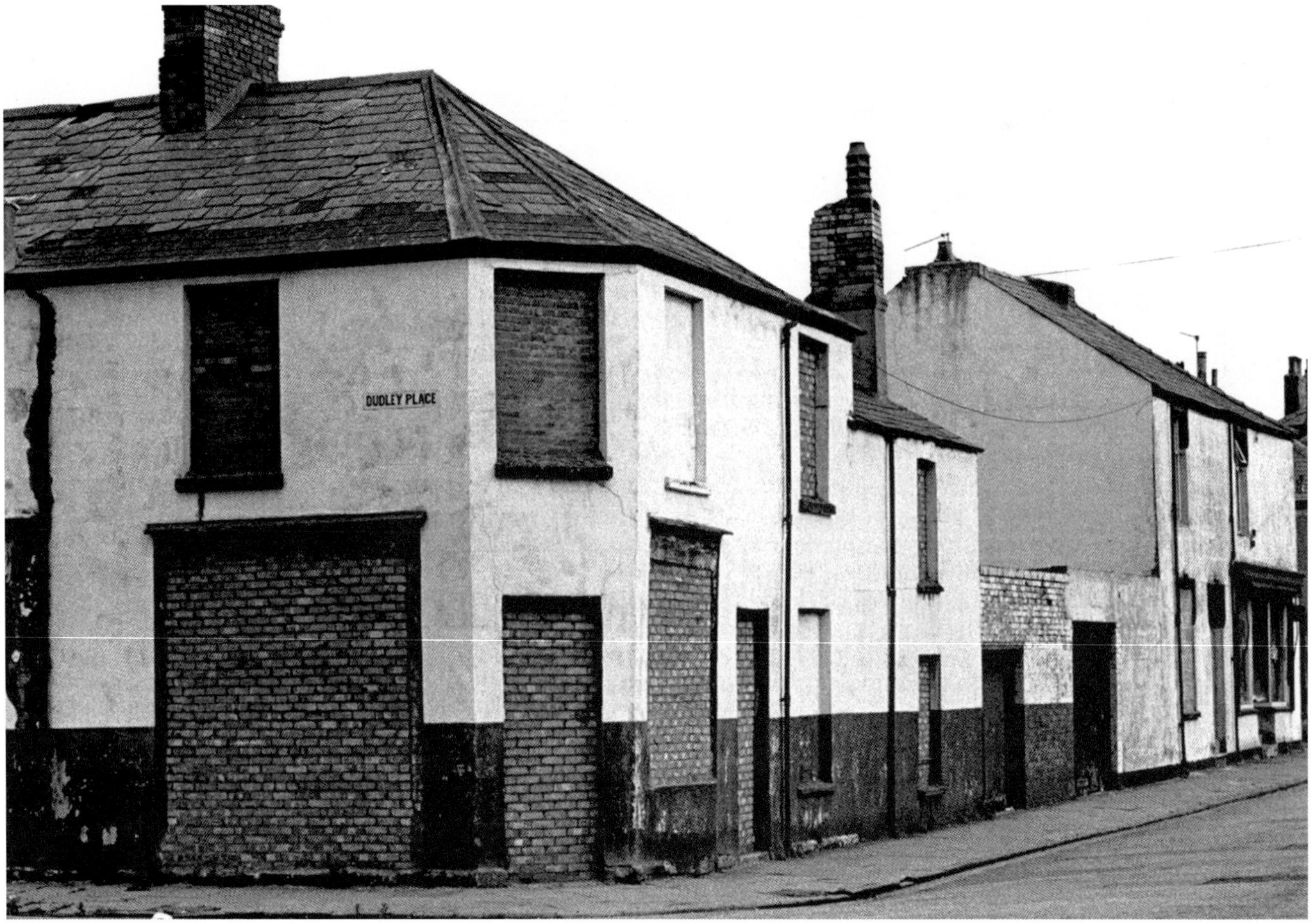
DUDLEY PLACE

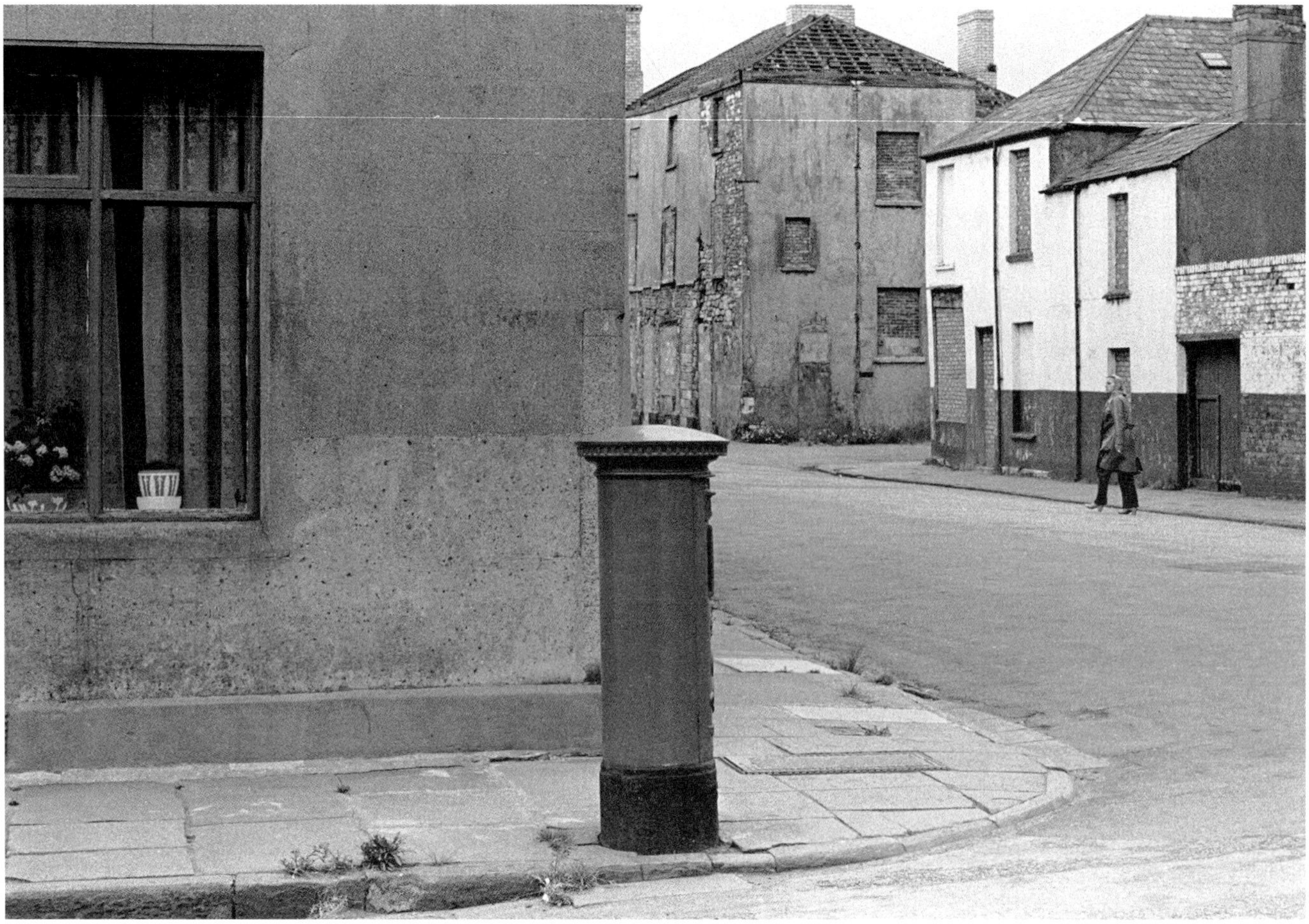

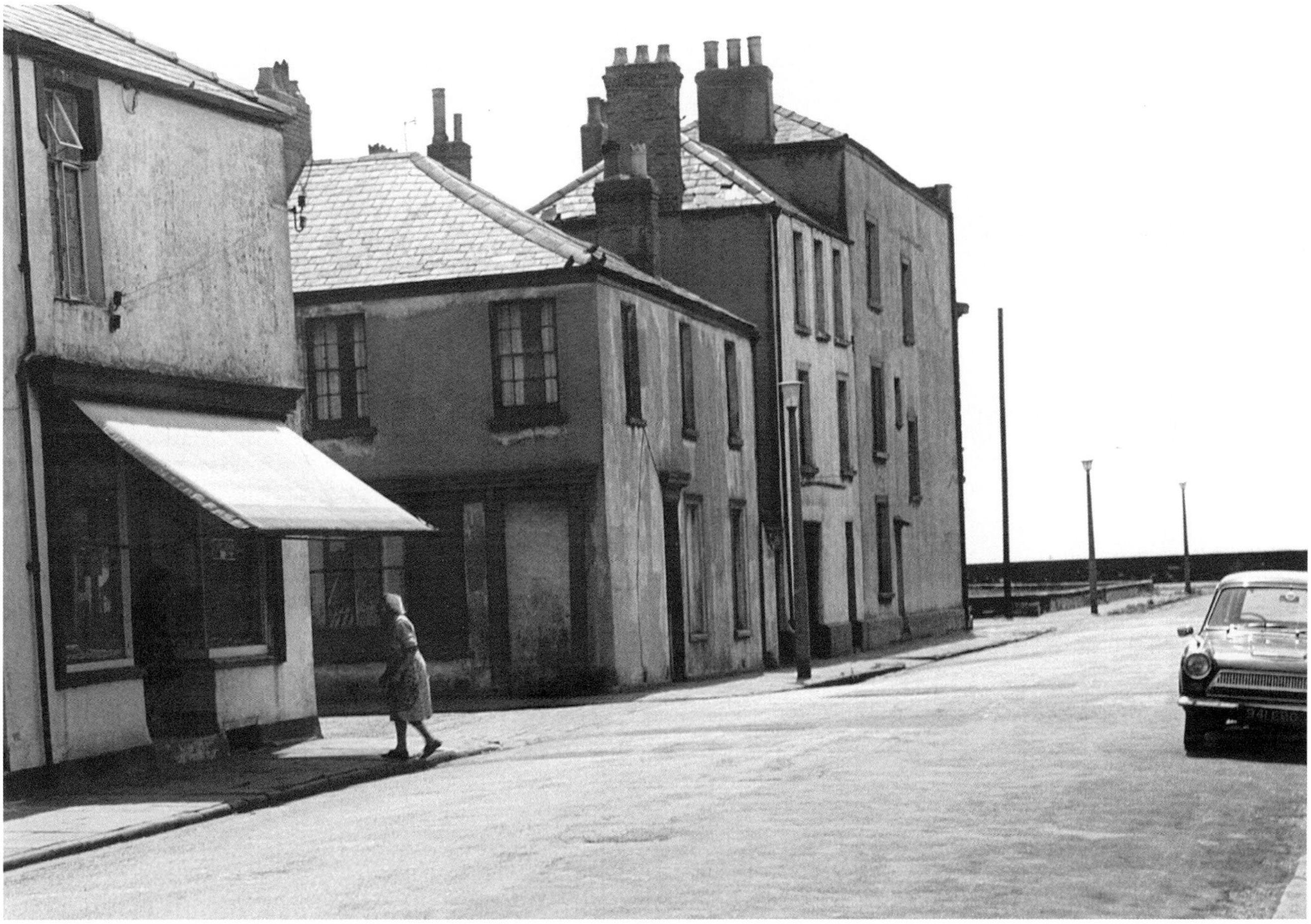

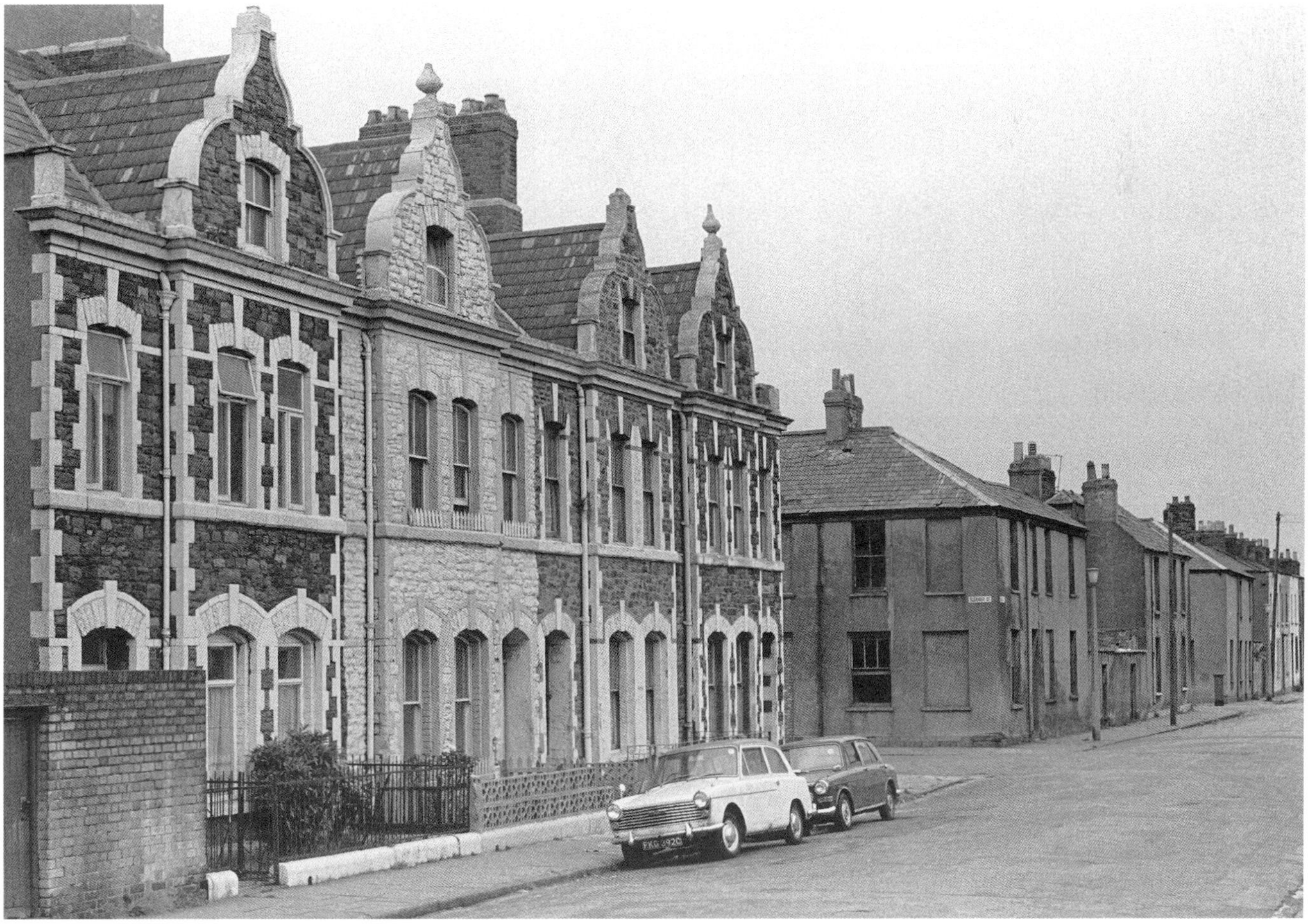

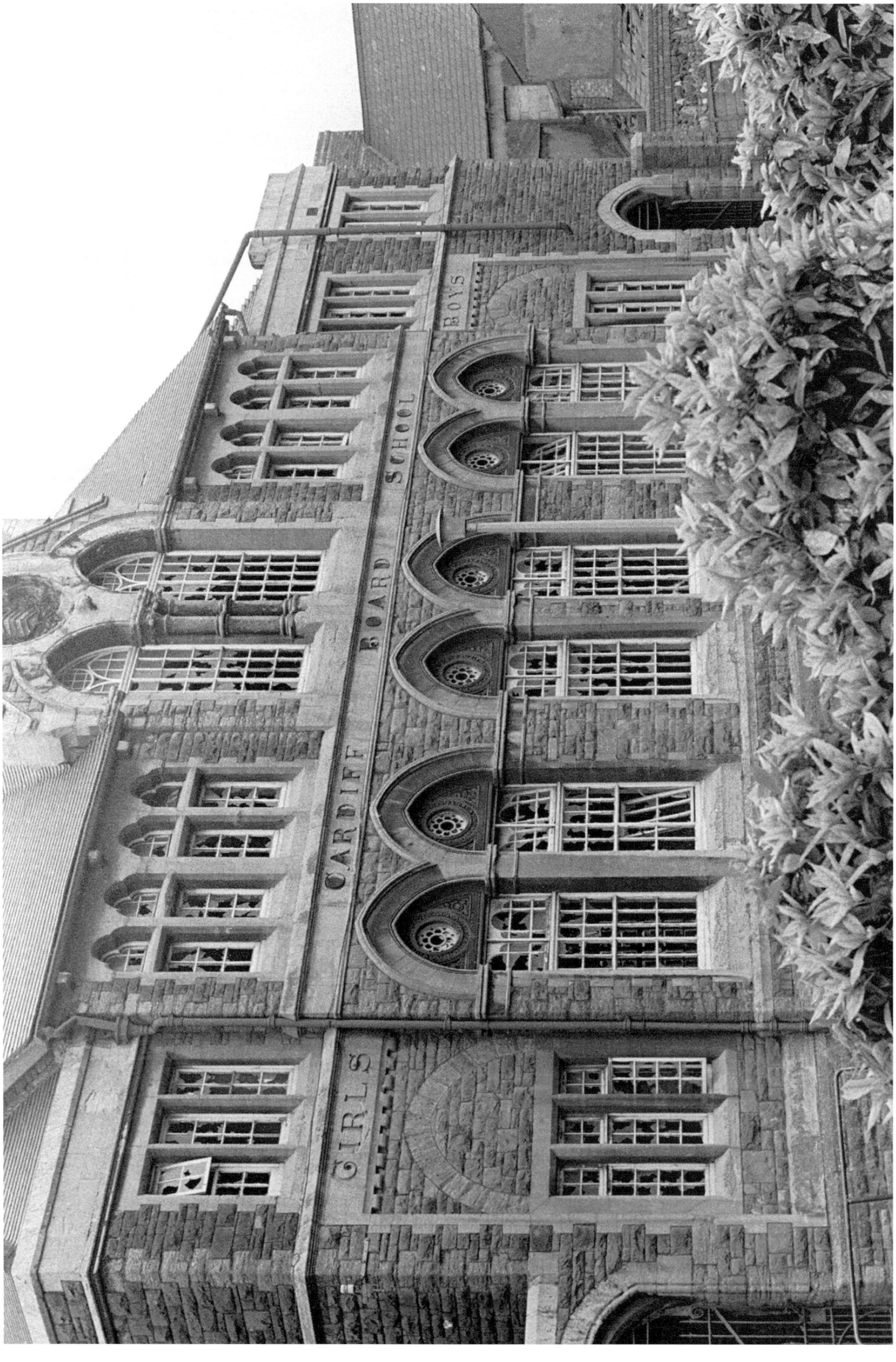
CARDIFF BOARD SCHOOL
GIRLS
BOYS

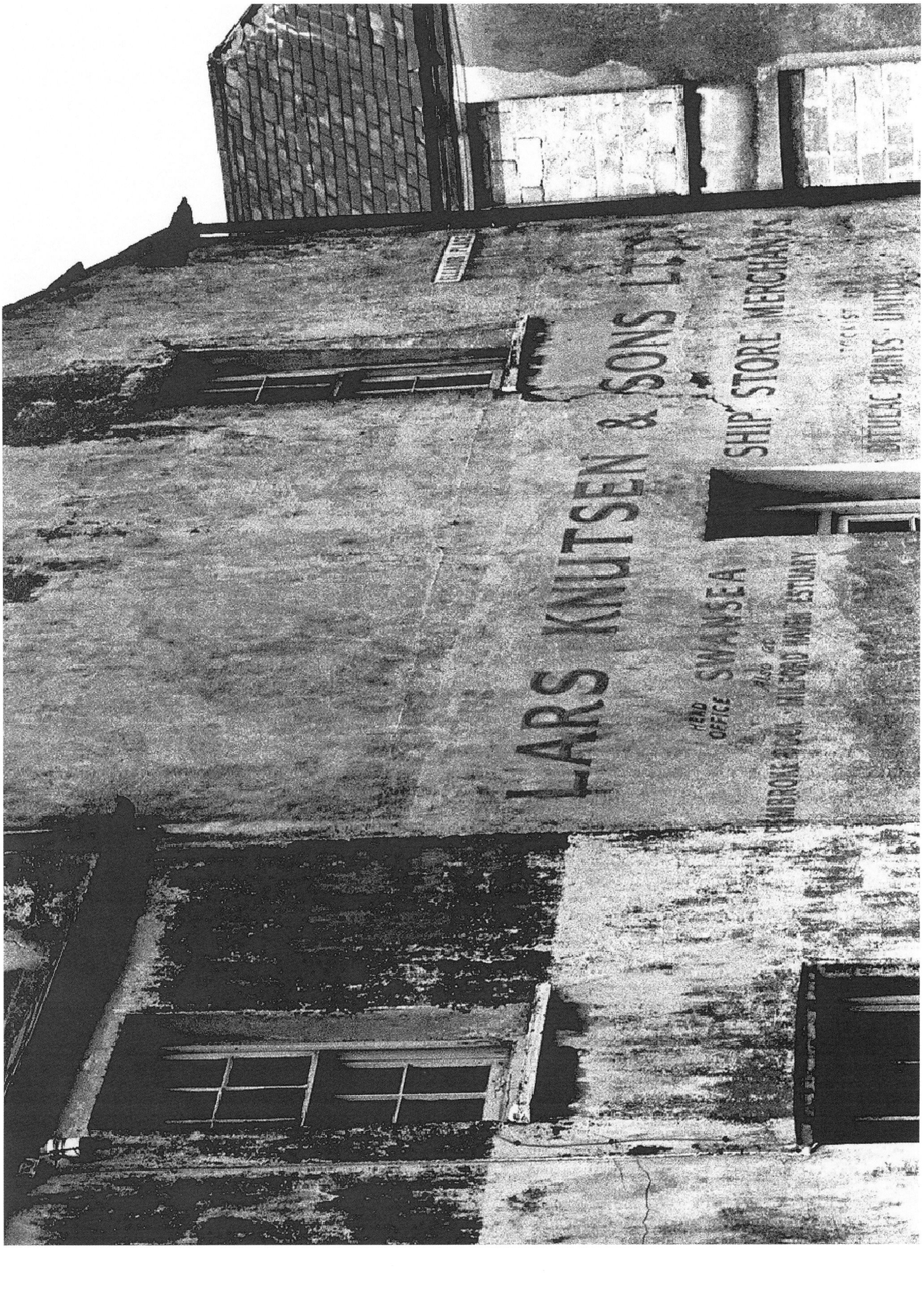
LARS KNUTSEN & SONS
HEAD OFFICE SWANSEA
SHIP STORE MERCHANTS

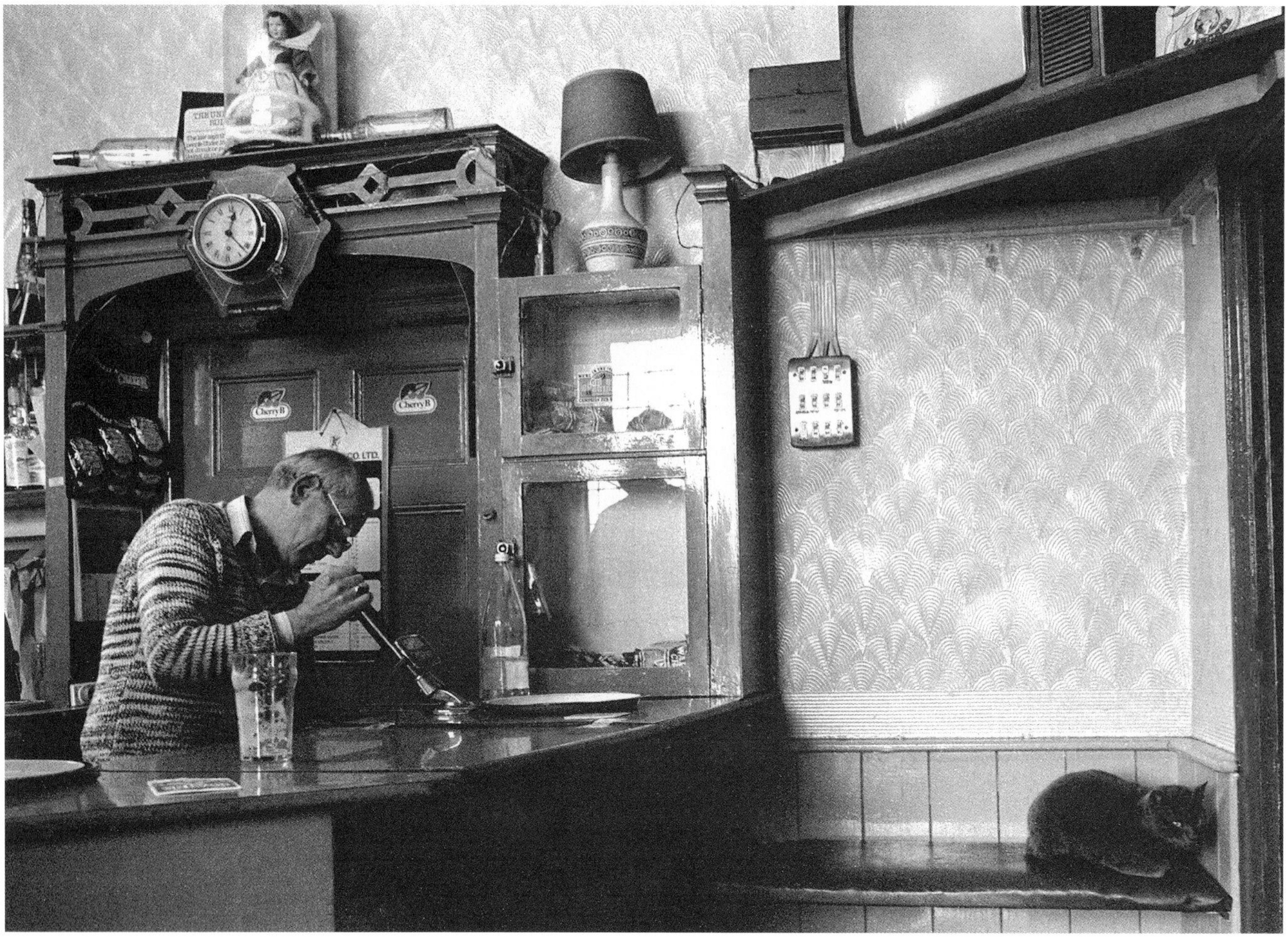
Cherry B
Cherry B

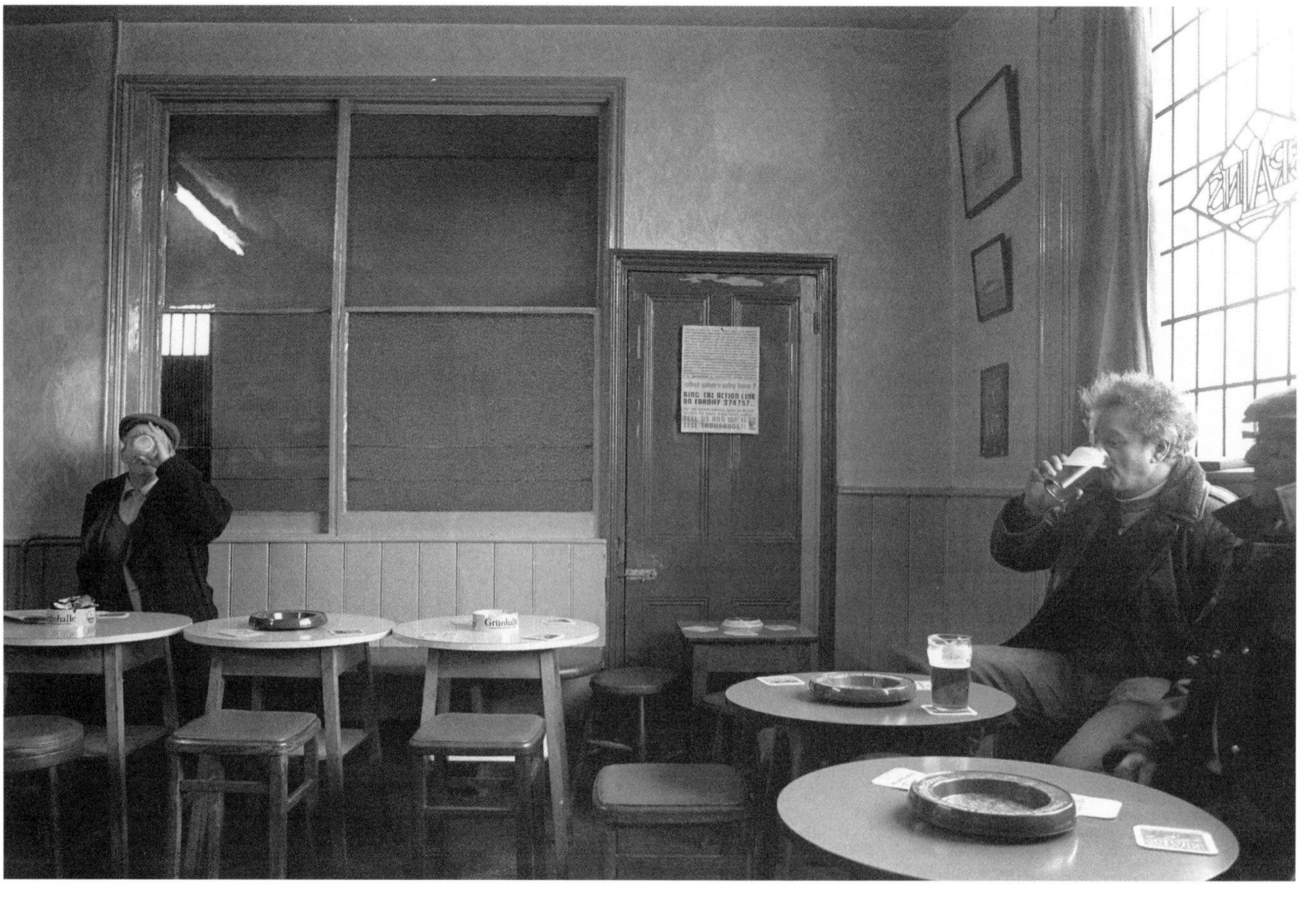

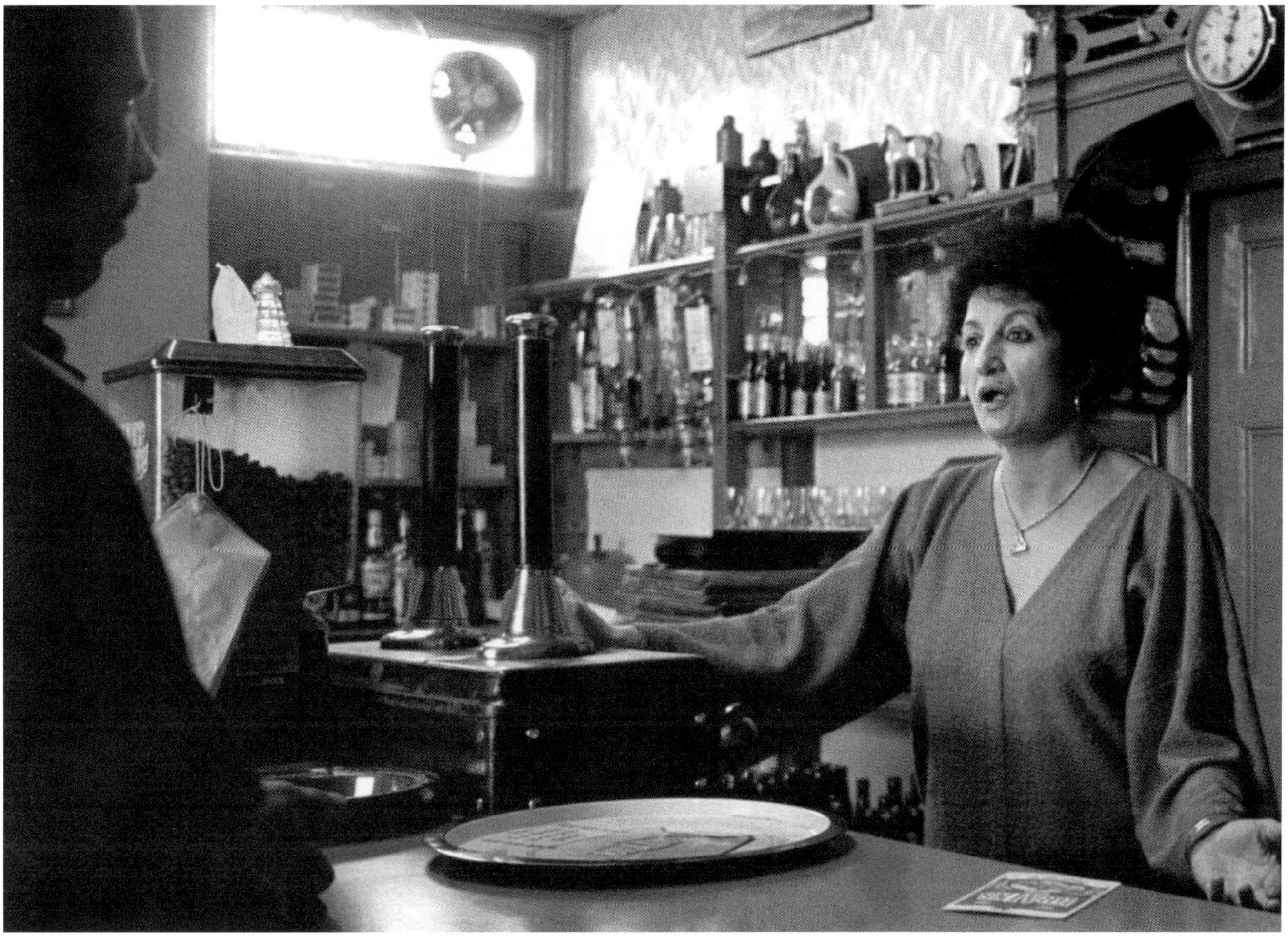

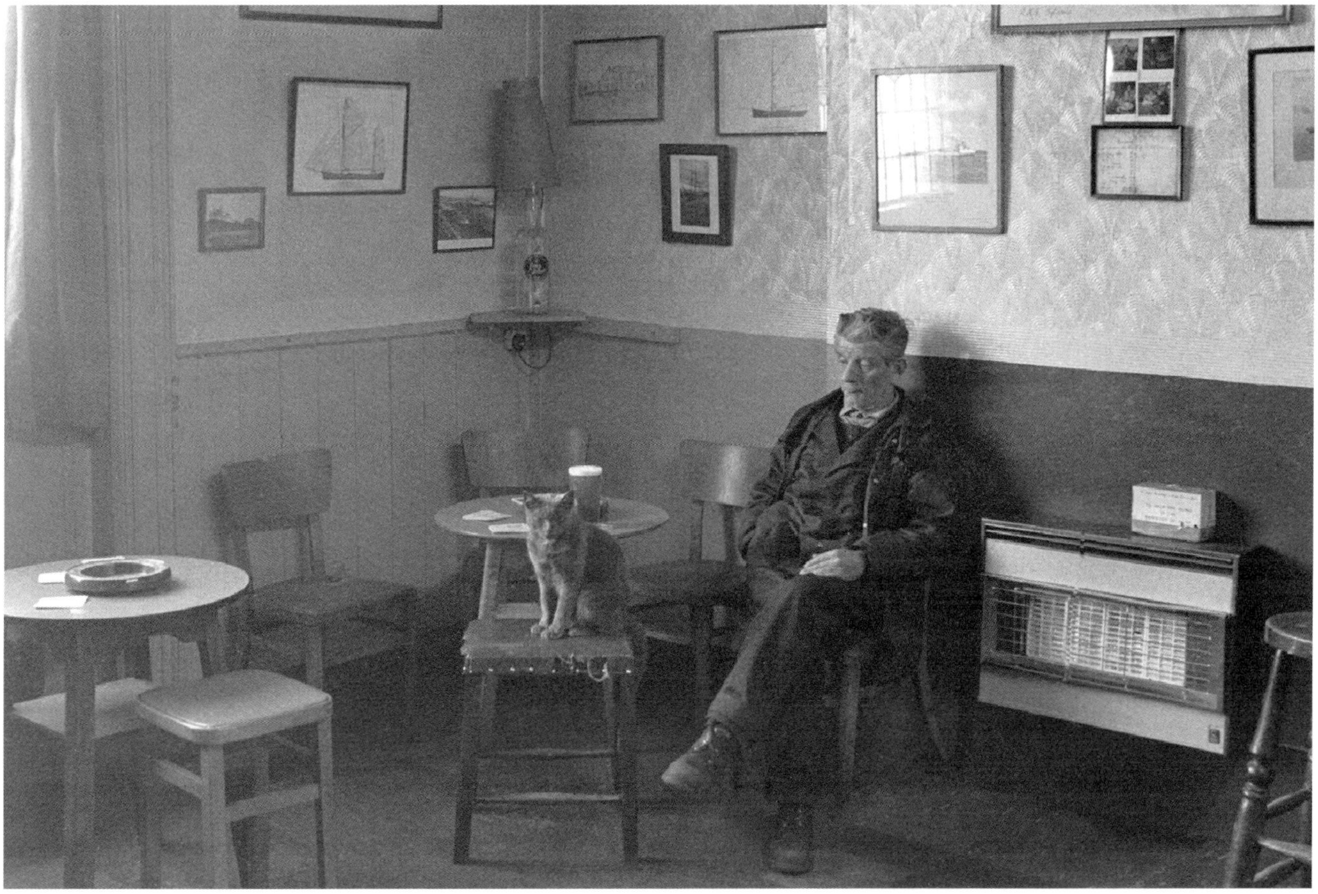

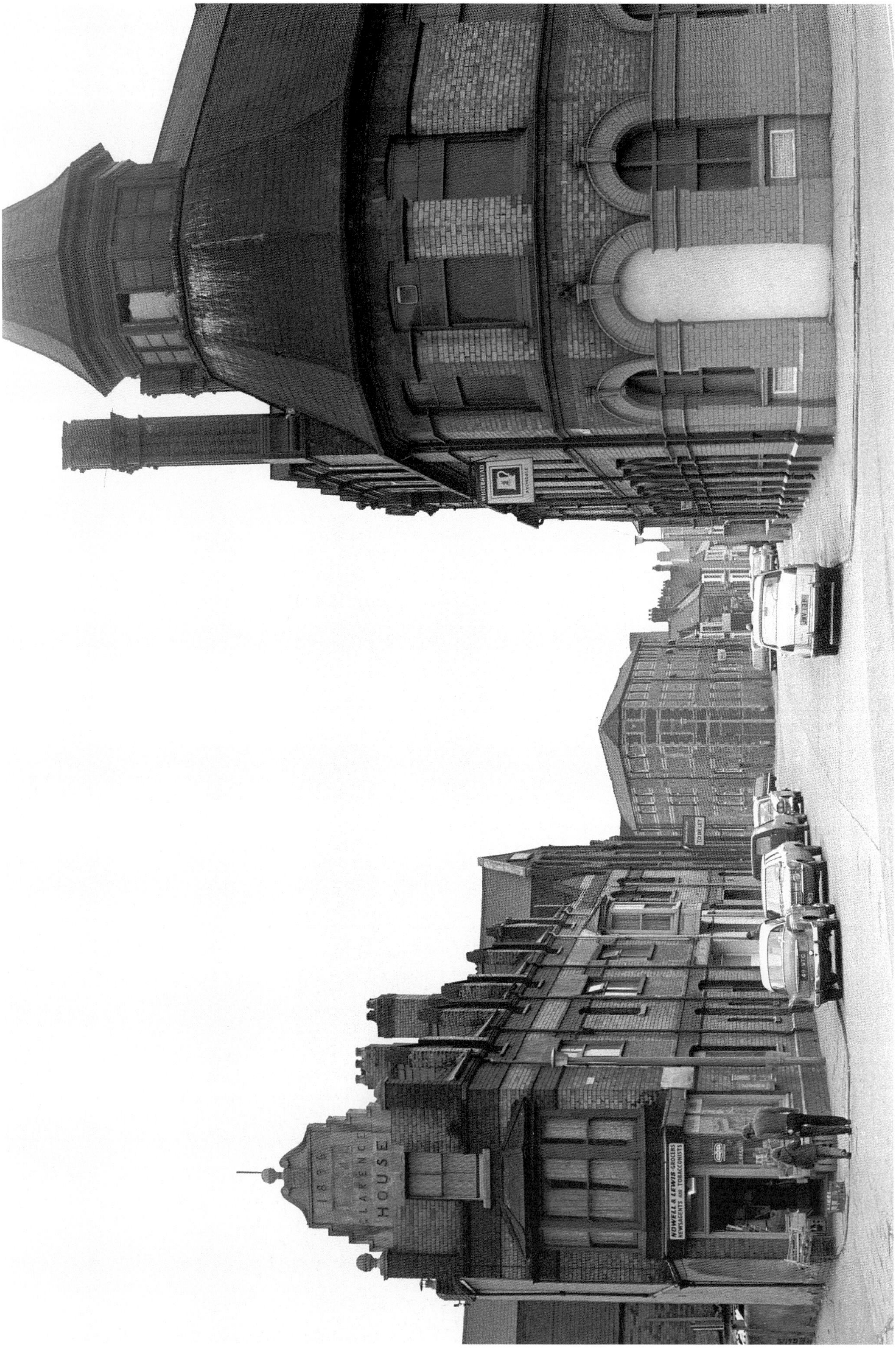
1896
CLARENCE HOUSE
NOWELL & LEWIS GROCERS
NEWSAGENTS AND TOBACCONISTS
WHITBREAD
TO BE LET

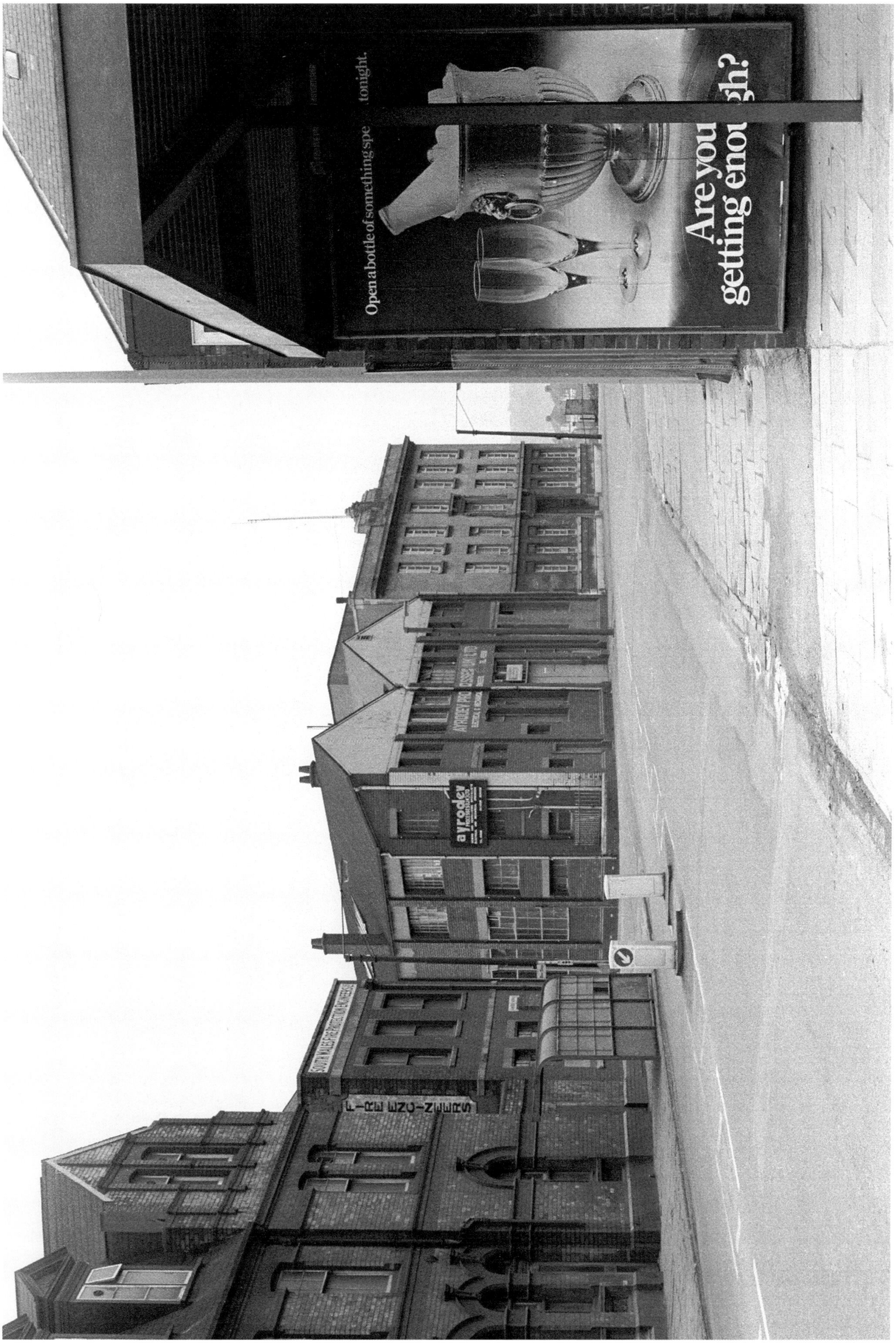
Open a bottle of something spe
tonight.
Are you
getting eno
gh?
ayrodev

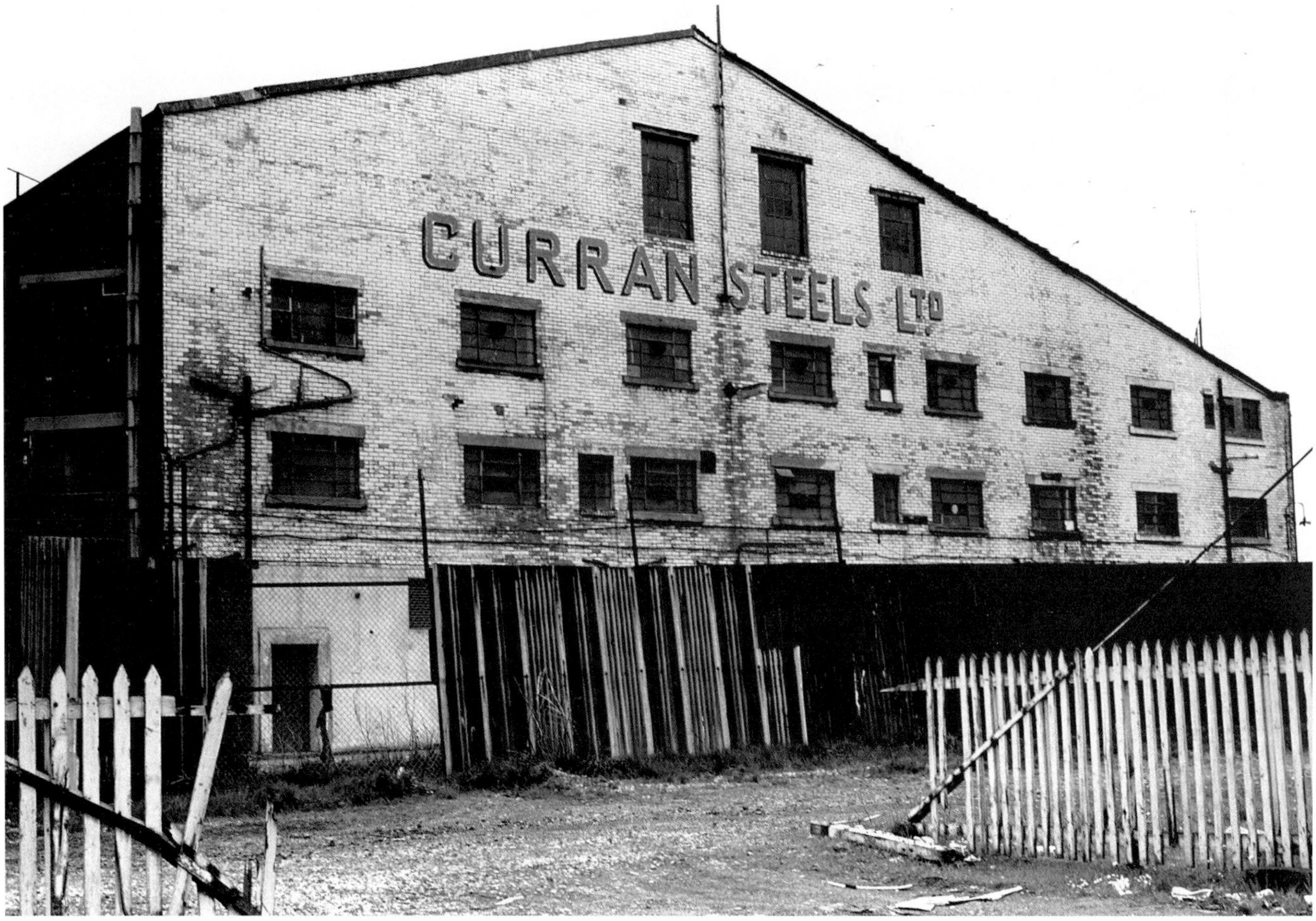
CURRAN STEELS LTD

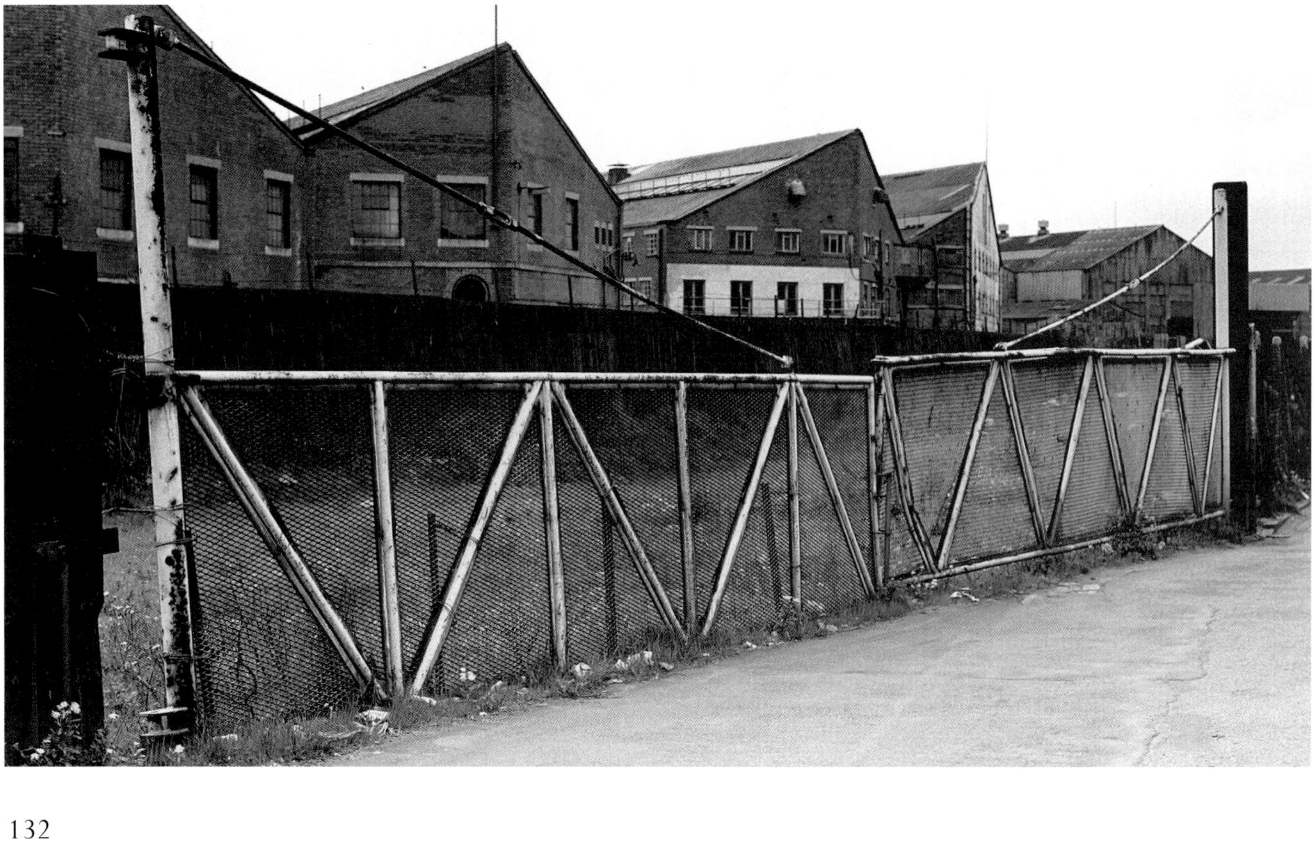

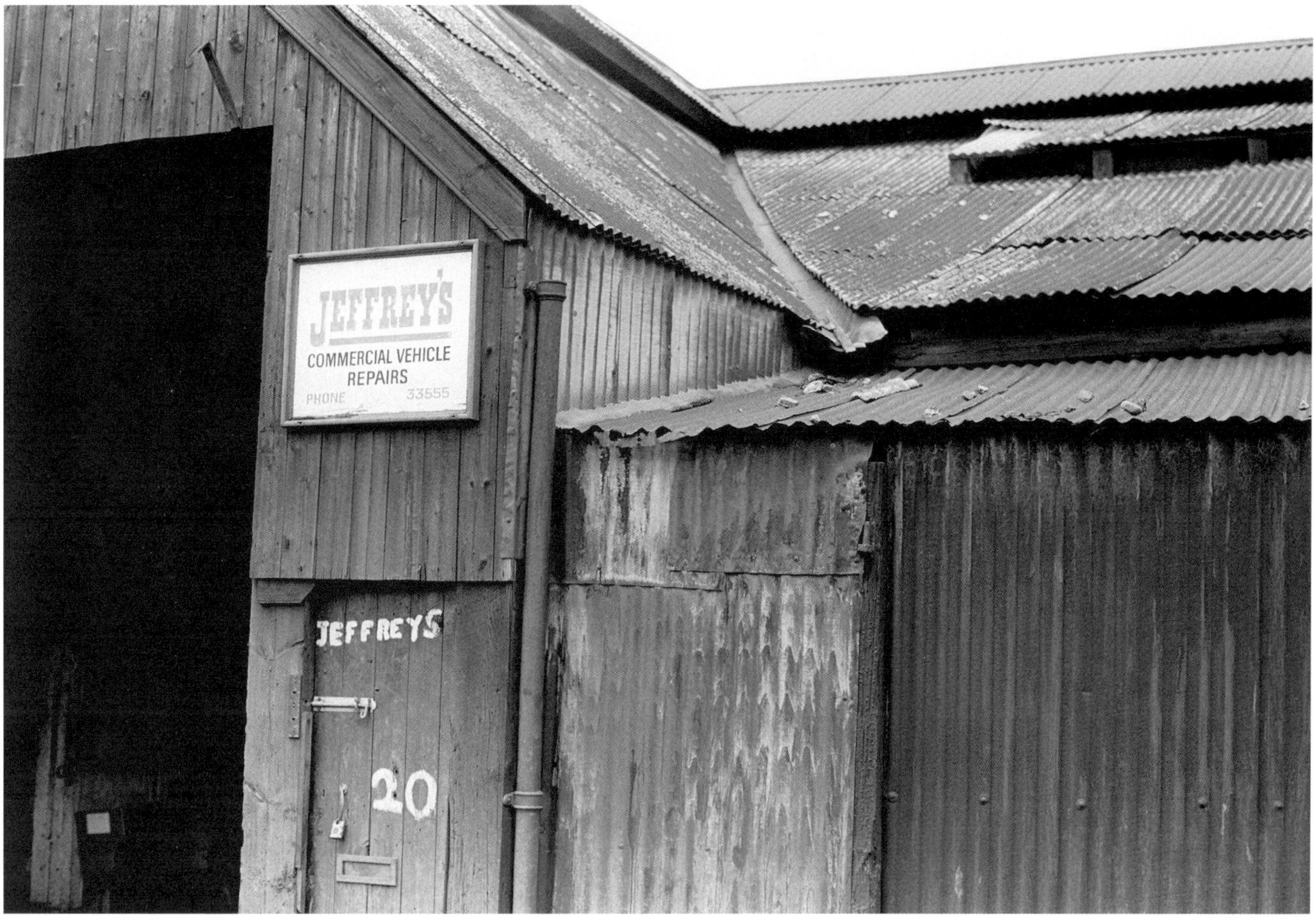
JEFFREY'S
COMMERCIAL VEHICLE
REPAIRS
PHONE 33555
JEFFREYS
20

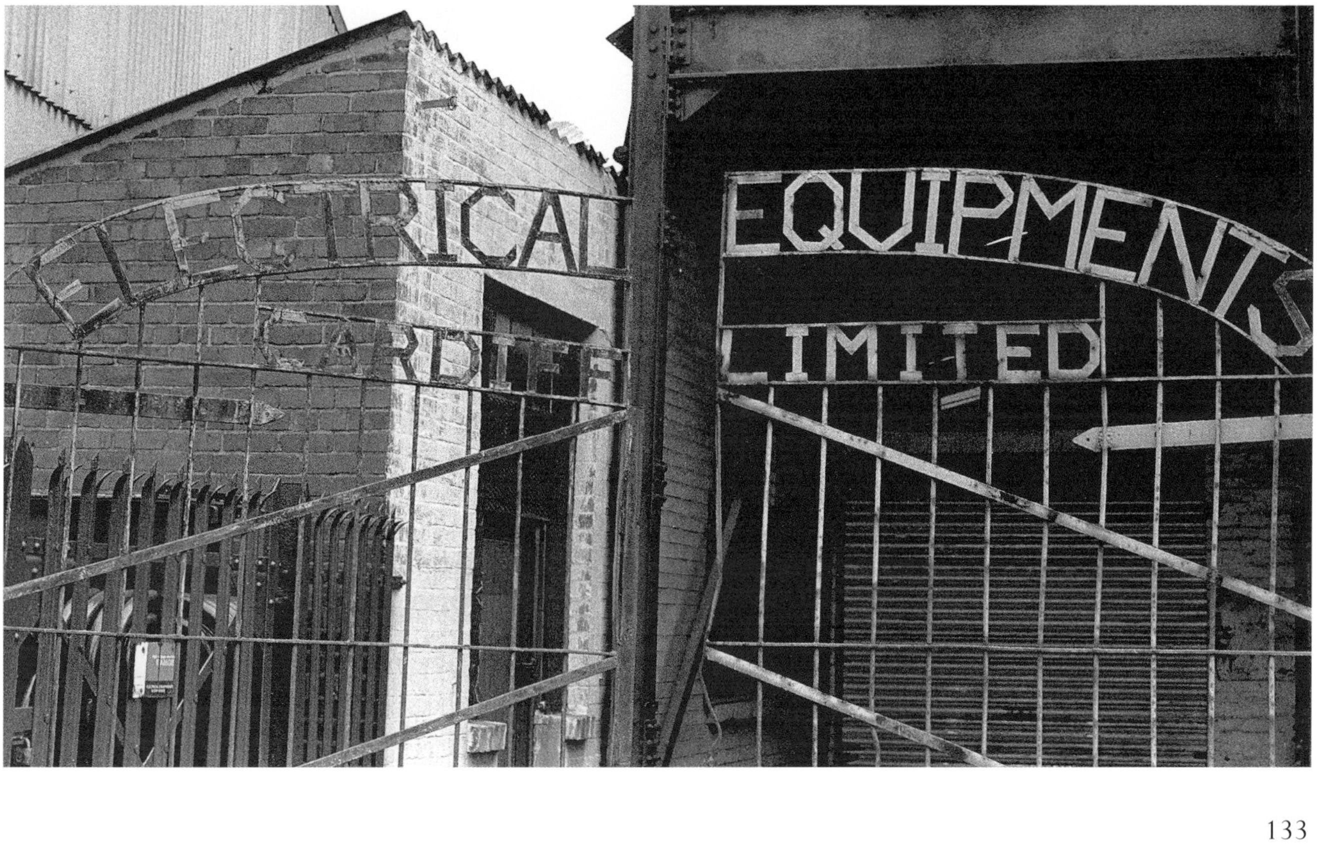
ELECTRICAL
CARDIFF
EQUIPMENTS
LIMITED

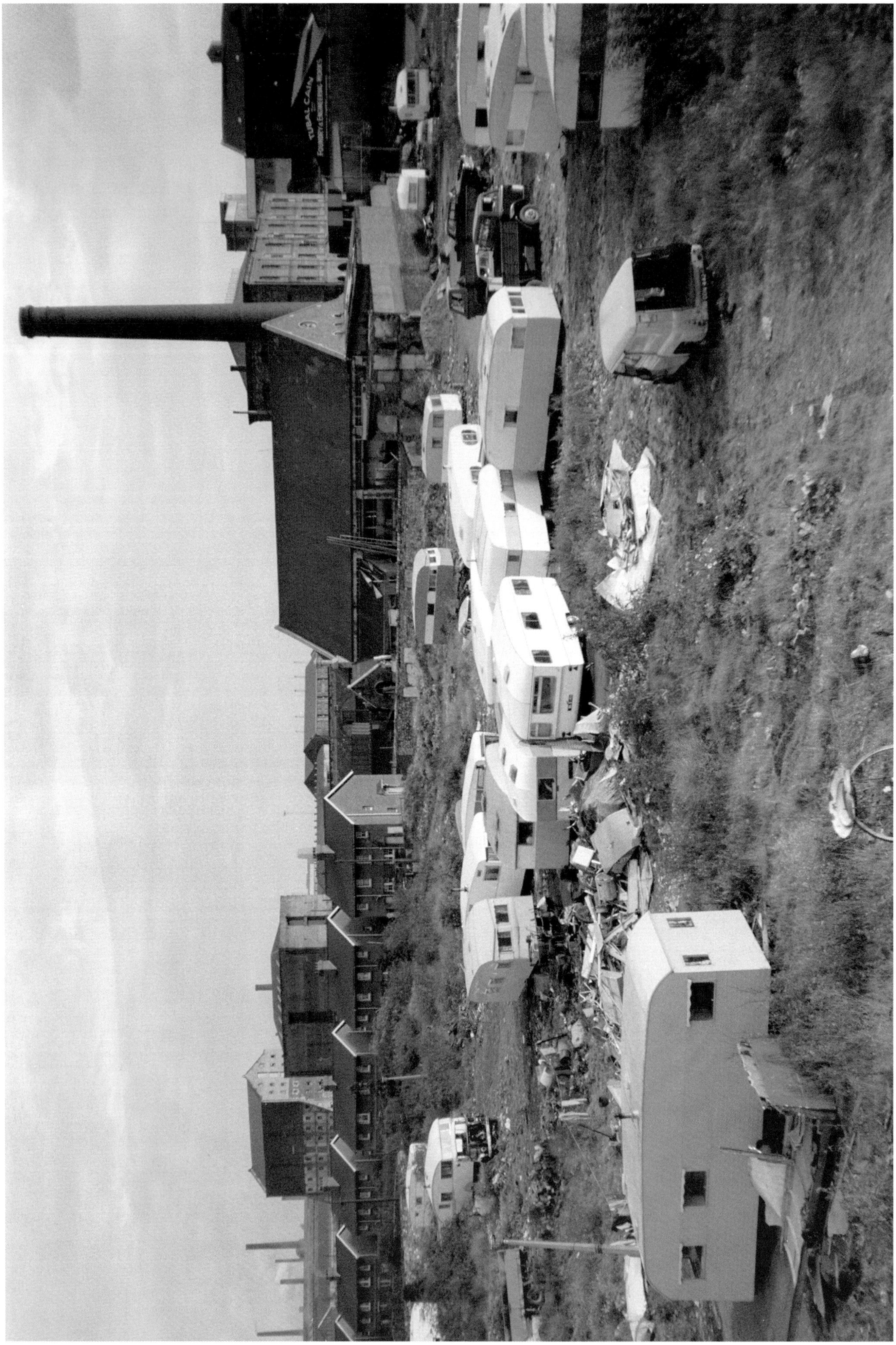
TUBAL CAIN

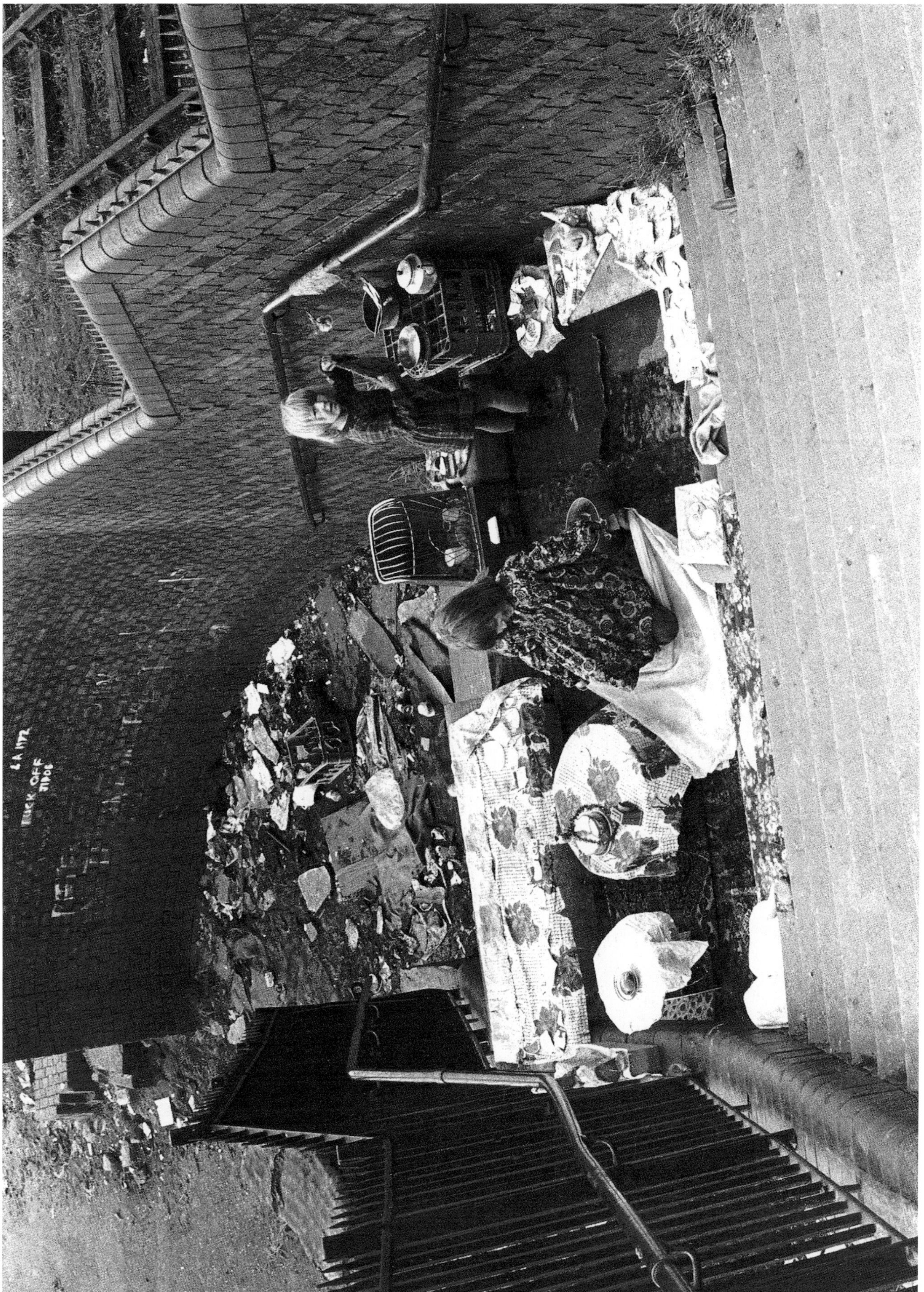

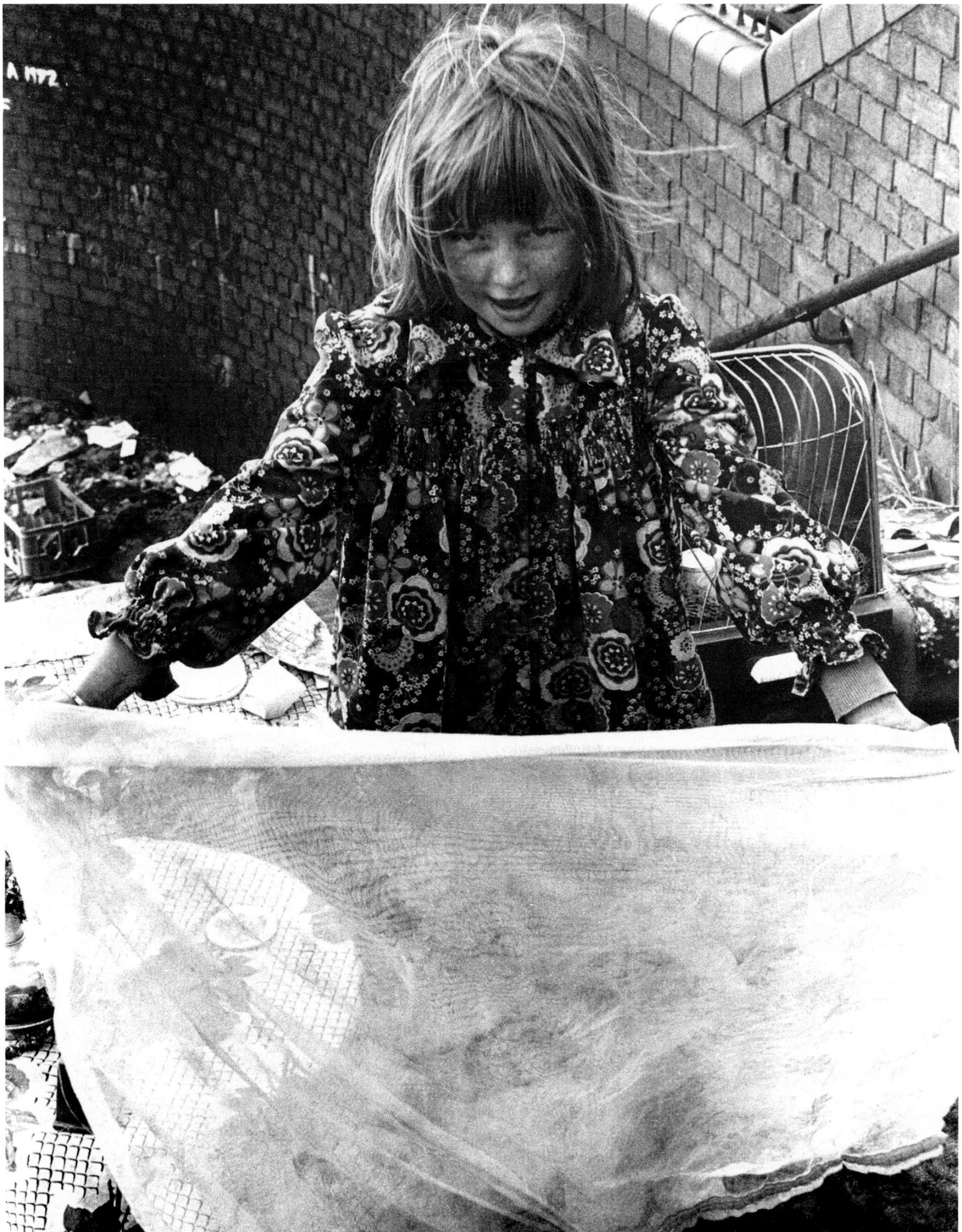
A 1972

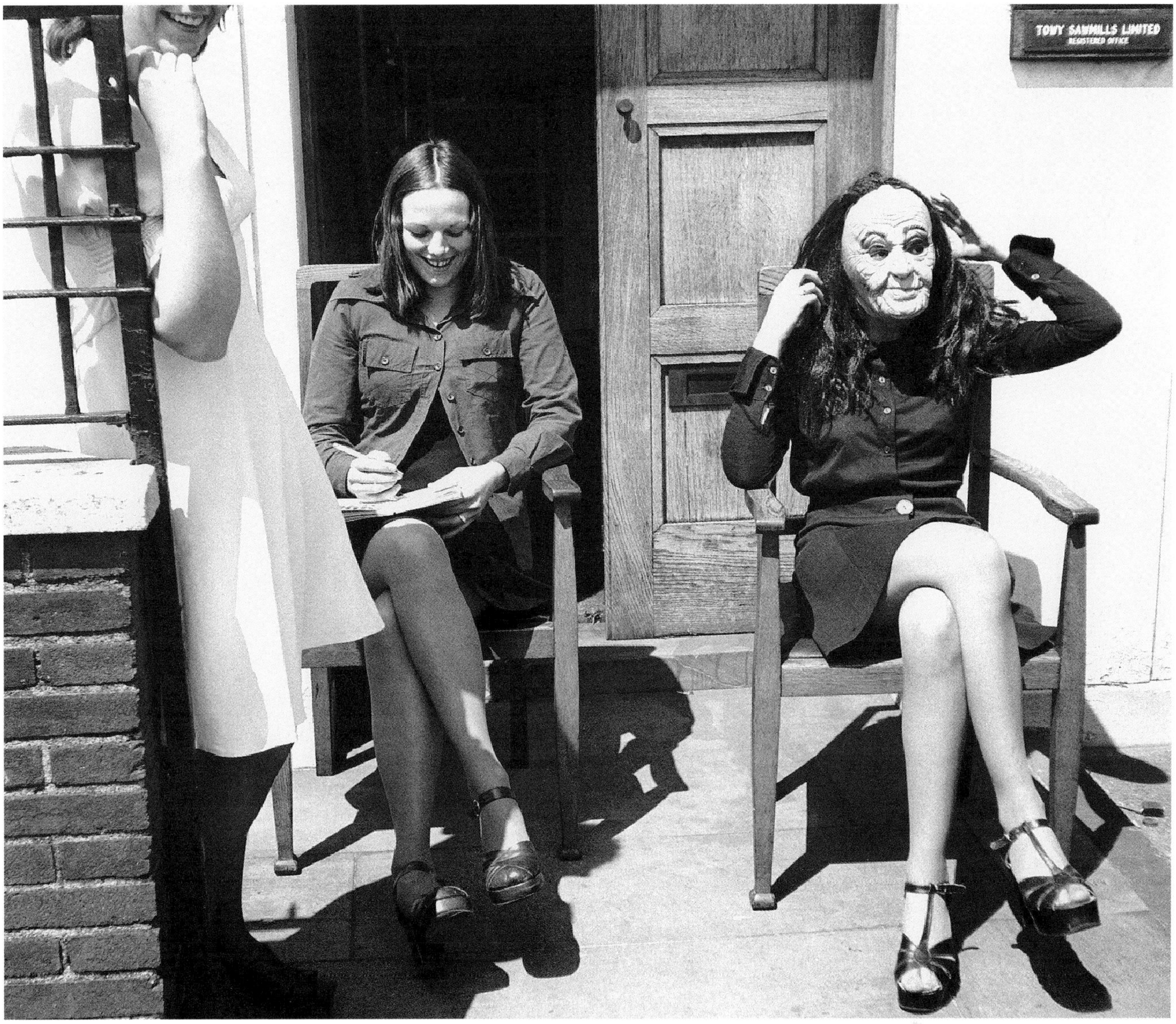
TOWY SAWMILLS LIMITED
REGISTERED OFFICE

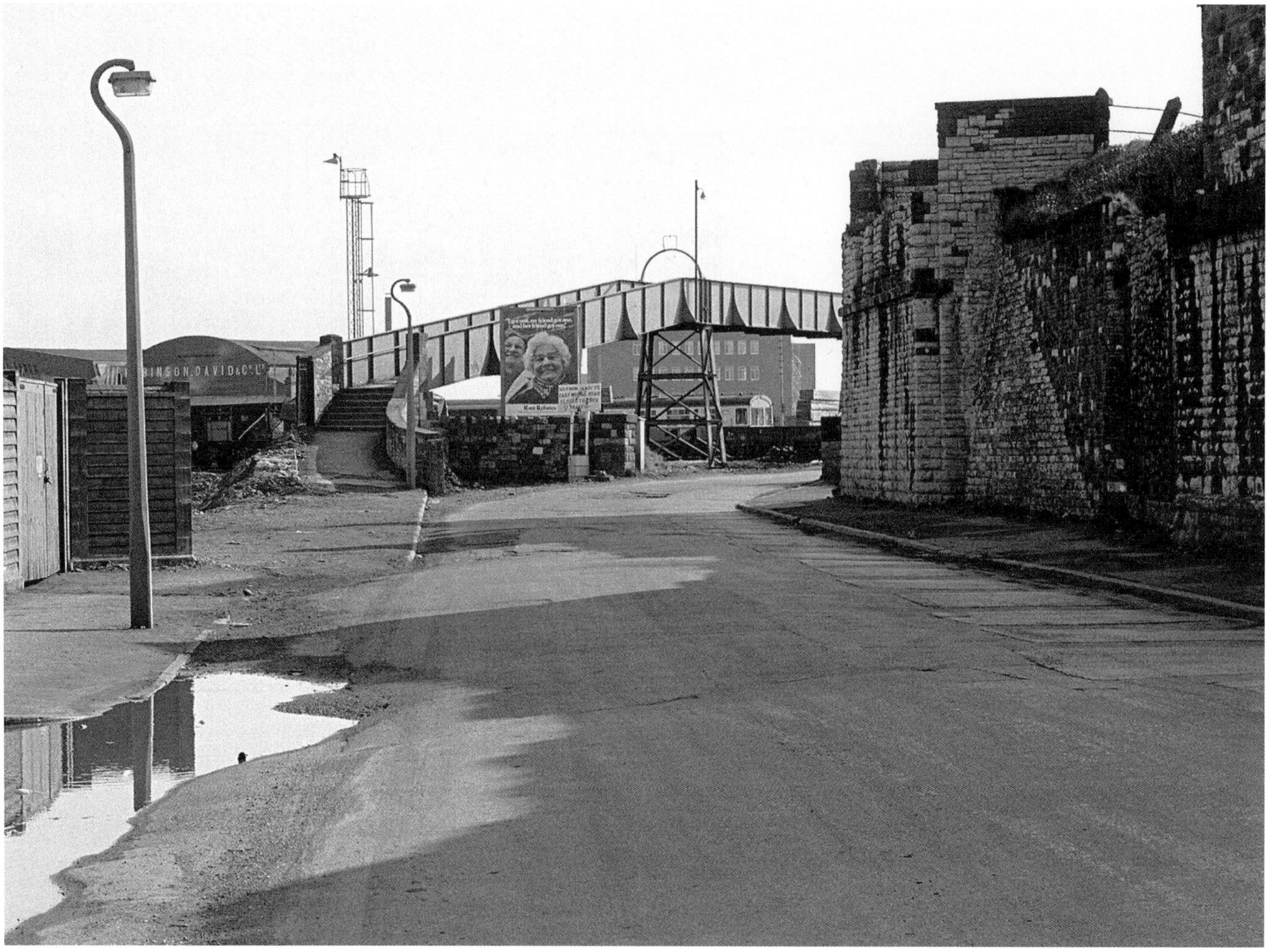
DAVID&C°

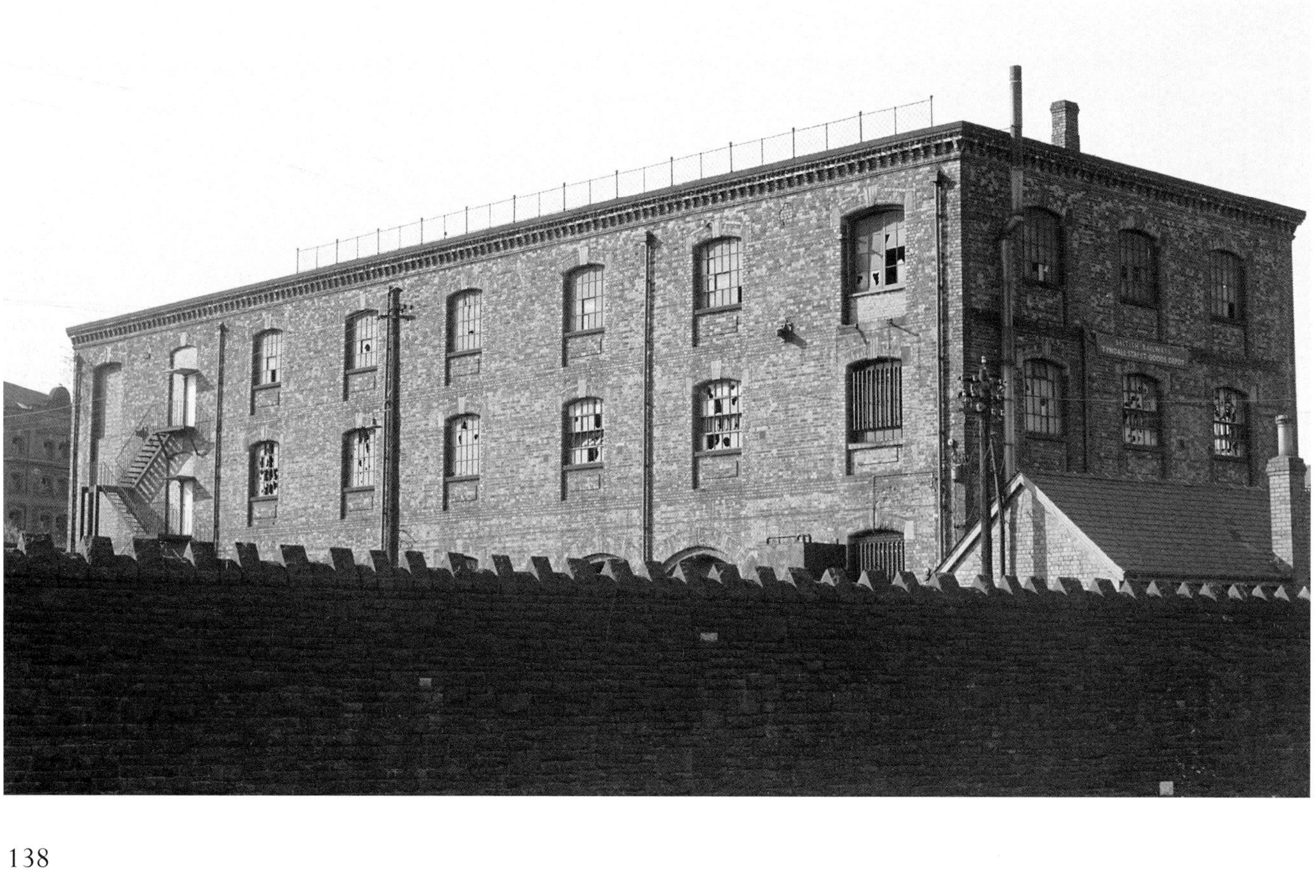

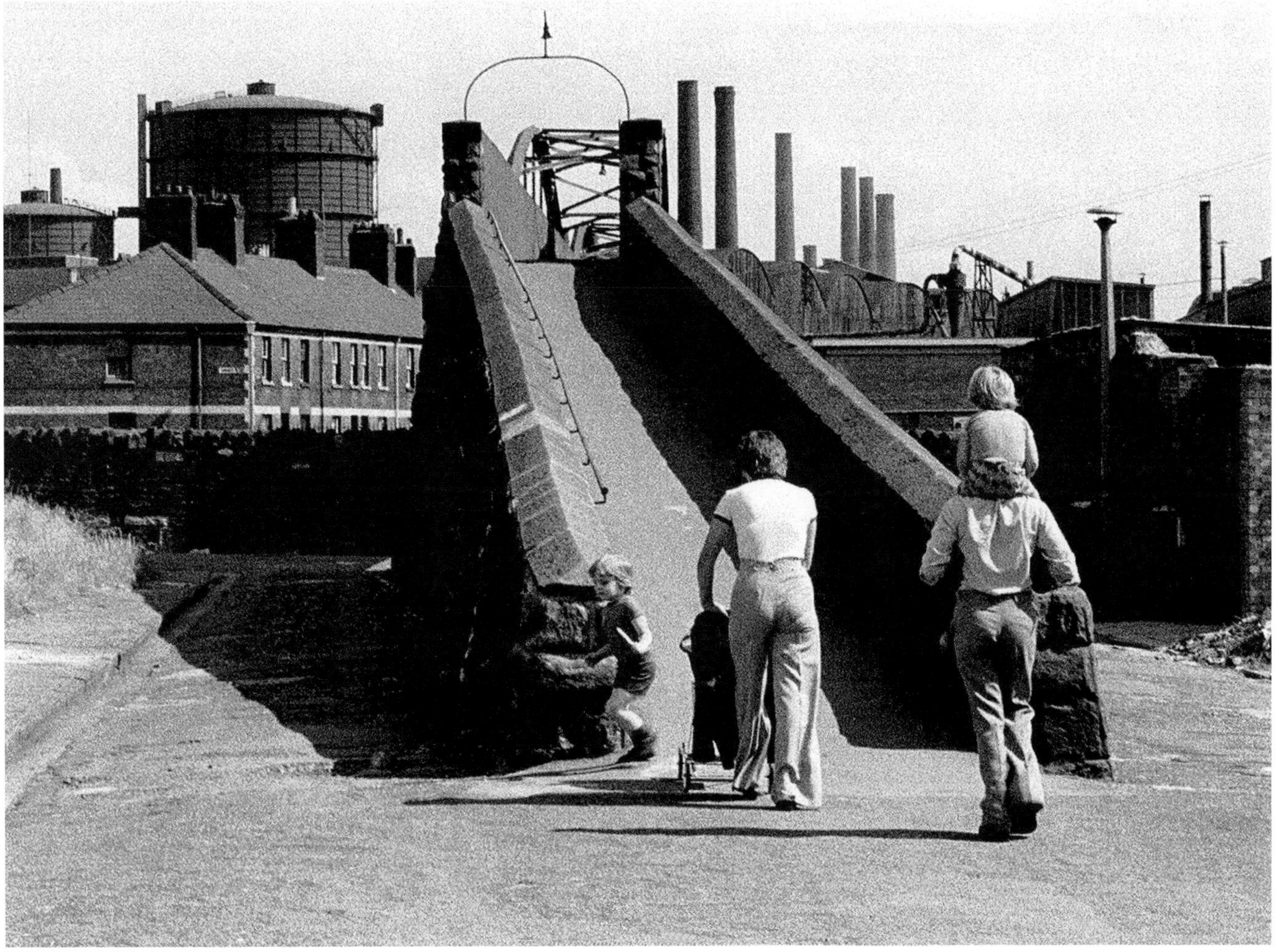

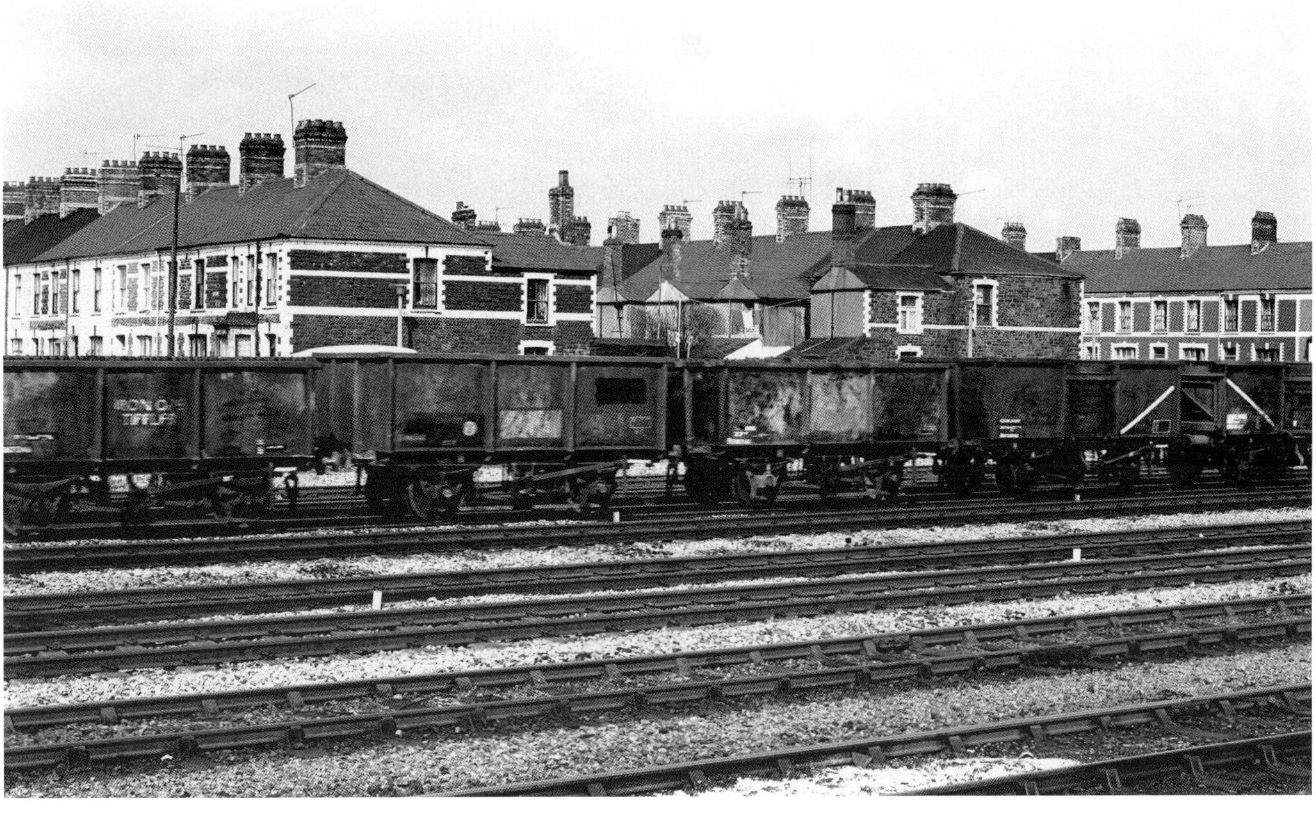

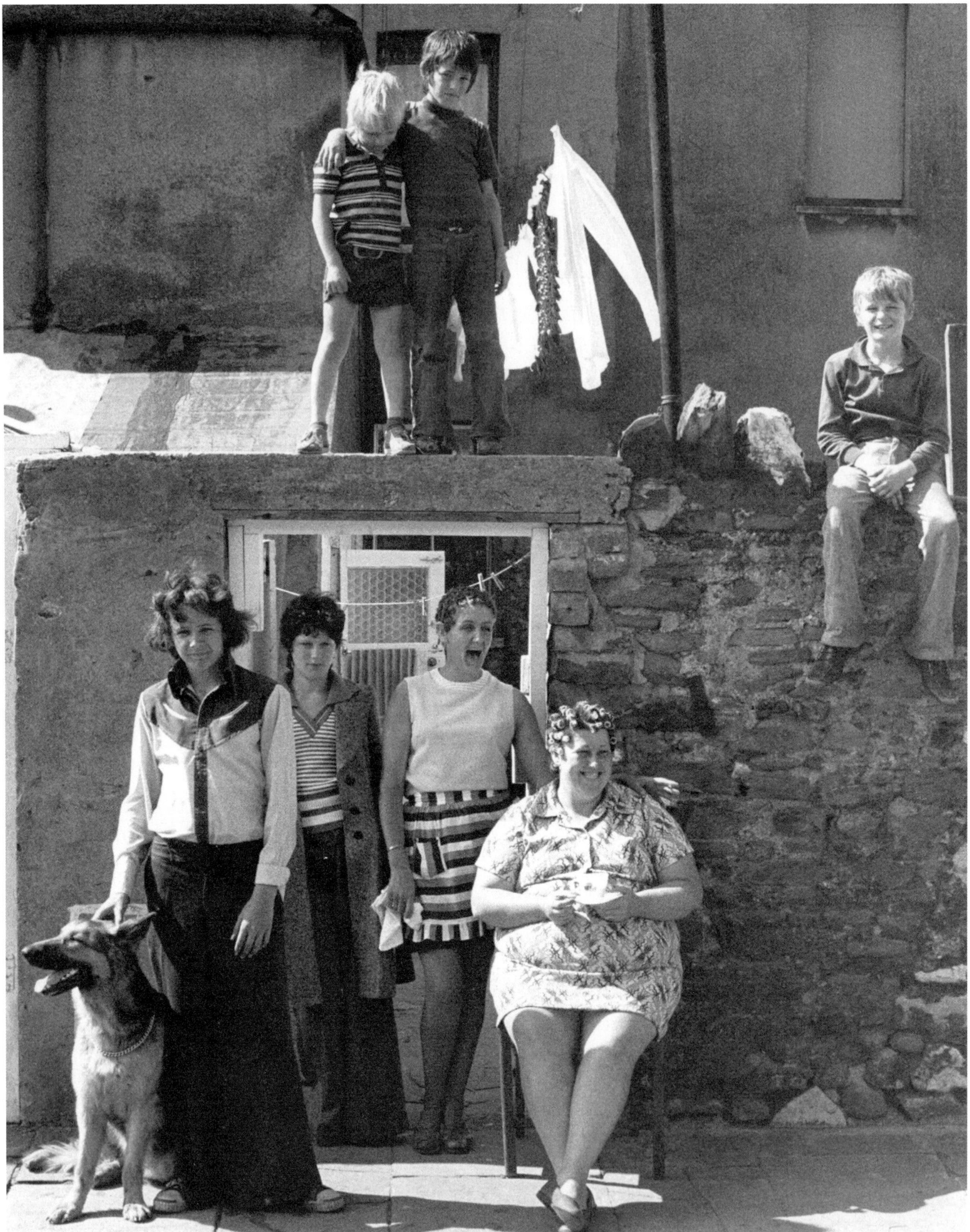

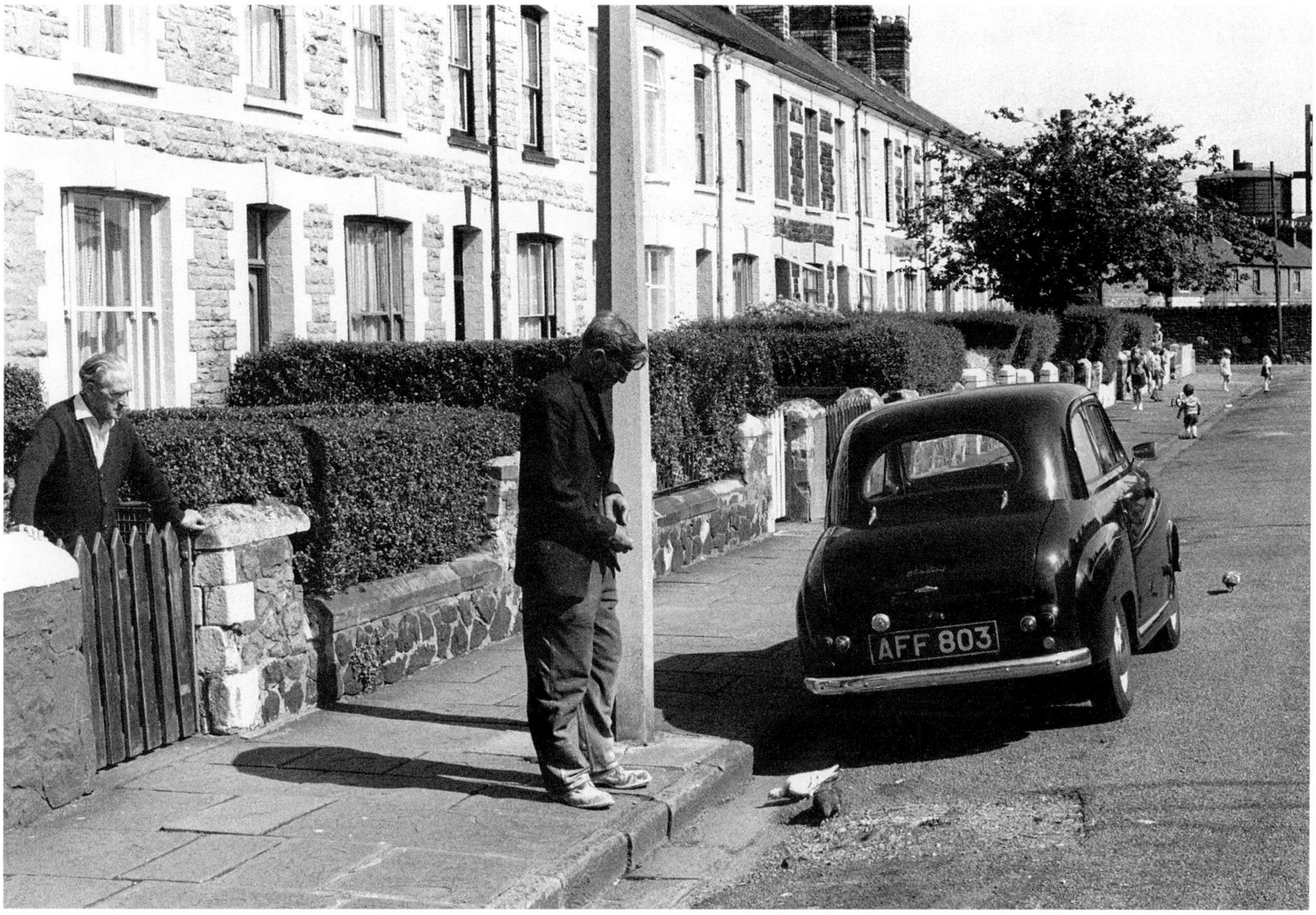
AFF 803

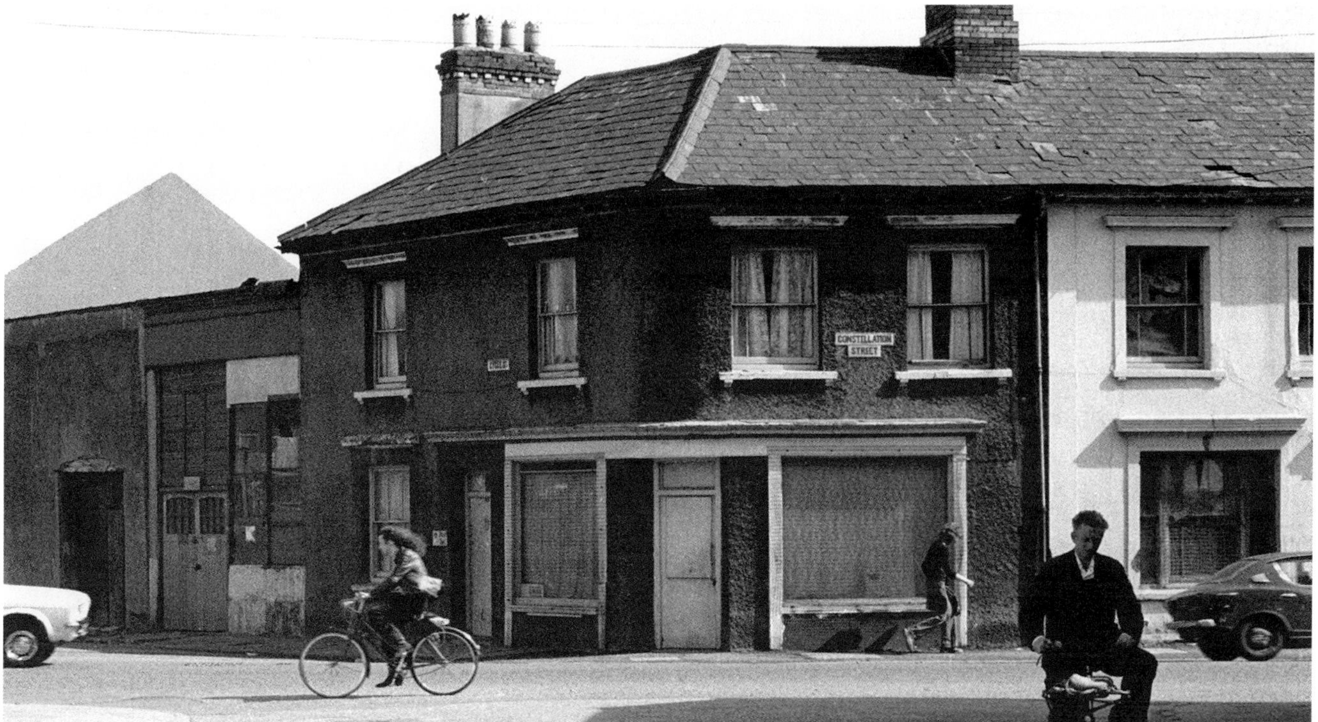
CONSTELLATION
STREET

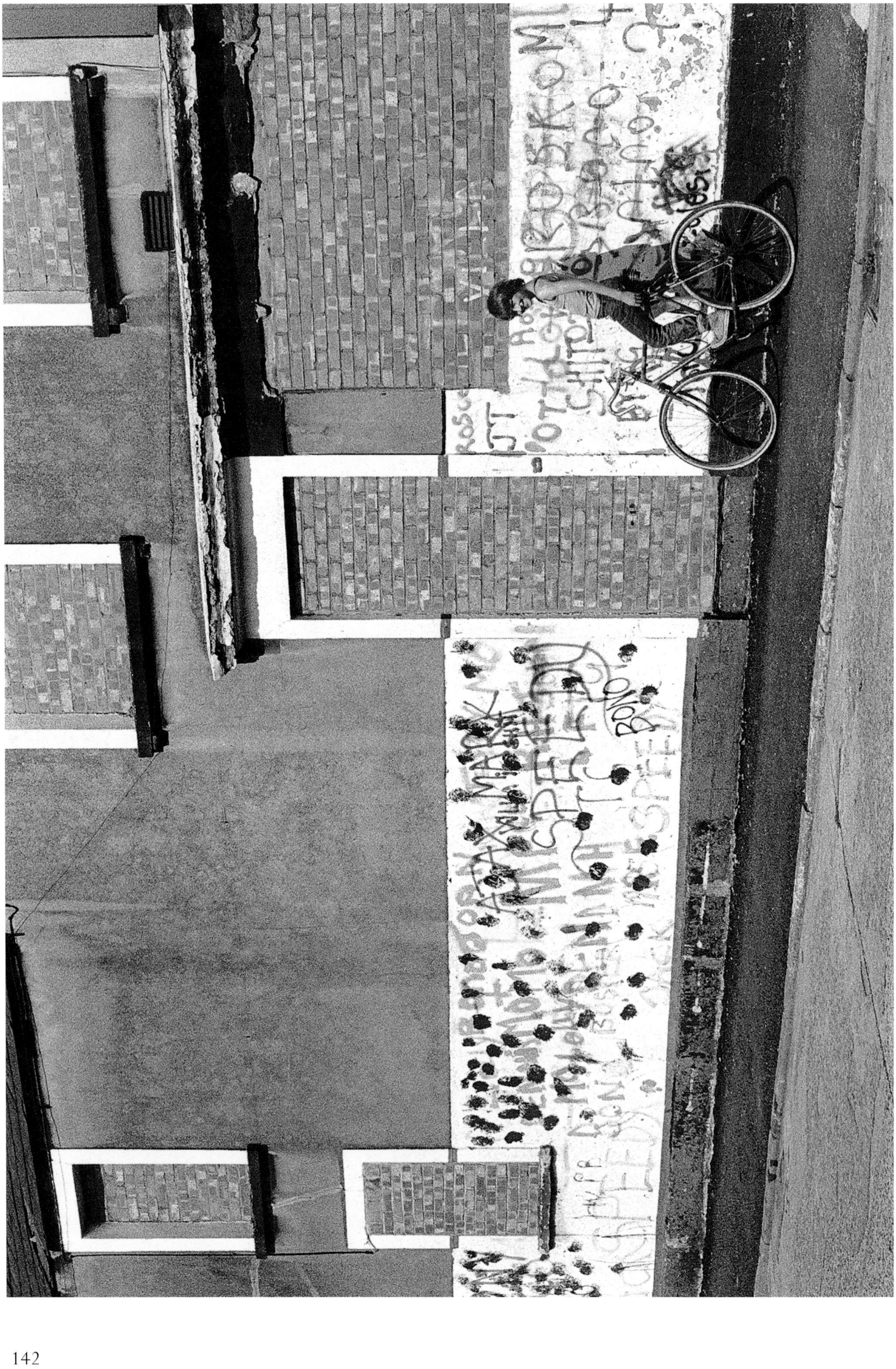
SPEEDY
BONO

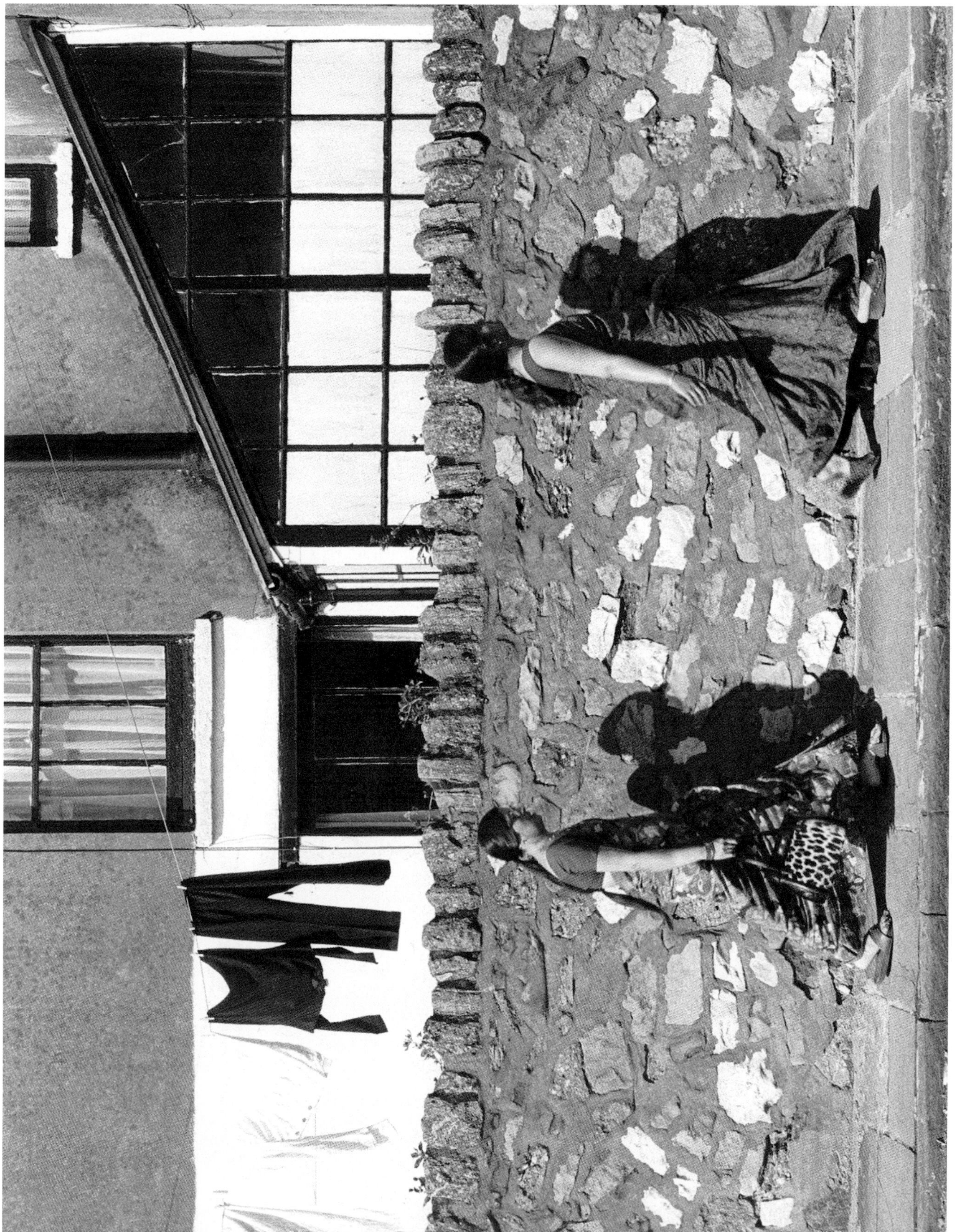

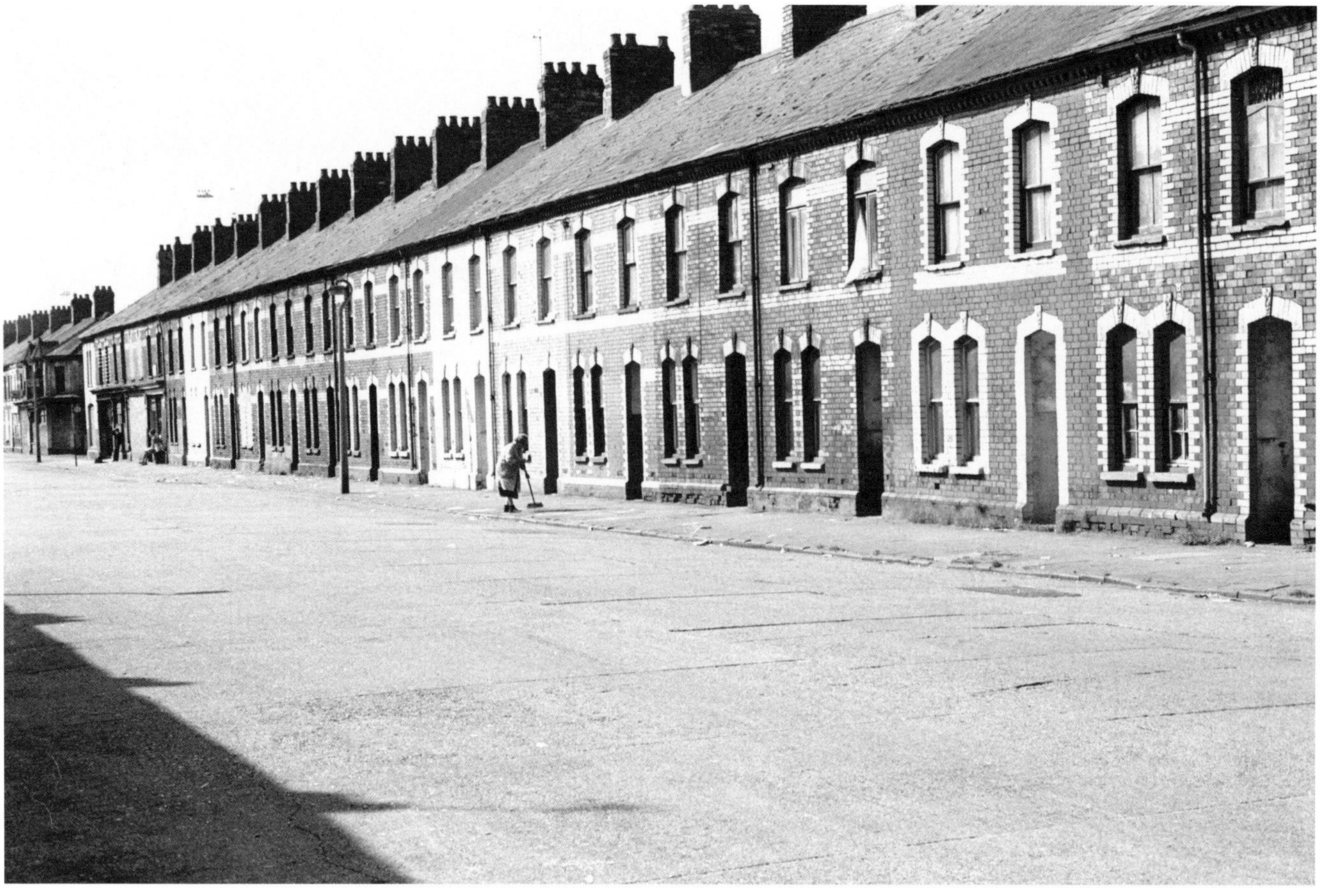

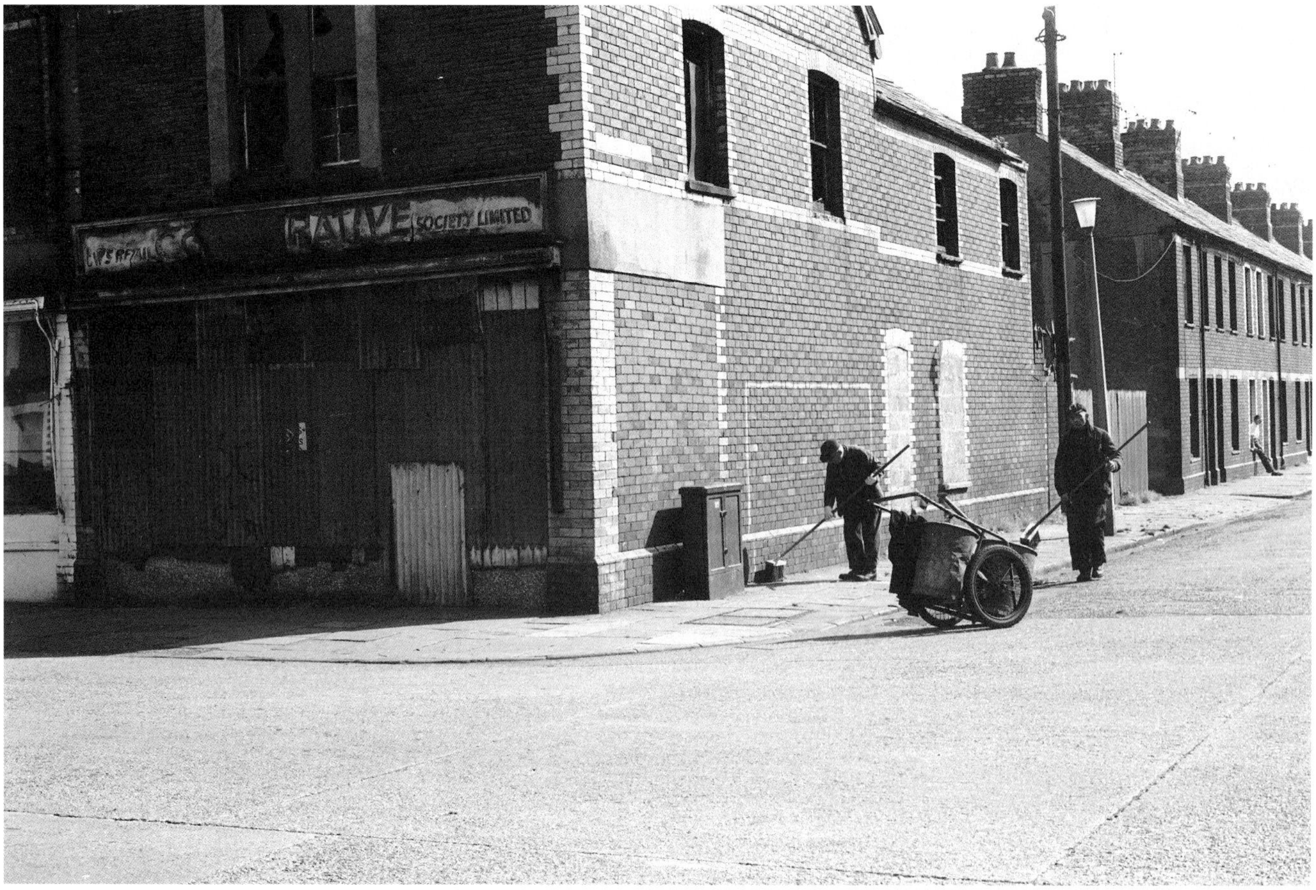
RATIVE SOCIETY LIMITED

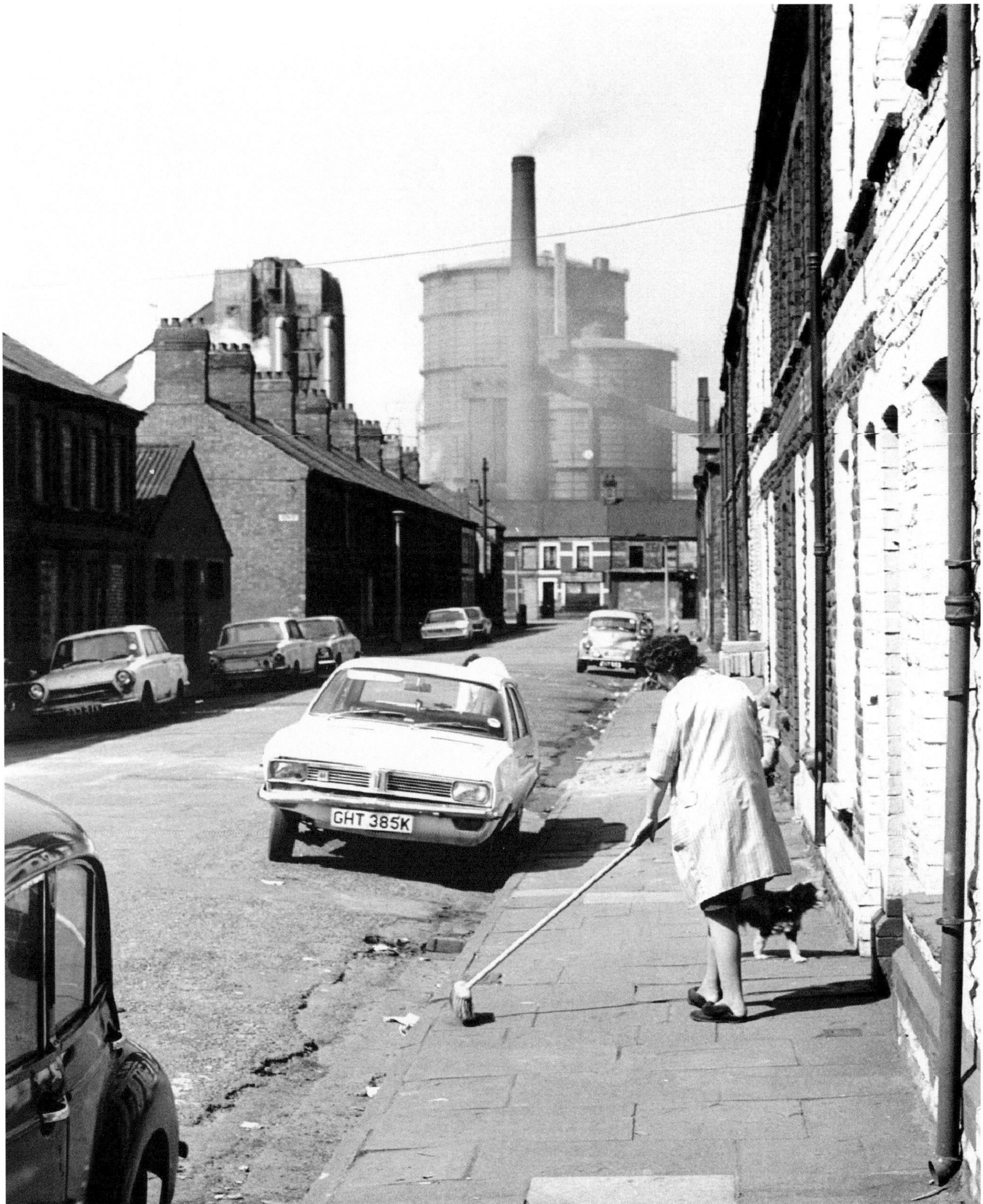
GHT 385K

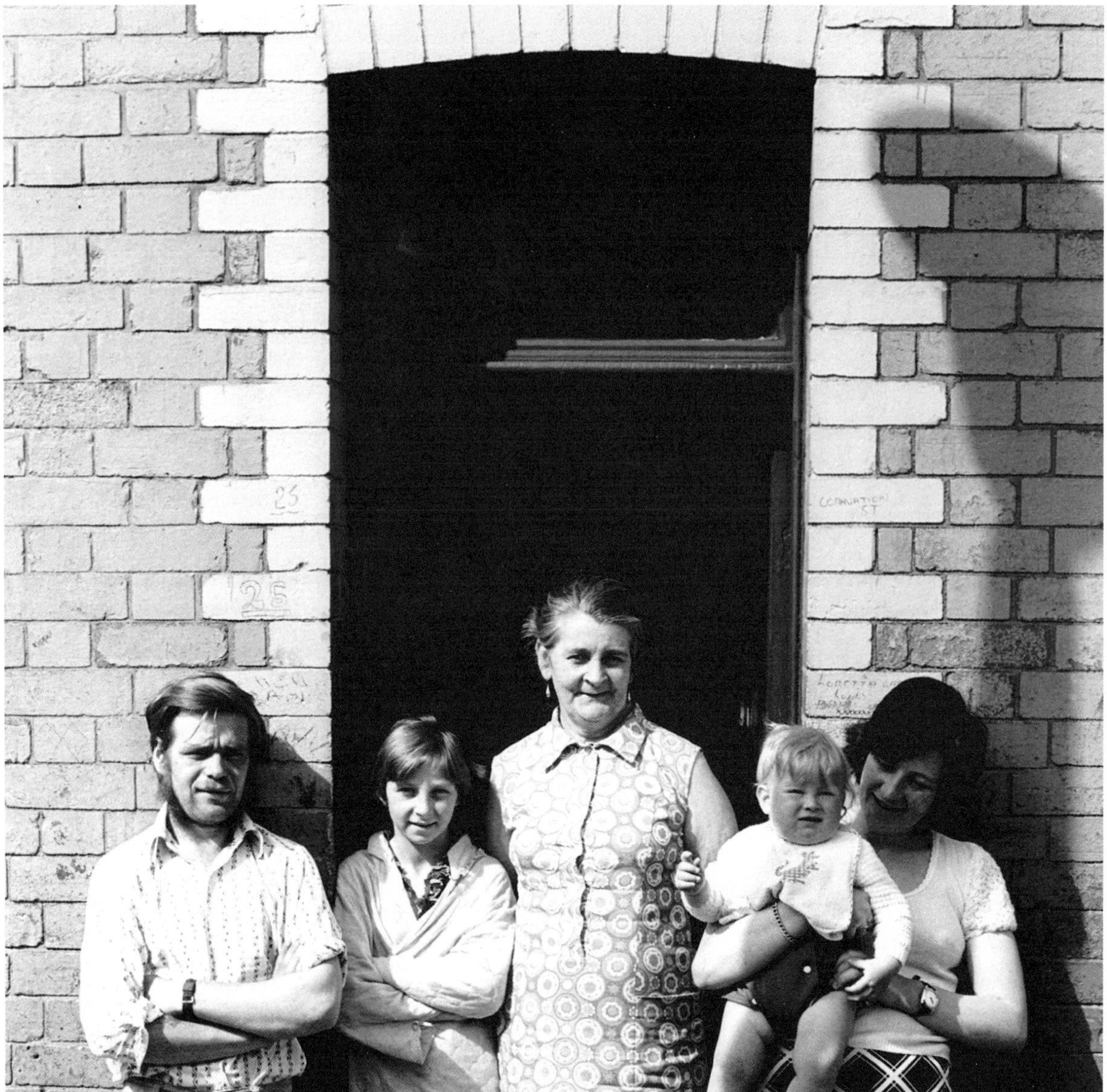
25
25

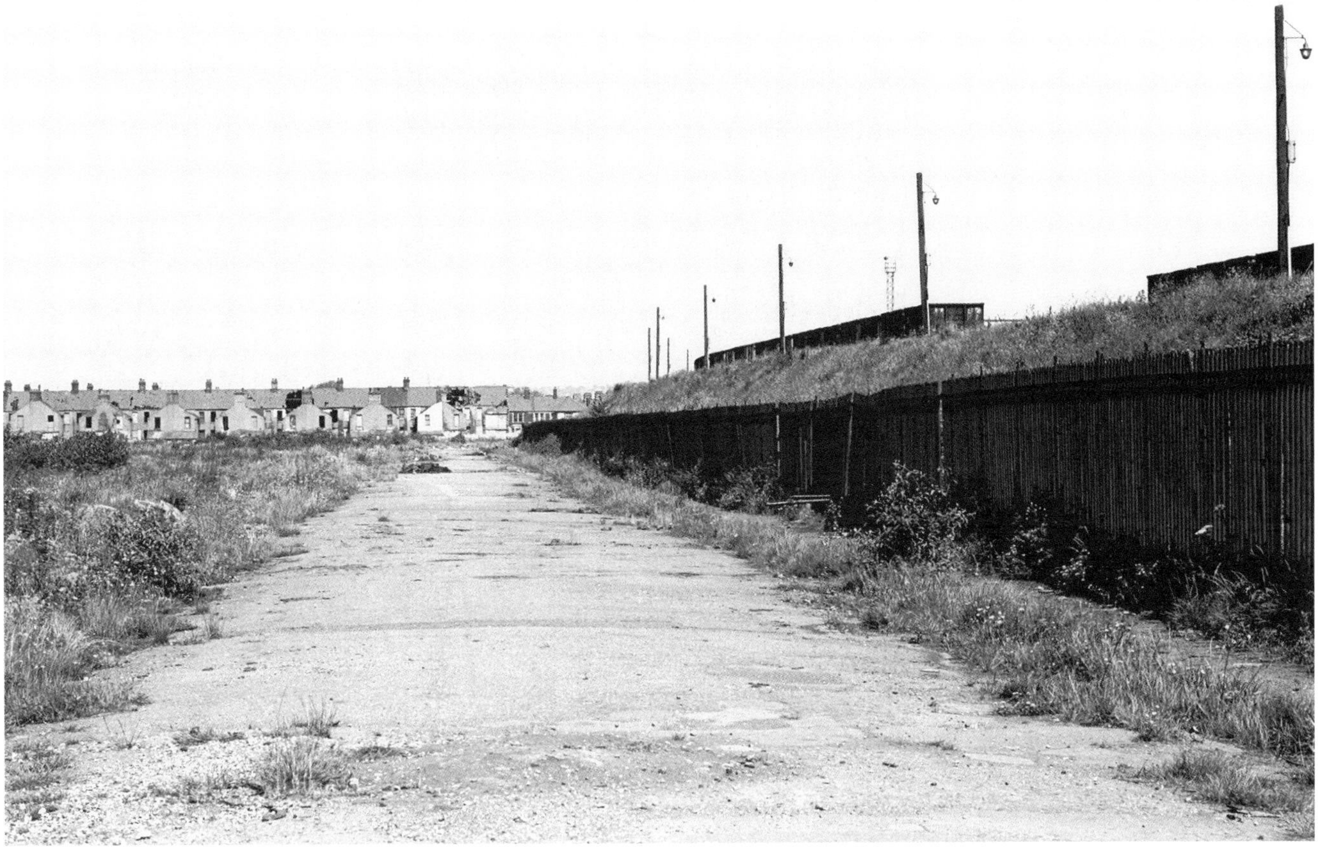

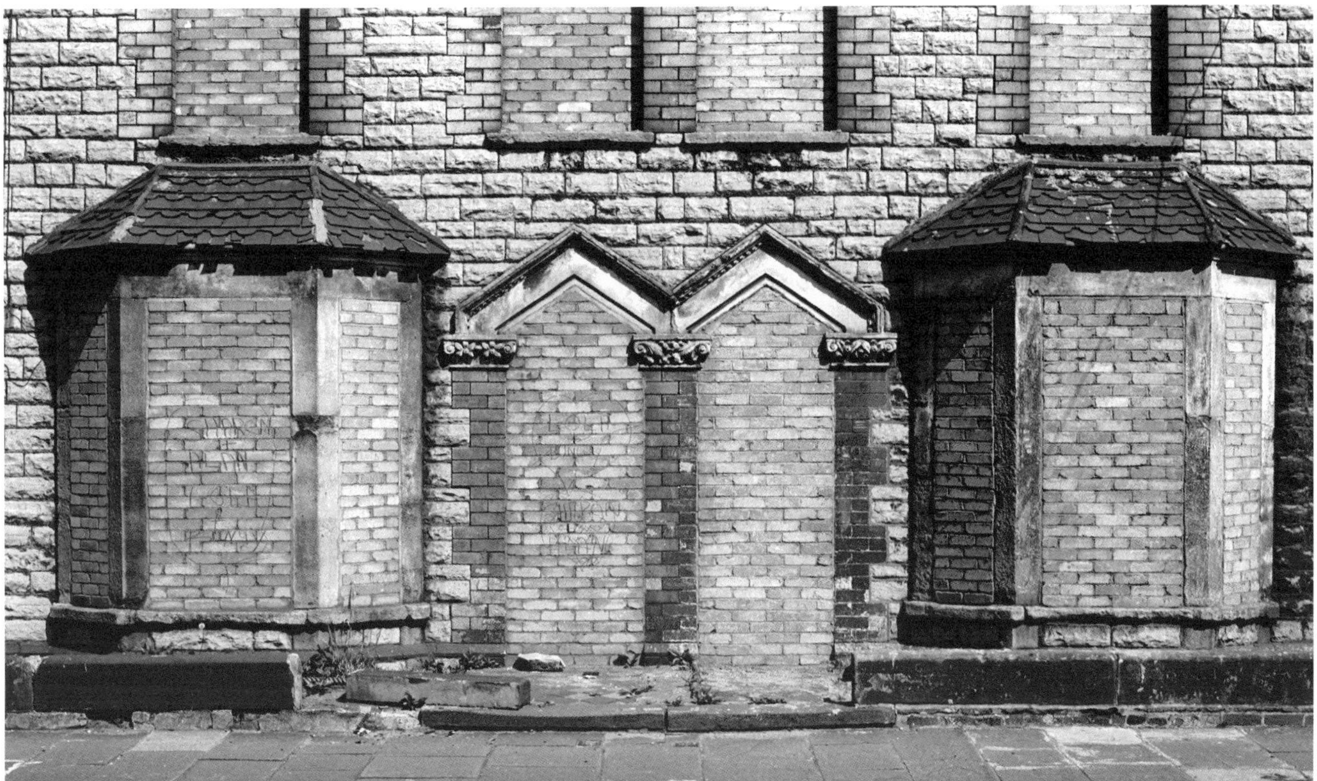

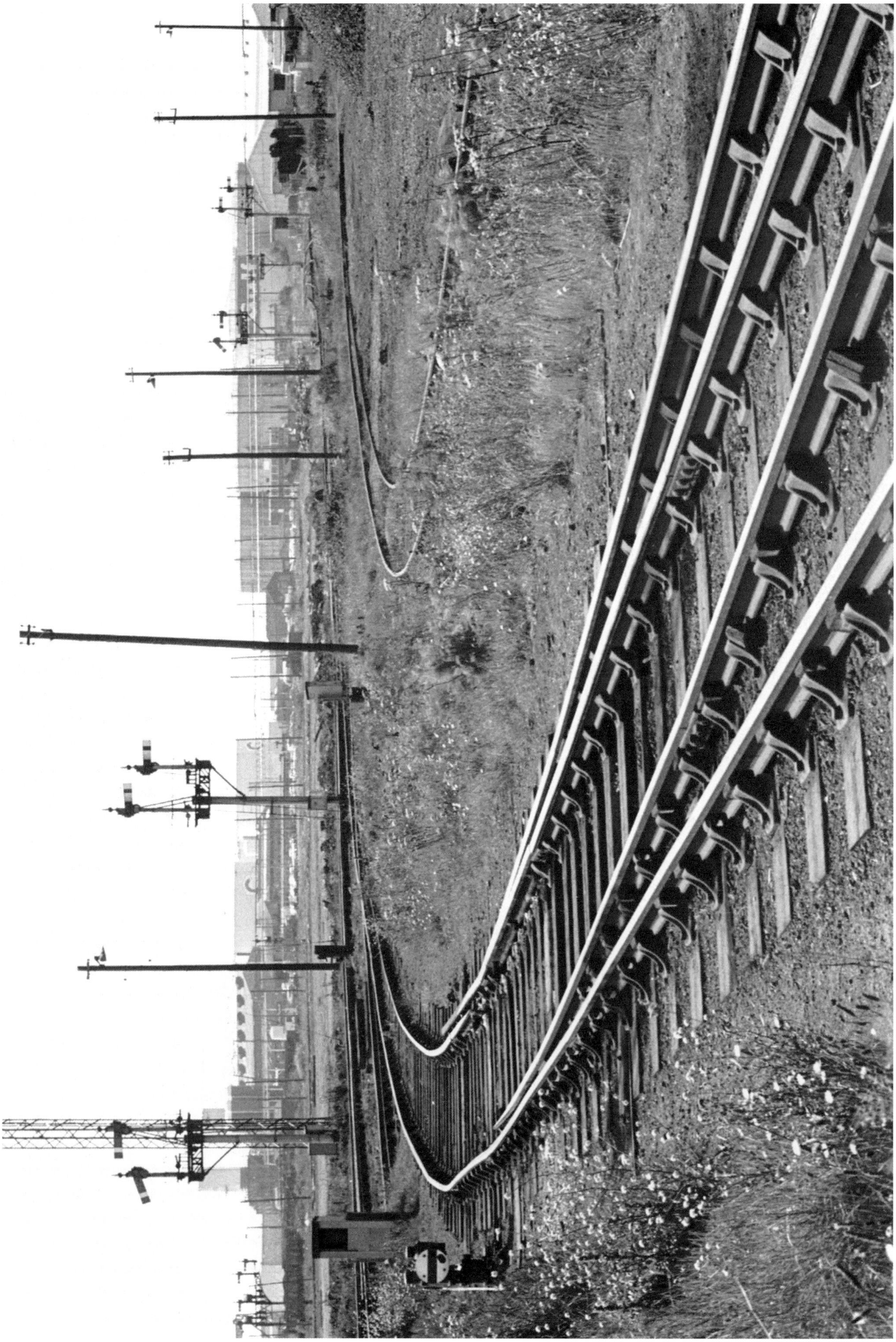

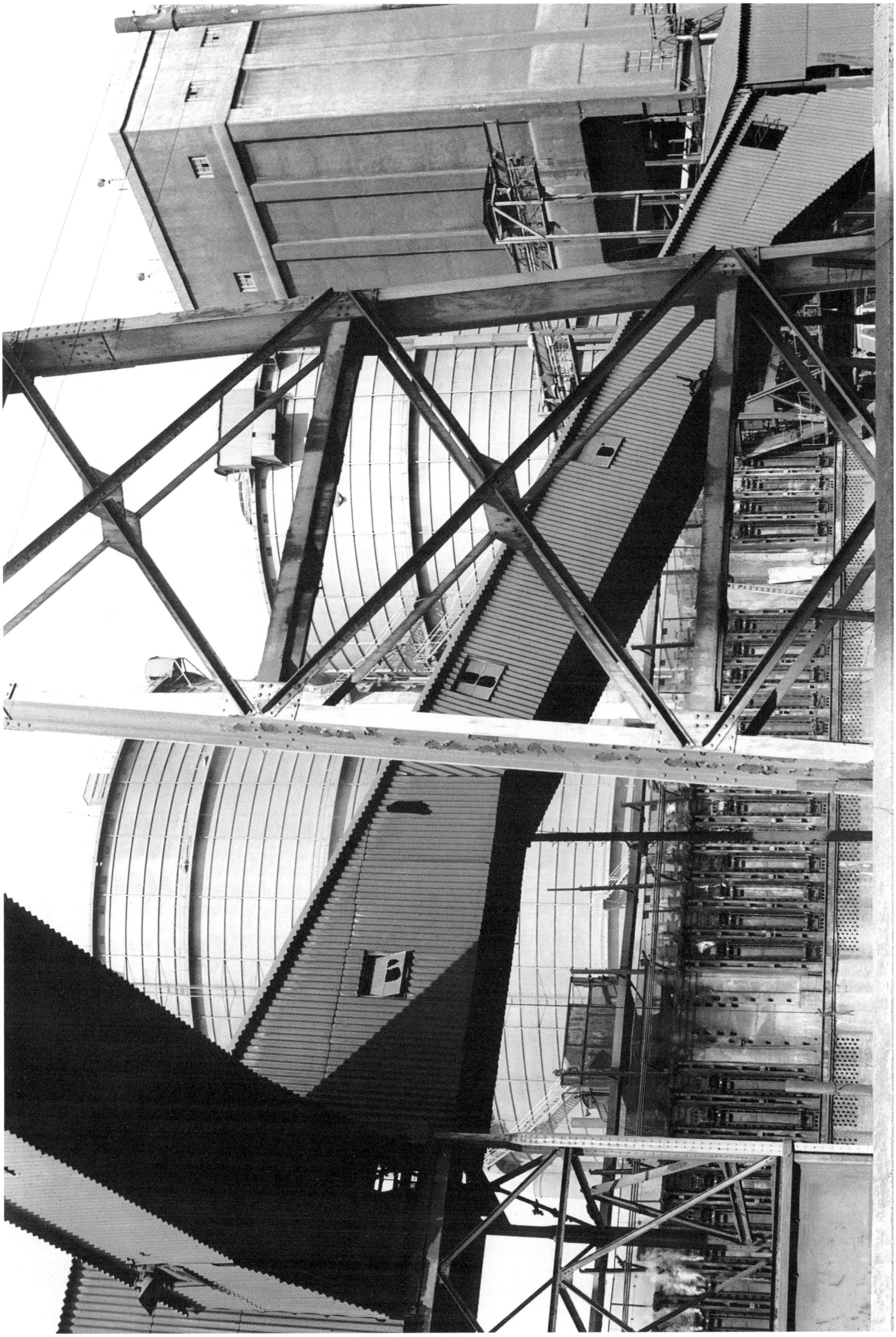

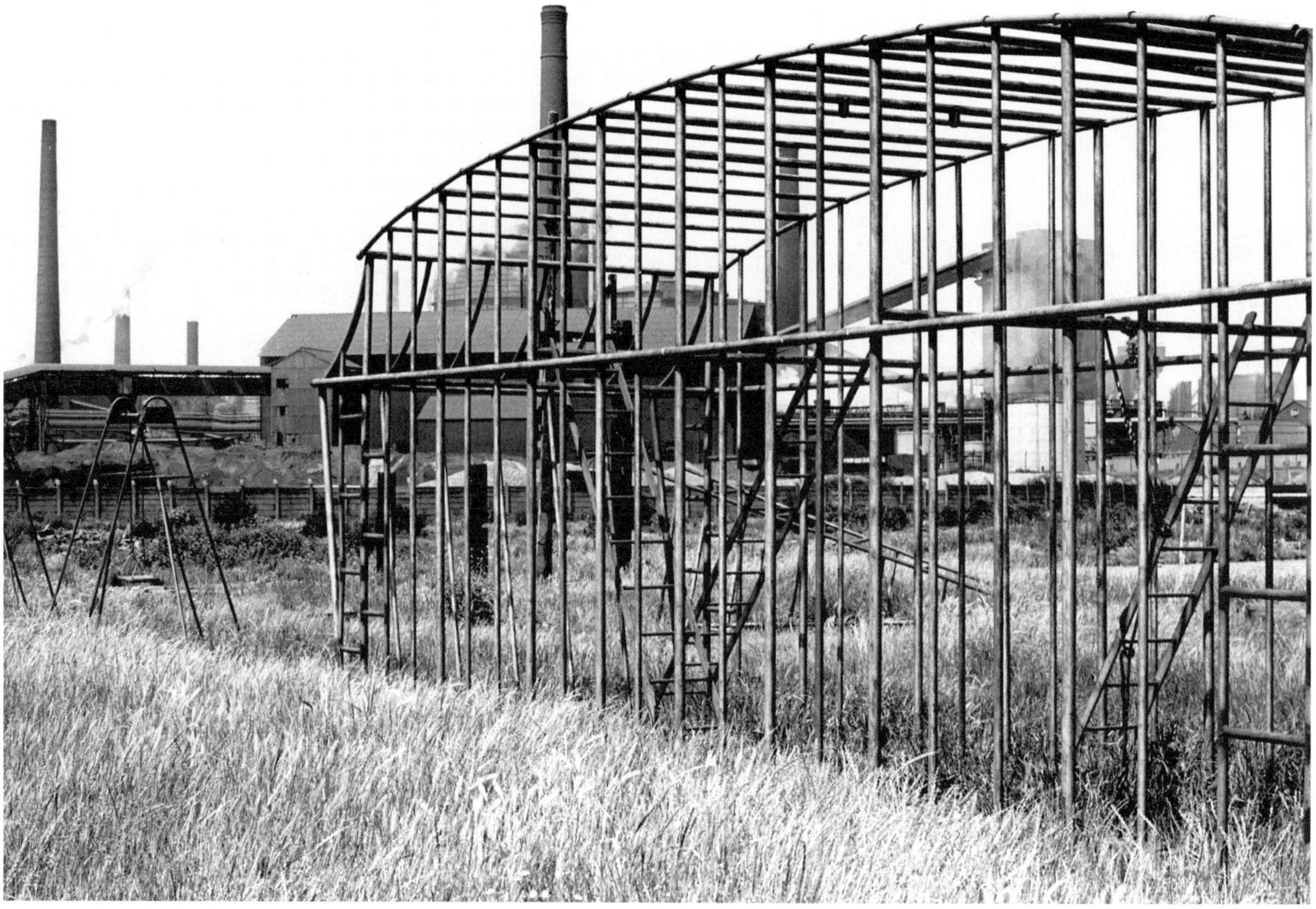

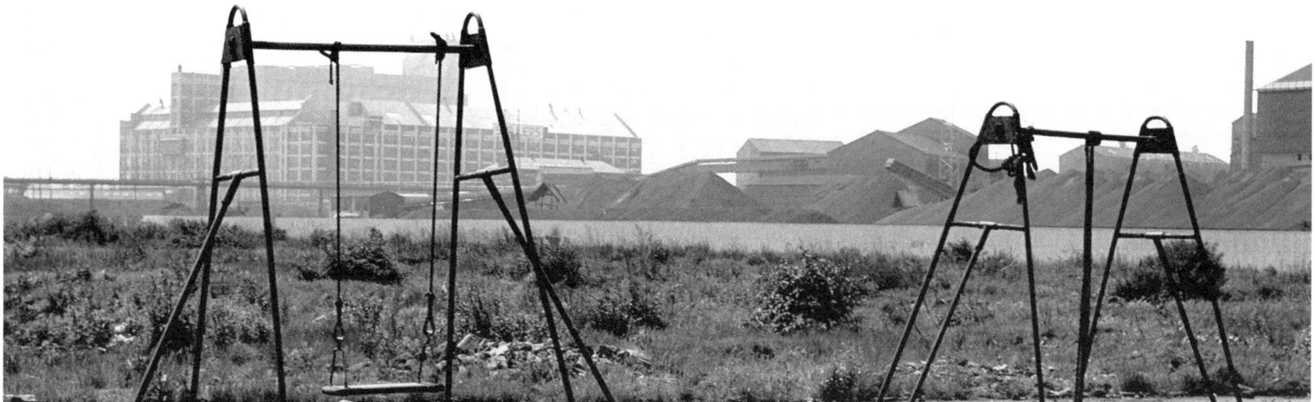

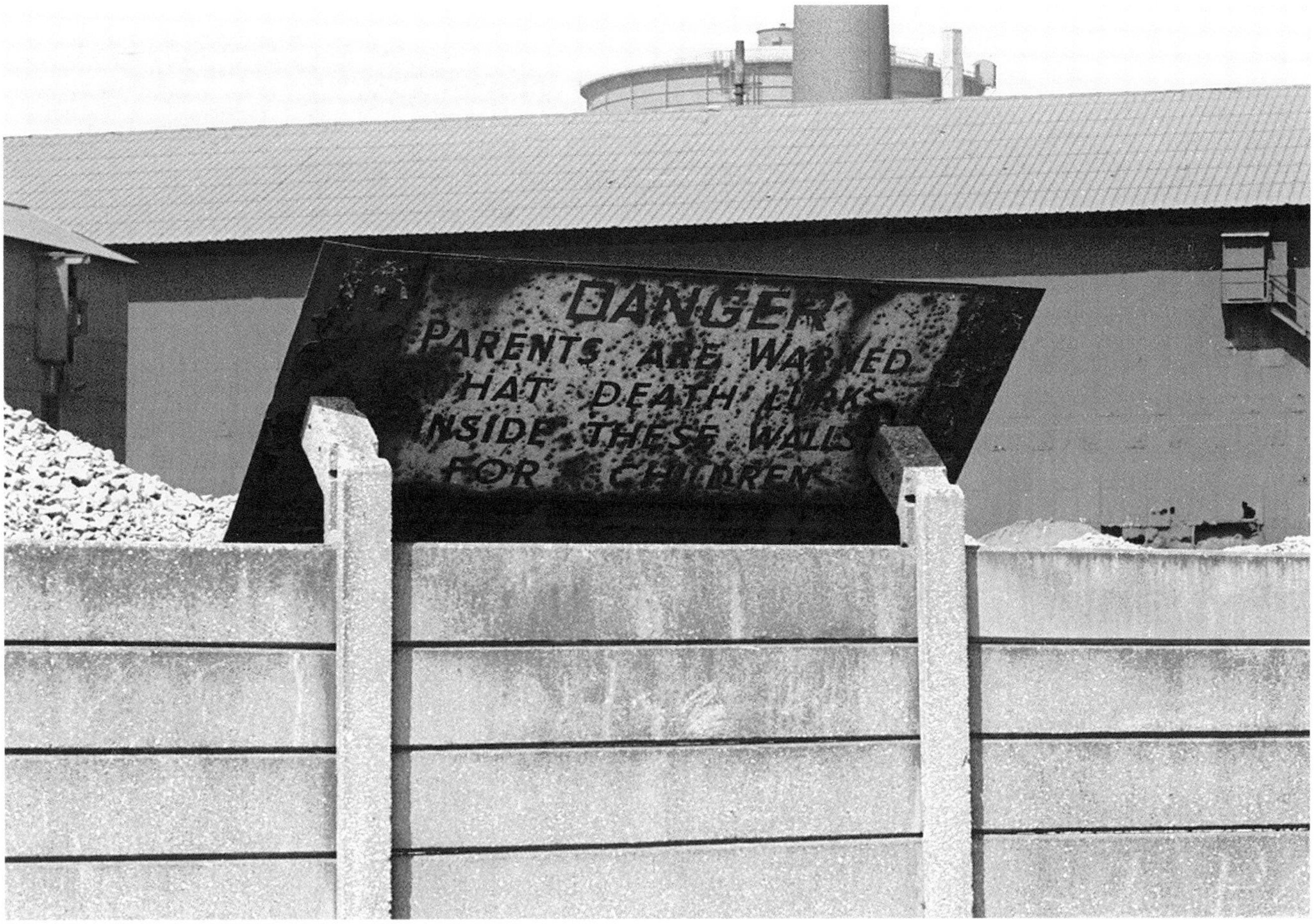
DANGER
PARENTS ARE WARNED
HAT DEATH LURKS
INSIDE THESE WALLS
FOR CHILDREN

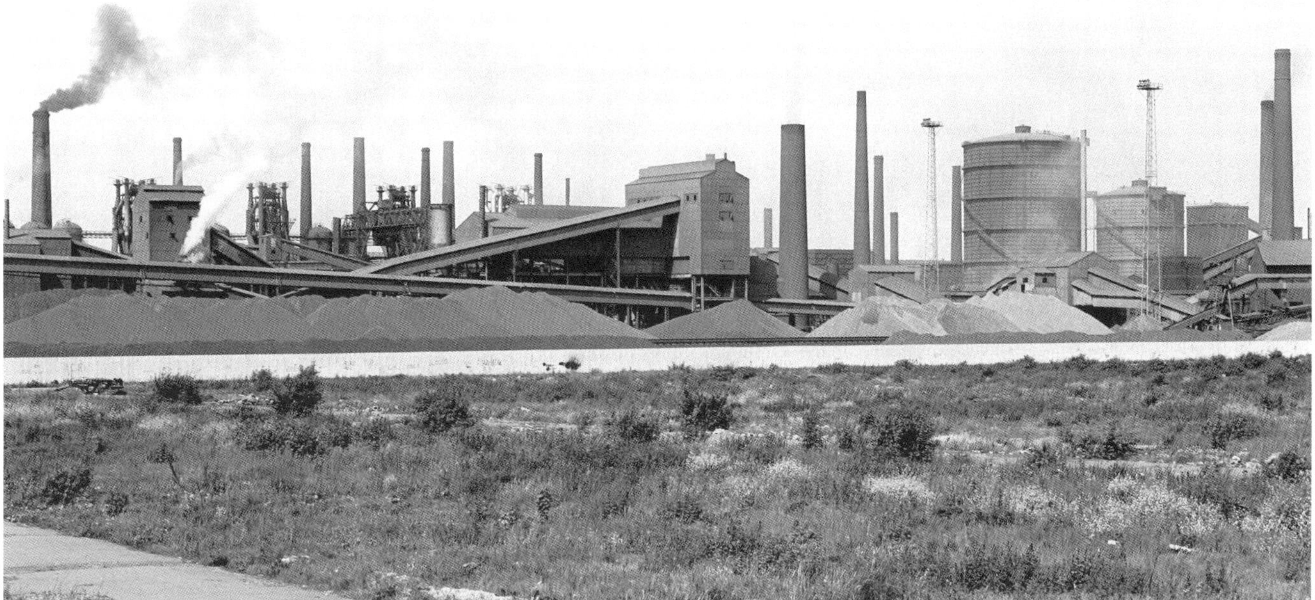

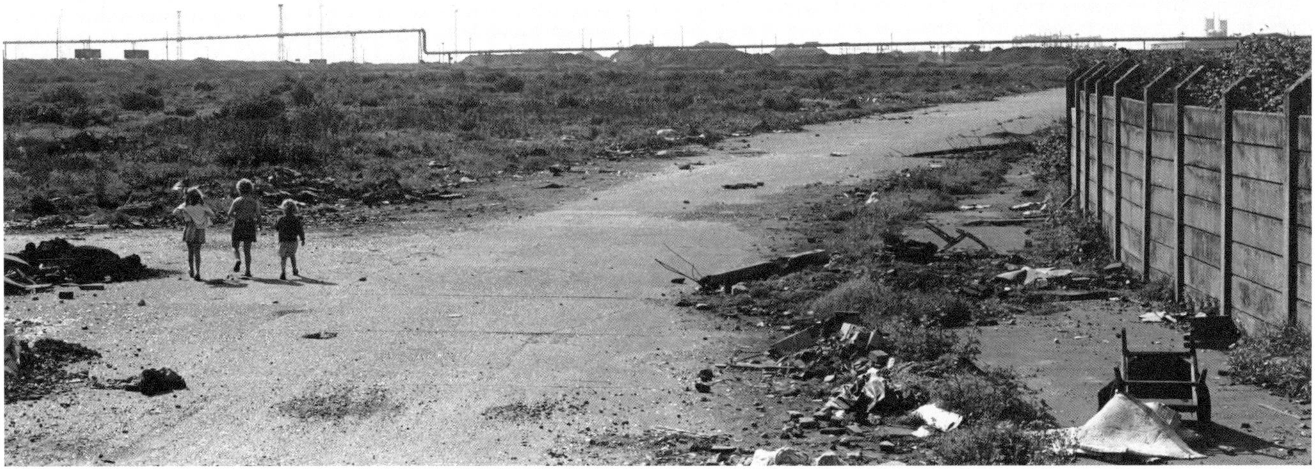

Acknowledgements

The publication of these photographs, the personal and historical details which accompany them and my ever-increasing familiarity with the people and the places of the communities represented in this book could not have been possible without the consent, support and information provided by many individuals and certain organisations. To the following I owe a very large debt of gratitude and hope that this book will give them the same sort of satisfaction upon seeing it published as it gives me:

Butetown History and Arts Centre and those associated with its work both the past and present: Humie Webb, Julia Young, Neil Sinclair, Glenn Jordan, Chris Weedon, Molly Maher, Tony Lewis, Patti Flynn and Mavis Jackson.

Butetown History and Arts Centre Seafarers Group: Sidney Robinson, Cyril Actie and others.

Danny and Reita Whelan, former licensees and their patrons of The New Sea Lock and White Hart, present licensees and their patrons of The New Dock Tavern, Broadway, Splott.

Debbie and Paul Reid, licensees and their patrons of The Baltimore Arms, Bute Street, Docks.

Patrons of The Splotlands, Meteor Street, Adamsdown and licensees Jon and Julie Morgan, who kindly loaned me the photograph on page 86.

Jill and Faouzi Aboussanouga, former proprietors of The Tiger Bay Café, Bute Crescent, Docks.

Mr. Mac McGuiness and patrons of The Splott Conservative Club, Habershon Street, Splott.

Mr. David Furnish, licensee and his patrons of The Fleurs Club, Portmanmoor Road, Splott.

John Coles, Anthony Toppar, Jacky Williams (née Carlin), Susan Moseley (née Dale).

Mr. Alan Lewis, Warden, Missions to Seamen Seafarers' Centre, Roath Dock.

Peggy Holder, Education Department, St. David's Hall.

The Camilleri Family, Adamsdown.

The Chambers Family, Tremorfa.

Leslie Owen, Audrey Scott, David Jenkins, Ken Hodgson, Eurof Thomas, Graham Chamberlain, Richie Jones (who took the photograph of me in The Mount Stuart on page 88).

Part 1: Bute Street, from the Hayes to the Pier Head

11 **'Johnny Onion', The Hayes**
The Hayes (meaning hedge or enclosure) has since the mid-nineteenth century been associated with Cardiff's fruit and vegetable trade. It was a natural location for the sale of produce coming up to town from the docks, situated as it was at the town end of Bute Street, built in the 1830s. Onion sellers from Brittany have been coming to Cardiff since that time. Olivier Berthevas, the last of the older generation of 'Johnnys', or 'Sionis', was a familiar figure in The Hayes during the 1970s. He and his partner rented a house in Bute Street for accommodation as well as the storage and stringing of onions, making their last trip to Cardiff in 1978. Since their departure young mobile-phone carrying 'Johnnys' have taken their place and can be seen in The Hayes once again, especially during the festive season.

12 **Children listening to the Salvation Army Band, 1965**
In 1915, the Cardiff headquarters of the Salvation Army was at 1 Bute Street, just across from Hayes Island where the band used to congregate under the John Batchelor statue. Here two children are being entertained in 1965. Mr. Danny Whelan, former landlord of The White Hart in the Docks informed me that the band also used to play in James Street well into the 1970s, starting at the top of Harrowby Street, working their way towards Bute Street and stopping at intervals to perform for passers-by.

13 **The Hayes, Barrowboy**
Fruit and vegetables were sold from barrows until the late 1970s, even when permanent stalls in nearby Mill Lane were a prominent feature of the area. To this day there is a stall in The Hayes outside David Morgan's department store. In the background can be seen Howell's department store, St. John's Church and the old Central Library, converted in the 1990s into the short-lived Centre for Visual Arts.

14 **Mill Lane Market**
This classic fruit and vegetable market just south of The Hayes was a familiar sight from the 1940s until the late 70s. Used by townspeople and especially the residents of Tiger Bay and the Docks, the open-air markets of Mill Lane and Custom House Street were important to these communities as a perennial and accessible source of fresh produce. Today the market stalls of Mill Lane have given way to the tables and chairs of Cardiff's new 'café quarter'. Fresh produce is now sold from stalls near St. David's Shopping Centre, a pale imitation of the sight, sounds and smells of this, Cardiff's former fruit and vegetable district.

15 **Rear of Mill Lane Market**
Loading and unloading fresh produce from the lock-up sheds behind the stalls was a daily occurrence. Here the market worker faces the camera with, it appears, his boss the stall owner. Such workers sometimes came from the docklands communities, less than a mile away.

16 **Market Worker**
Sitting in a quiet place behind the market stalls, he is at his table – an upturned crate. In the background are buildings in Mill Lane and St Mary Street. The rear of the Costa Rica Coffee House is visible above the stalls and the roasting of coffee beans there afforded a rich aroma to this part of the city, often mingling with the pungent odour of beer brewing in the nearby Brain's brewery.

17 **Stairway to the New Moon Club, New Street**
Famous for its jazz in the 1960s, the New Moon was at the top of a textiles warehouse in New Street and overlooked Mill Lane market. By the mid-70s it was a night club and student haunt where the phrase 'cheap, sweaty fun' could well have been coined. Swigging Newcastle Brown from the bottle while enjoying the live music was *de rigueur*. Entrance to the club was at the top of this Victorian iron staircase, its curve mirrored by that of the lamp post in New Street below.

18 **The Salutation, 14 Hayes Bridge Road**
This handsome pub exterior was to be found at the junction of Homfray Street with Hayes Bridge Road, originally the part of Bute Street nearest the Hayes until it was renamed in 1937. It was one of several pubs in the area including The Fishguard and The Golden Cross which welcomed seamen coming into town from the Docks via Bute Street. It was also one of many examples of Cardiff's fine pub architecture, Georgian in style, sacrificed with many other small businesses to the featureless facade and huge floorspace of the city's first American megastore, Toys 'R' Us.

19 **Landlady, The Salutation**
Taken in 1974 this photograph was of perhaps the last landlady the pub had before it was demolished. First licensed in September 1847, The Salutation closed on 21 January 1982 as the character of this part of the city was radically transformed by comparatively gargantuan new buildings and multi-storey carparks.

20 **Huxley's shop window detail**
The fascinating array of signs for cures and remedies merited closer inspection. The sign of Saunder's toy and party goods shop opposite Huxley's was also reflected in its window from across Barrack Lane.

21 **Huxley's Medical Stores, 71 Bridge Street, junction with Barrack Lane**
Huxley's was an old Cardiff business which suffered the same fate as many others when redevelopment of the Bridge Street/Hayes Bridge Road/Mill Lane area took place in the early 80s. Antique and second-hand book shops were also typical of the area.

22 **Grocer, 16 Bridge Street**
Shops catering to the tastes and culinary needs of Cardiff's multi-racial community were found in the sidestreets near the Hayes not far from Mill Lane Market. There were several others like this one, the Asian Spice Box, but all were demolished to make way for the new Bridge Street multi-storey car parks.

23 **Watch material suppliers, Caroline Street**
This Gothic-arched frontage, so typical of Cardiff's Victorian architecture, belonged to L. Nevies' shop at 9-10 Caroline Street near Mill Lane Market.

24 (t) **CWS shirt factory, Bute Terrace**
Its Cardiff factory reminiscent of those of northern mill towns, the Co-operative Wholesale Society at one time had a network of businesses across Cardiff including a biscuit works, bakery division, funeral service, travel service, soft drinks factory and motor trade department. The shirt factory was active in the 1960s but disused by the 1970s. With many other buildings in Bute Terrace it was cleared to make way for the Cardiff International Arena.

(b) **The Tredegar Arms, Bute Terrace**
Also sacrificed to the CIA was this facade with its stylishly-crafted wood framing, enamel border tiles and mosaic work. It stood in Bute Terrace near Mary Ann Street, not far from the CWS building.

25 **Baby and Young Mum(s), Custom House Street**
The Golden Cross provides the backdrop for these ladies, obviously very pleased with the new arrival, as yet too young to take notice of the pub's impressive exterior.

26 **The Golden Cross, window detail**
This style of pub window, with its distinctive diamond-shaped Brain's panel in blue and white, was typical of the craftsmanship found in most Brain's pubs until the 90s. Such leaded-glass windows were a familiar feature of the city's architecture for many years. Thankfully this one and a few others survive; most old windows were taken out and replaced by clear glass devoid of any aesthetic appeal.

27 **Girls in the Saloon Bar of The Golden Cross**
A cluster of pubs at the top of Bute Street were the centre of the shadier side of Cardiff's streetlife in the mid-1970s, especially at night. People I encountered in these pubs were usually quite easy about being photographed and these two girls in their 70s fashions were no exception.

28 **Gent in the Public Bar, The Golden Cross**
The public bar is remarkable for its late-Victorian tiling. After the brewery made known its intentions to knock the pub down in 1978 it was made a listed building. Now the only one of its era remaining at the top of Bute Street, The Golden Cross is an oasis in the desert of modern American-style structures surrounding it.

29 (t) **Newsboy in flat cap, The Golden Cross**
When selling the *South Wales Echo*, he would shout 'Echob!', a good friend of mine told me. He was a one-eared newsboy and this photograph is undoubtedly of his best side.

(b) **The Golden Cross Saloon Bar entrance**
The etched glasswork was typical of more opulent Cardiff public houses. The Golden Cross was and still is the most ornate of those dating from the early twentieth century.

30 (t) **The town end of Bute Street in 1974/75**
'The bridge' has always been an important symbol for docklands residents, marking the point of passage into or out of Butetown. On the left are the entrance to Cardiff Central Boys Club, Sherman's betting shop, a Greek café/boarding house and The Glastonbury Arms Hotel. The tall building beyond the railway bridge had a Greek delicatessen on the ground floor and after it came John Street with The Crown on the corner. On the opposite corner across Bute Street was The Custom House pub at the entrance to Crichton Street. The building to the immediate right of the bridge and opposite The Glastonbury harboured the once-popular CIACS – Cardiff International Athletics Club. Today, the only structure still remaining in this scene is the bridge itself.

(b) **Bute Street, town end, man lighting his pipe**
Having placed his bet at Sherman's bookmakers a punter pauses to light his pipe.

31 (t) **Bute Street, town end, man smoking his pipe**
The camera focuses closer again to the two buildings and the man in 30(b). Unperturbed, he calmly enjoys his pipe.

(b) **Top of Bute Street, man waving his arms**
I was told the man was a Docks carpenter, Mr. Meletti, from Louisa Street. By the time this photograph was taken in 1976, the exterior of The Glastonbury had been repainted.

32 **Rear of The Glastonbury Arms**
Taken from School Lane which ran behind the pub up to St. Mary's National School in Bute Terrace. The style of lettering and advertising slogan were particular to Brain's who displayed their name prominently on pub facades and other structures such as bridges throughout Cardiff.

33 **The Glastonbury Arms, 26 Bute Street**
A famous Victorian Cardiff pub, it was known to seamen all over the world as 'The White House' because of its white, enamel-tiled exterior. Dating from 1840 it was bought by Brains in 1922 and remodelled in this style by the distinguished Welsh architect Sir Percy Thomas. A vigorous preservation campaign was mounted in the mid-80s to save it, but the pub was demolished in 1990 to make way for the AXA insurance building, the car park to which now covers the site.

34 **Four ladies in Bute Street**
About to walk under the railway bridge on their way into town, I am told that they were possibly gypsies from the nearby site in Tyndall Street.

35 **Three ladies in Bute Street, near John Street**
Behind these three smiling women stand Johnston Buildings, a Victorian warehouse which eventually became the headquarters of the Welsh National Opera. 1970s fashions were in evidence as these girls gleefully paused to pose for the photographer.

36 **The Custom House, corner of East Wharf and Custom House Street**
The Cardiff Custom House (Customs and Excise, not the pub), built in 1845, stood beside the Glamorganshire Canal just beyond St. Mary Street where the canal widened into a mile-long basin before reaching the sea lock. In the canal's heyday, and before the opening of the Bute Docks in the mid-1800s, this building was at the hub of Cardiff's sea-going trade. But by 1882 the Custom House was, according to the *Cardiff Guide* for that year, 'inconvenient' due to the fact that it was no longer located near the centre of maritime activity. It is now the headquarters of the Welsh Rugby Union.

37 **Man walking towards Tiger Bay, East Wharf**
Running parallel to Bute Street, this was also a frequently used thoroughfare between Tiger Bay and the city centre. On the left is The Glendower pub at the entrance to Crichton Street, once the heart of the Greek Cypriot community. On the right are the well-known Snowden's Fish Wholesaler's, and Frixa House, a government training centre, which were built on the site of the Glamorganshire Canal when it was filled in in the 1960s. Loudoun House and Nelson House, the two tower blocks in the distance, stand in Loudoun Square at the centre of the old Tiger Bay area. Except for these two buildings, all of those nearest the camera have been bulldozed to make way for a new road extension called, as the planners would have it, St. Mary Street.

38 **Crichton Street from East Wharf**
In the 1972 *Cardiff Street Directory*, 34 houses are listed in the street belonging to inhabitants who had Arab, Greek or Irish names. The north side of Crichton Street seen here in 1975 was demolished soon after and became the site of new business units whilst the south side survived for a few more years. Cardiff's Greek Cypriot community was centred here, not far away from the Greek Orthodox Church. Towards the end of its existence it was better known for its nightlife and the street where one went to The Quebec to hear Tiger Bay jazzman Victor Parker play the guitar. The people on the pavement emulate the 70's fashions on the posters. Nowadays, Crichton Street has been obliterated by the new Bute Square development and there is no trace at all of the buildings and streets which made up this old community.

39 **Tinker's Barrow, Hope Street and Crichton Street**
Barrows such as these were normally used for moving fish and fresh produce along the streets. Many older Tiger Bay residents will remember Tommy 'The Fish' Letton selling fresh fish from his barrow. This one seems to belong to a scrap merchant or perhaps a 'rag and bone man' from the nearby Tyndall Street gypsy camp. The iron tyres indicate this was a very old barrow.

40 **Ornate Building in Bute Street between 'the bridge' and John Street**
I have not been able to find precise details of this building's origins but the 1882 *Cardiff Street Directory* lists The Bute Arms in this location. In 1975 the ornate facade was a derelict but striking reminder of Bute Street's former glory. The ground floor belonged to the Greek grocer S. Lambros during the 1960s and early 70s, apparently one of the last shops serving the local Greek Cypriot community.

41 **The Crown and The Custom House, Bute Street**
The last two pubs to survive the advance of redevelopment before it completely engulfed the town end of Bute Street in the 1990s. John Street still exists but Bute Square now covers Crichton Street and others. These two streets originate in the first two Christian names of the Second Marquess of Bute, John Crichton Stuart.

42 **Old Gent in the Custom House**
Photographing in this pub was not always welcome because of its colourful clientele and the barman, Mr. John Scully, looks none too pleased. The old gent didn't seem to mind though.

43 **The Custom House, Public Bar, 1979**
The decor may have been basic but the atmosphere was always lively. The man in the woolly hat is Mr. Owen Leslie whom I was to photograph again by chance twenty years later in The Crown across Bute Street (see 44).

44 **Owen Leslie in The Crown, 1997**
Having photographed Les in The Custom House (43) I happened to be showing the photo to people in The Crown shortly after The Custom House had closed for good. I was told he would be in shortly and when he duly arrived I asked him to sit for a more formal portrait. He willingly obliged and then told me about his life as a merchant seaman. There are many others like Owen Leslie still in Butetown with stories to tell but fewer and fewer places to tell them as the old pubs disappear.

45 **Greek Priest and friend, Bute Street**
Near the junction with Herbert Street and not very far from St.Nicholas Greek Orthodox Church, a priest converses with a Greek hairdresser, Mr. Panayi, whose barber shop was on the other side of Bute Street. The suitcase probably belongs to the priest who, I am told, travelled by train to Cardiff at weekends from the Greek monastery in Wrexham to help with masses at St.Nicholas Church on Sunday.

46 **Five young 'Bay Boys' in Bute Street**
There are at least three nationalities amongst these young lads attesting to the racial diversity of Tiger Bay. The photograph dates from 1975 and was taken in Bute Street near the canal bridge. Some of the boys, now in their 30s, still live in the area. (From left to right: Leroy Stone, Wayne Tullock, Ian Fleming, Marvin Maynard and Christopher Welsher.)

47 **Buildings in Bute Street, west side, near Bute Bridge (canal bridge)**
Leading up to the Bute Street canal bridge are, from left to right: Mahfood's Cafe, the Rainbow Club (where Shirley Bassey made a well-publicised appearance in support of their work with local children), Panayi's Barber Shop, Davies's Ironmongers, the Charleston Club and Abraham's Betting Shop (formerly Bute Street Post Office) at the corner of Wharf Street. Before the dismantling of the Tiger Bay community began in the early 1960s such premises were characteristic of Bute Street's commercial life along its entire length. Here they stand just before being swept away after having served Bay people for over a century.

48 (t) **Signs old and new along the West Junction canal path**
Taken along the canal footpath a few yards from the steps leading down from Bute Street, these examples of Brain's advertising were perhaps unique and certainly the best of their kind to survive into the 1970s when these building were being demolished. The enlarged AI was the idea of Albert Jones, an artist employed by the brewery in the early 1900s. The crude racist graffiti is in direct contrast to the past and present character of Tiger Bay, one of Britain's most diverse multiracial communities where racial harmony has been the rule throughout its past.

(b) **Hand-painted Brain's sign, off Bute Street**
Close-up detail of a delightful example of the sign painter's art as well as a clever pun.('Brains on his tail', if you haven't yet got it), shown in the previous photograph.

49 (t) **52 Bute Street, east side next to Bute Bridge**
This crumbling structure stood opposite the buildings in photograph 48(t) and in the late 1800s was a dyer's and a photographer's. In the 1960s and 70s it finished life as a café and boarding house. The poster in the window advertises a dance. Despite its near-derelict state and signs of various incarnations during its lifetime, the building was another example of the architectural style typical of this famous avenue. The four posts to the right protected the steps leading down to the canal.

(b) **Figure in the window, 52 Bute Street**
A close up shows that the building was still occupied when I photographed it in 1975. The 1972 *Cardiff Street Directory* verifies that several people resided here, a throwback to the days when a multitude of lodging houses and hostels in Butetown provided temporary accommodation for seamen on shore leave in Cardiff.

50 (t) **Course of West Junction Canal looking east**
Just beyond the canal bridge parapet the canal bed passes under the railway bridge towards West Bute Dock. In 1880, this location inspired the author of *The Book of Cardiff, No.1* to write:

> When we reach the bridge spanning the canal...without much trouble we can see at this end of the docks a potato boat from Ireland or Brittany being unloaded by a gang of strong-looking women who carry the heavy sacks on their backs from the vessel to the warehouse

In the background are those potato warehouses belonging to the Edward England family, dating from 1842 and still in use today. The stone railway arch also still exists and a pedestrian walkway has been laid under it leading to the newly enlarged Lloyd George Avenue, formerly Collingdon Road. The Bute Street canal bridge and steps leading down to the canal have been levelled.

(b) **Course of West Junction Canal looking west**
This short waterway connected the Glamorganshire Canal to the West Bute Dock and its main purpose was to provide a steady flow of water into the main canal. In this view looking west towards Bute Street, the empty canal bed makes its way under the railway bridge and then under Bute Street as it heads towards the former Glamorganshire Canal basin. In the film *Tiger Bay* Hayley Mills played under this very bridge whilst the leading man, Horst Bucholz, hid a gun in the pub above. Today, on that site, and in place of the row of buildings with the Brains chimney (see 49[t]), stands the recently completed Salvation Army centre Ty Gobaith.

51 (t) **Man walking along Bute Street**
This well-dressed gentleman is crossing Maria Street as Bute Street proceeds past Tiger Bay towards the Docks. He and a small tree are the only signs of life in an otherwise bleak landscape. In the background beyond the elevated Bute Road railway line are the last remains of industry stretching along Collingdon Road.

(b) **Gents' convenience, Bute Street**
A no-longer existent facility in Bute Street. In busier times, no doubt a welcome stop for sailors calling in Bute Street pubs or for local people working and living in this once-thriving area. The cement facades on the stone wall were at one time lettered and painted with advertisements for local businesses.

52 **Old Gent at the Salvation Army Hostel, 240 Bute Street**
Carrying on past the canal bridge and along Bute Street, one came to the Salvation Army Hostel. Its elegant facade was originally that of Spillers & Bakers, grain merchants and flour millers whose first Cardiff mill stood in Collingdon Road. Erected in the 1880s and demolished in the 1970s it is another example of the stylish structures built by Victorian industrialists, now lost. The landmark towers of the Church of St. Mary the Virgin stand out boldly in this 1975 view as they still do today. In front of its stone boundary wall lies the course of North Church Street and the vacant site of St. Mary the Virgin School, now rebuilt. The lane where the old gent is standing led down into Greek Church Street, part of which still exists. He is no doubt a resident of the hostel and perhaps the same person I photographed earlier in The Custom House pub (37).

53 **Man beneath railway embankment**
Reclining in a sun-drenched, peaceful spot above Bute Street is perhaps another patron of the Salvation Army Hostel. Visible above

him is the railway line overlooked by a pristine white cottage and a corner of the former Elliot Wire Ropes works.

54 (t) **Girls waiting at the bus stop, Bute Street**
The low stone wall on which the schoolgirls are sitting extended down from the canal bridge on the east side of Bute Street. They seem to be a school group from nearby St. Mary's waiting for the bus. Behind them is the disused Elliot Wire Ropes works, closed down in the early 60s and originally built as the Spiller's Flour Mill. The faded lettering on the building includes words from both industries. Behind the retaining wall the railway line runs towards its terminus at the Bute Road (now Cardiff Bay) station. The striking white-washed stone cottage was apparently inhabited until 1975 by the local Cadogan family who were its last residents. Only the road and the railway wall remain, in stark contrast to what was written about this part of Bute Street in 1880 in *The Book of Cardiff, No.1*:

> As we continue to walk down Bute Road we find that the street is much wider, flanked on one side by a railway embankment and huge warehouses, flour mills and other industries, and on the other side by shops, many of which are kept by foreigners from all parts of the world.

The fence in the foreground is that of St. Mary the Virgin Church.

(b) **Girls waiting at the bus stop – close up view**
Local residents will perhaps recognise faces in this group or perhaps even find themselves as children amongst their number.

55 **Girls in front of Nur Ul Islam Mosque, Peel Street**
A delightful group in front of the mosque which was an Islamic cultural and educational centre as well as a place of worship. Some of these girls still live in the Tiger Bay community. (From left to right: Audrey Scott, Cynthia Jeffers, Kathleen Blades and Monica Jeffers).

56 **Nur Ul Islam Mosque, Peel Street**
The mosque was built in 1947, the first of its kind in Britain. Set in amongst the terraced houses of Peel Street, here it now stands on its own in 1975 following the demolition of houses on both sides in the late 1960s. The pointed towers of St. Mary's church in the background make an interesting comparison to the rounded turrets of the mosque which was itself hauled down in the late 1970s and replaced by a new one in nearby Alice Street, to serve a growing Muslim community. Although Peel Street no longer exists a new centre for Islamic worship has been erected on the original site of the old mosque.

57 **Nur Ul Islam Mosque and modern house**
The mosque contrasts sharply with a Ford Anglia and the late-60s style houses typical of those built to replace the original Victorian houses in the Loudoun Square neighbourhood of Tiger Bay.

58 **Ship Lane**
Coming out of Tiger Bay and into the Docks area of Butetown ('down the Docks and up the Bay', as the saying goes) there was ample evidence that one was entering a district that had the trappings of much past if not present commercial activity. This lane still exists and looks very much as it did in this 70s view. It runs behind the west side of Bute Street just before it meets James Street where it used to end at a pub called The Ship.

59 **Jaguar in lane off West Bute Street**
In spite of so many ghosts of the past, there was the occasional sign of present affluence. A cobbled lane leads off West Bute Street into a courtyard where warehouses and storerooms once stood. At the time this photograph was taken in 1974, F.D.Searl of 17d West Bute Street was a rug cleaner and polisher. The premises are still in use.

60 (t) **The Pier Hotel during demolition**
As they were about to disappear, the details of Victorian chimney pots, brickwork and lettering made one wonder if anything with similar architectural character and variety might replace these beguiling old docklands buildings.

(b) **The Pier Hotel frontage**
Once the nearest pub to the Pier Head and the end of Bute Street, it was demolished not long after this photograph was taken in 1974 (see 60[t])

61 **Pier Head Chambers, 75-82 Bute Street near the Pier Head**
This row contained several offices serving the shipping and coal industries and was one of many stylish frontages demolished during the 1970s and 80s to make way for new development at the bottom end of Bute Street. The area is now covered by the sprawl of the recently completed Mermaid Quay complex. A Victorian pub, The Packet, does survive, however, and with the old long-disused Coal Board office standing nearly opposite, is the last nineteenth-century building still in use before the millennium monoliths dominate the last hundred yards of Bute Street. In fact Bute Street at this point is little more than a pedestrian alleyway hardly worthy of the former avenue which used to end at the Pier Head and the sea with a sweeping view of Cardiff's Inner Harbour.

62 **The Pier Head, outlook towards the sea at high tide, 1974**
Following Bute Street to its very end at the Pier Head I was finally confronted by this quiet yet inspirational panorama, framed by a century-old wooden structure, encompassing Cardiff's harbour, part of its docks, Penarth Head and the Bristol Channel beyond. This was the outlook for generations of seamen and their families, and the setting which, for a century and a half, witnessed the arrival and departure of an infinite number of sea-borne travellers to and from all points of the globe: the crews of countless sailing vessels and tramp steamers carrying Welsh coal to foreign ports, nineteenth-century émigrés bound for America with hopes of starting a new life, immigrants escaping from potato famine in Ireland to help build the Bute Docks, Captain Scott on board *Terra Nova* departing on his ill-fated expedition to Antarctica, trawlermen of the Neale & West fleet fishing off the south coast of Ireland, thousands of American troops disembarking from troopships in the build up to the D-Day invasion, hordes of daytrippers bound for seaside resorts on a 'Campbell steamer', sailors from foreign lands working on ships of all shapes and sizes bringing timber, grain, oil and iron ore to help turn the wheels of Welsh industry. Now 'going to sea' from this point can never exist again, not even in the imagination, as the new barrage holds the sea at bay and away from Cardiff's harbour forever.

Part 2: The Bute Docks

65 **Pier Head Building and the dolphins**
William Frame's masterpiece was built for the Bute Dock Company in 1896 at the peak of Cardiff's success as a port. Naturally, it was sited in a commanding position overlooking the Inner Harbour and the approach by sea to the Bute Docks. Fashioned of red terracotta brick and tile with a majestic yet delicately ornamented clock tower in the French Renaissance style, it remains a stunning example of late-Victorian architecture. The Pier Head Building stands virtually unchanged, but its pre-eminence as docklands icon and symbol of Cardiff's maritime past is now challenged by the monoliths of the Millennium Waterfront development.
The Inner Harbour moorings, known as 'dolphins' and built of African hardwood, have been in place for at least a century. Serving as platforms for the repair and cleaning of smaller ships, the vessels tied up to them would settle at low tide onto gridirons resting in the mud in front of these well-known structures. They continue to be a prominent feature of Cardiff's waterfront. On either side of the Pier Head Building piles of rubble have been pushed up against the lock gates of the East and West Dock basins after their closure in the 1960s. Just visible in the distance beyond the first 'dolphin' and to the left of the Pier Head Building are the All-Souls Mariners' Church (Missions to Seamen) and the fine Georgian-style buildings of Bute Crescent.

66 (t) **Paddle Steamer landing stages in front of the Pier Head**
A second group of 'dolphins' standing on the west side of the Inner Harbour are these, the docking facilities for the pleasure steamers so common in Cardiff's maritime scene from the 1870s to the 1960s. When operational, these structures were linked by floating pontoons, and a swinging gangway to them from the shore allowed daytrippers to board. From here vessels from P&A Campbell's 'White Funnel' line with evocative names such as *Devonia, Glen Usk* and *Cardiff Queen* took their passengers to popular Bristol Channel seaside destinations: Ilfracombe, Minehead and Weston-super-Mare amongst them. Now that such pleasures have disappeared from Cardiff, these wooden frames, complete with the semaphores which signalled the inward and outward movement of the boats, have been restored for the appreciation of visitors to the new Millennium Waterfront. Those wanting to sail on a paddle-steamer can still do so from Penarth Pier in the summer months aboard the *Waverley* (70[b]).

(b) **Mooring stumps: Pier Head and the Inner Harbour**
Standing near the Pier Head Building at the former entrance to the Oval Basin, this was the view in 1976 looking towards the west side of the Inner Harbour. Small craft rest on the sand near the defunct paddle steamer moorings while the workshops of the Mount Stuart Dry-docks dominate in the background behind them. To the right is Landsea House (now Harry Ramsden's), the Q Shed (formerly the 'White Funnel' paddle steamer ticket office) and the Pilotage Building. On the extreme right is Stuart Street which still leads to the residential part of the Lower Docks.

67 **The Pier Head Building viewed from Stuart Street**
The building's elegant architectural detail is highlighted in this photograph of its west facade. The white chains at the front protected the edge of the West Dock Basin, recently refurbished under the auspices of the Oval Basin Project. The curve of the iron railings in the foreground probably followed that of Bute Crescent when it originally met Bute Street at the Pier Head. They were presumably removed when the Industrial and Maritime Museum was built on this site in 1975, that insitution giving way in its turn to the Mermaid Quay development.

68 (t) **Boats at rest on the mudflats**
Survivors of the many changes in and around the Inner Harbour are the craft of Cardiff Yacht Club seen here in 1975 at rest on the mudflats, a unique feature of Cardiff's maritime landscape until the beginning of the twenty-first century. In the background are the remains of paddle steamer landing stages and also their floating pontoon in front of the walls and entrance channels to the Mount Stuart Dry-docks. The former rise and fall of the tide, the second highest in the world, was a spectacular natural phenomenon and will be missed by many. Not the least of the losers are the many varieties of sea birds for whom the mudflats were a plentiful feeding ground.

(b) **Sunken skiff at low tide, Inner Harbour**
It is probably a linesman's boat, traditionally manned by a crew of two whose job it was to bring ships' mooring ropes to the quayside. Beginning in the 1870s they performed this service for the paddle steamers approaching and leaving the pontoons at the Pier Head. Although this skiff must have lain for years in the sand and mud of the Inner Harbour, covered and uncovered countless times by the tide, it seems to have remained basically intact – testimony to the materials and craftsmanship of the men who made it. Many such craft were made at the Cardiff Boatbuilders yard off nearby Roath Basin but it is doubtful whether any have been preserved. Thanks to two septuagenarians, undoubtedly the last two of their kind in Cardiff Docks, who still ply their skills in the workshops there, the industry survives today.

69 (t) **High tide at the Pier Head**
This view of the paddle steamer 'dolphins' at high tide and the docks two years later in 1976 shows noticeably fewer cranes in the distance on the Queen Alexandra Dock. The BCSR (Bristol Channel Ship Repairers) Channel Dry-dock works is clearly visible, as is the ship being serviced in its bay. The entrance to the Channel Dry-dock was from the Inner Harbour at the stern end of the ship and at the opposite end, it was nose-on to the Bute Dry-dock chamber.

(b) **Mount Stuart Dry-docks jetty**
A short distance further along the harbour wall are the lock gates to the Mount Stuart Dry-docks and just beyond them was this wooden jetty some 300 feet long. It was used to inspect and make floating repairs to ships not requiring maintenance in the dry-dock bays. The bays have been saved and are now part of the Techniquest complex but the jetty was removed prior to the construction of the St. David's Hotel at this point. Across the harbour can be seen the working docks as they were in 1974 – with other 'dolphins' at the approach to Roath Basin lock, the Cardiff Boatbuilders works and slipway, the funnel of a ship being serviced in C.H. Bailey's Bute Dry-dock and cranes standing in their numbers behind the transit sheds of the Queen Alexandra Dock. The Queen Alexandra and Roath Dock (not visible in this photograph) constitute the active part of the Port of Cardiff today.

70 (t) **P&A Campbell Paddle Steamer Ticket Office**
This kiosk was located off Stuart Street behind the Pilotage Building. Passengers purchased their tickets here before descending the gangway to the paddle steamer pontoons but when the kiosk fell into disuse, tickets were instead bought on board ship as indicated by the weather-worn sign. The last of P&A Campbell's paddle steamers *Bristol Queen* and *Cardiff Queen*, sailed from Cardiff in the late 1960s and then were scrapped.

(b) **Paddle Steamer *Waverley***
Although she spent most of her sailing life on the Clyde in Scotland, the *Waverley* is similar to the graceful 'White Funnel' vessels that sailed from the Cardiff pontoons. In fact it is thanks to the *Waverley*, restored by the Paddle Steamer Preservation Society in the late 1970s, that the tradition of steam-powered excursions in the Bristol Channel has been successfully re-established. Now the last sea-going vessel of its type in the world, *Waverley* regularly operates an annual excursion programme, sharing the task during the summer months with the M.V. *Balmoral*, a graceful motor vessel, the very last to sail for the P&A 'White Funnel' fleet, also restored by the Paddle Steamer Preservation Society.

71 (t) **Fire-damaged *King George V* in Roath Basin**
Another pleasure steamer, also reminiscent of Cardiff vessels but belonging to the Caledonian MacBrayne line in Scotland, the *King George V* was being refurbished in the Bute Dry-dock in 1976 as a floating restaurant when it was severely damaged by fire. The project was abandoned and the ship moored in Roath Basin (as seen here) before being towed to the ship-breakers. While being towed out of Cardiff harbour it began to sink and was hauled onto the foreshore near Queen Alexandra Dock where it was cut up.

(b) **Decks of *King George V***
The fire-gutted decks of the unfortunate *King George V*. The heat was so intense that the metal sections of the superstructure twisted and buckled into fantastic shapes.

72 (t) **Junction Dry-dock, proprietor C.H. Bailey Ltd**
The Junction Dry-dock started life in the nineteenth century as a ship-building yard as well as a ship-repair facility before devoting its full-time efforts to the latter. With Cardiff's career as a port growing ever busier after 1850, the need for ship repair and servicing facilities became acute, giving rise to an industry for which Cardiff became widely renowned. Junction Dry-dock, so named because of its location between the Bute West Dock and the East Dock Basin, was one of the largest and busiest. Here the name of C.H. Bailey, its last owner and one of the last ship-

repairing companies to survive in Cardiff Docks, is still in bold evidence in 1974.

(b) **Dockside and lock, Junction Dry-dock**
The disused lock was full of water and debris. The steps on the left gave access to the lock bottom when it was empty. Its gates to the East Dock basin were to the right.

73 (t) **Junction Dry-dock shed**
Taken from inside the empty shed, this view looks towards the eastern end of the lock and the working part of Cardiff Docks beyond. The funnel in the distance is that of a ship in the still active Bute Dry-dock also under C.H.Bailey ownership in the mid-70s.

(b) **Telephone box near Junction Dry-dock**
Disused, vandalised and seeming to stand in the middle of nowhere, this public telephone would have been much used when the area was thriving, situated as it was in a once-busy part of the port not far from the Junction Dry-dock and next to the East Dock Basin. Beyond it a few of perhaps the last remaining docks' workers are making their way past the dry-dock buildings towards Bute Place, former site of the old dock gates. This area is now covered by the southern extremity of the new Lloyd George Avenue. The tops of the Georgian houses in Bute Crescent at the edge of Butetown are visible in the distance.

74 **Ship's bows in Roath Basin**
The vessel towering over the quayside is moored adjacent to the sand wharf where a lone worker tidies the area. Sand dredged from the Bristol Channel has long been an important commodity in the docks and at present is still discharged at one of the Roath Basin wharves, though probably not for much longer. In the distance and partially hidden behind the C.H. Bailey office, a ship is in repair at the Bute Dry-dock. The dockside of Roath Basin, opened in 1874, has now been redeveloped as Britannia Quay, a residential and recreational area with mooring space for the restored lightship *Goleulong* as well as for vessels paying courtesy visits to the Port of Cardiff such as *H.M.S. Caerdydd.*

75 **Docker next to ship in lock**
By the mid-1970s there were few men like this working in Cardiff Docks to hint at its great maritime past or the massive amount of manpower required to carry out the great number of jobs on the docks. The workforce at one time numbered in the thousands: men were needed to work on the quaysides, in the dry-docks, at the locks, in the transit sheds, at the coal hoists, as stevedores, labourers, boatmen, nightwatchmen. By the 1970s there were perhaps less than a hundred men working on the docks.

76 **Bute Dry-dock, southwest corner of Roath Basin, C.H Bailey Ltd**
Originally the site of the Bute Shipbuilding, Engineering and Dry-dock Company, dating from 1886, where the first steam vessel constructed in south Wales was launched. But shipbuilding never flourished in Cardiff and the company turned its attention to the ever-present need for ship repair by the end of the century. In 1975 the facility was near the end of its life and this Norwegian ship was perhaps one of the last to be repaired there. The Bute Dry-dock offices, under the ownership of C.H. Bailey Ltd., were up for sale as indicated by the sign on the building. These offices and dry-dock chamber remained in their disused state at the end of 2000, presumably awaiting clearance. In the foreground are the inner gate of Roath Basin lock and an 'inner' linesman's boat for picking up mooring ropes from ships in the docks as opposed to an 'outer' linesman's boat seen in 68(b) which worked in the harbour.

77 **Commercial Dry-dock off Roath Basin**
This chamber and its gates were located at the east end of Roath Basin, diagonally opposite the Bute Dry-dock. With its lock gates removed, the old Commercial Dry-dock has been transformed into a maritime feature bordered by waterside apartments.

78 **Buoy and crane**
Of the hand-operated variety, the crane raised and lowered heavy mooring buoys, such as the one pictured here, into the waters of the dock. This piece of docks history was one of the outdoor exhibits at the Industrial and Maritime Museum before it closed to make way for the Mermaid Quay development at the Pier Head. It is incredible that for all its seafaring history, Cardiff no longer has a museum to celebrate it and provide an attraction for visitors.

79 **Cranes in Roath Basin**
Still standing in 1975 along the quayside of Roath Basin, these cranes were among the last relics of former dockside activity. They were removed some time afterwards to make way for the Britannia Quay dockside walk which visitors enjoy today. Preservation of one or two of these ancient cranes could have provided a bold reminder of the dock's former life.

80 (t) **Ship at its moorings, Roath Basin**
Fortunately access to the docks in 1974-75 was not restricted and it was possible to observe quite easily the shipping activity there. This ship, the *Ocean Energy*, was a general cargo vessel with passenger accommodation. It is quite similar in appearance to the vessels of the South American Saint Line, a Cardiff shipping company known for the revolutionary design of its ships built in the mid-1940s.

(b) **Ship at iron ore terminal, Roath Dock**
Coal, the 'black gold' from the Rhondda, was Cardiff's famous export but imports such as pit wood, grain and iron ore, vital to the well-being of the Welsh mining, milling and steel-making industries, also came in through the port in very large quantities. The steelworks, built purposely at dockside in 1891 so that it could receive imported ore directly, had its discharging facility at Roath Dock. Most of the iron ore was imported from Spain. Here in 1975 a bulk carrier, the Singapore-registered *Mahanda*, unloads at the iron ore terminal, an activity which had ceased by 1978 with the closure of the steelworks. Today piles of scrap metal ready for export to foreign steelworks line this side of the dock.

81 (t) ***Bowfleet* sandboat and tanker**
The discharging of sand dredged from the Bristol Channel was once by a fleet of sandboats which had the prefix 'Bow' attached to their names, in this case the *Bowfleet*. Here the vessel is berthed temporarily at the Spiller's grain terminal jetty in Roath Dock, perhaps to make room for another vessel unloading at the sand wharf on the other side. This activity continues in the present-day Port of Cardiff with larger, more modern vessels.

(b) **Gulf Oil Tanker, Roath Dock**
It was the replacement of coal by oil as the fuel for transport and industry that contributed inexorably to Cardiff's decline as a coal port. Where once upon a time rows of hoists lined the quayside of Roath Dock tipping Welsh coal into ships' holds, in 1975 (and to this day) storage tanks stand there and are filled with oil brought into Cardiff Docks by tankers such as the *Belgulf Union*. Having discharged her cargo she rides high in the water in readiness for having her tanks flushed out.

82/83 **Damaged ship *Redfern* in Channel Dry-dock**
This bulk carrier with 'contact damage' is being repaired in the Channel Dry-dock of Bristol Channel Ship Repairers, the last facility of its kind in the Port of Cardiff. The damage might look serious, but was nothing compared to the *Tafelberg*, a Norwegian whaling ship which struck a mine and was blown in two in the Bristol Channel in 1941. The two halves of the ship didn't sink, were beached at Barry, made watertight and then floated to the Channel Dry-dock in Cardiff. In this very dry-dock they were joined together and the ship was put back into service. When this photograph was taken in the mid-1970s, the remaining dry-dock sites were: Channel Dry-dock (BCSR), Junction, Bute and Hills' Dry-docks (C.H. Bailey's), Mount Stuart and Commercial Dry-docks.

84 **Anchor and Cable of the *Redfern***
These lie at the bottom of the dry-dock next to the keel supports as the ship undergoes repairs. Ship's anchors can still be seen and bought at the Lloyd & Beal's anchor yard in Dumballs Road, surely one of the very last maritime businesses remaining in Cardiff's docklands.

85 **Docker on bus**
Our destination was the same on the no.8 bus to the Pier Head, he perhaps to work on the docks and me to photograph them. The notice might be seen as a cryptic comment on the fate of his job or that of the docks themselves.

86 **The Mount Stuart Hotel**
A famous pub which stood at the corner of Bute Crescent and Bute Place at the entrance to the docks, The Mount Stuart was the first (or last) port-of-call in Butetown used by countless seamen and dockers until it closed in 1985. In more prosperous times admittance to the Luncheon Bar, through the left-hand door in Bute Crescent, was reserved for shipowners, brokers, colliery proprietors, docks entrepreneurs and their office staff whilst entrance to the Public Bar, meant for seamen and dockers, was through the right-hand door in Bute Place.

87 **The Mount Stuart Hotel, inside entrance to the Luncheon Bar**
The exterior of the building (86) was quite plain compared with many other docklands pubs and the Public Bar was functional. In another part of the pub, however, the more lavishly decorated Luncheon Bar was worthy of the men of power and wealth in Cardiff Docks. A strict apartheid was observed and no docker or ordinary seaman ever dared set foot here. The single-door entrance to the Public Bar is across the passage from the double-door entrance to the Luncheon Bar. The leaded-glass figure and moulded panelling hint at the plush style of what was for many decades the wealthy docksman's 'inner sanctum'. The pipe-smoking Dutchmen were, to my knowledge, a unique set of stained-glass figures not used in any other Cardiff pub.

88 **In the Public Bar of The Mount Stuart Hotel**
Large and basic, it seemed designed to accommodate a good many thirsty seafaring men. It was said of this bar that in the days of coal-fired ships barmaids would sweep a hundred-weight of coal dust from the floor after the 'trimmers', men whose job it was to shovel and keep coal evenly distributed in a ship's bunker, had been drinking on a Saturday night. These patrons from the year 1977 are by no means coal trimmers, but the surroundings and atmosphere left one feeling one was in a hallowed institution from a bygone era.

89 **Clock and Welsh Dresser, the Luncheon Bar**
When this picture was taken in 1977 the Luncheon Bar had not been used for years. There was barely enough light to photograph the exotic embossed wall-frieze, the ornate woodwork, bronze-faced clock and cut-glass detail of the Welsh dresser which graced this unique room. These examples of Victorian craftsmanship were presumably lost when the pub was replaced by a car park.

90/91 **Victorian iron railings, handrails and bootscraper, Bute Crescent**
The top railings are still in place Bute Crescent in front of what were the first residences financed by the Second Marquess of Bute in the 1840s for the top echelons of Butetown society. They were home to sea-captains and the Butetown gentry, built to face the entrance basin to the first Bute Dock thereby allowing the occupants to witness sailing ships arriving and departing. Such folk did not remain long in noisy, coal-dirty Butetown and their homes became coal and shipping offices. With the final demise of coal exports after the 1950s, parts of Bute Crescent fell into disuse and dereliction. Now, since the 1990s, the two properties behind these railings have been refurbished as a pub and a cafe/restaurant facing the newly-renovated Oval Basin. The stylish curved handrails and bootscraper were lost, however, with the demolition of some of these lovely former residences.

92 **Thisbe House, Missions to Seamen Hostel, Bute Crescent**
This seamen's hostel, whose architecture shows an Art Deco influence, was also located in Bute Crescent between fine Georgian houses and the Mount Stuart Hotel. The well-known 'Flying Angel' figure is displayed on the top balcony. The hostel offered cheap accommodation to seamen, an alternative to the often-notorious boarding houses of Bute Street, as well as dayrooms for watching television, playing cards, reading and letter-writing. On the other side of Bute Crescent opposite Thisbe House was the All Souls' Missions to Seamen Church. Both buildings succumbed to new development plans and had gone by 1980. The present-day Missions to Seamen Centre is in a much smaller building at the west end of Roath Dock.

93 **Unknown Figurehead, Bute Crescent**
Attached, as far as I remember, to the exterior of Thisbe House, was this beautifully sculpted figurehead. In spite of extensive research I have not been able to identify it.

94 **Norwegian Church, Bute West Dock**
The church was built in 1868 to serve the Norwegian community in Butetown and the Scandinavian community throughout south Wales. The Scandinavian timber industry, vital to the production of wooden pit-props for the coal mines of the Welsh valleys, sent many ships to Cardiff. Located at the edge of Bute West Dock and behind Junction Dry-dock, the Norwegian Church served its congregation until the 1960s. With the closure and filling of the West Dock the structure and surrounding area were derelict by the 1970s. An exact replica was built in the Nineties by the Roath Basin lock where it operates as the Norwegian Church Arts Centre.

95 **Site of filled-in West Bute Dock**
In 1974 this bleak landscape was all that remained of Cardiff's (and south Wales') first dock, constructed in 1839 by John Crichton Stuart, the Second Marquess of Bute. It was the cornerstone of Cardiff's growth and prosperity as the world's leading coal port in the late-nineteenth and early-twentieth centuries. Soon of limited use because it was too narrow to accommodate larger vessels and its facilities became inadequate for loading the enormous quantities of Welsh coal demanded throughout the world, it survived until the 1960s. Other well-known industrial concerns such as Spiller's the flour merchants and grain importers, Edward England the potato importers and the trawler fleet of Neale & West used its quaysides. The course of the West Dock is now covered by Lloyd George Avenue and new, upmarket housing units.

96 (t) **Junction Canal looking west**
The canal was crossed by a number of bridges, the taller carrying railway lines into the Bute Docks and the lower providing a road link between the Bute Docks and canalside industries. The smaller stone bridge in the centre is the only one to survive 1980s re-development, but in the 1970s it also marked the location of another well-known docklands industrial concern, the Tubal Cain Foundry and Engineering Works.

(b) **Victorian industrial remains, West Wharf Road**
The Spiller's building predominates but is surrounded by a fascinating array of other Victorian stone structures. On the left is the former London & Northwestern Railway (then LMS, finally BR) Goods Station and on the right are other buildings on Atlantic Wharf. In the foreground are the remains of Bute Viaduct, the bridge supports for the high level railway lines to the docks and West Wharf Road threading a way to the quayside industries of the East Dock.

97 (t) **Tubal Cain Foundry and Engineering Works, detail**
The curious brickwork in the gable was perhaps a ventilation

device. Tubal Cain began in the 1860s as a boatbuilders but then established itself as an engineering works making castings and parts for the shipping industry. It was one of the last docklands industries to survive before closure in the early 1970s.

(b) **Hill's Dry-docks buildings, Bute East Dock**
A general view of the derelict Hill's works shows one of the flooded dry-docks without its gates, and the waters of the East Dock to the left, with East Moors Steelworks in the distance. West Wharf Road in the foreground continues towards other former wharves. In the 1980s the Atlantic Wharf pub and restaurant was built on this site. The disused East Dock, isolated from the rest of the port, became a tract of open water surrounded by new developments including waterside apartments, offices, the Holiday Inn and at its southern end, the new South Glamorgan County Hall.

98 **Steps, Bute East Dock**
These lovely, vanished old stone steps led down from the top of the railway embankment between the East and West Docks to ground level somewhere near West Wharf Road.

99 **Junction Canal and Bute Viaduct**
The Junction Canal between the Bute East and West Docks passed in front of the Spiller's biscuit factory and under the Bute Viaduct, here minus its railway bridge and shortly to be completely demolished. But the canal and its path were given a new lease of life in the 1980s when they were restored and integrated into the redevelopment of the area.

100 **Railway arches near Spillers & Bakers**
These graceful stone arches in the Bute Viaduct, built in the 1860s for the Rhymney Railway, once bore the weight of coal trains in a sweeping curve around the north end of the East Dock, over the Junction Canal and down to the quaysides and hoists at dockside. The frontage of the Spiller's building behind the viaduct was designed to accommodate and attractively complement its curve. The arches of the viaduct were at some point bricked up to be used for storage and small industrial units. It was razed to accommodate the 1980s housing development.

101 **Steam locomotive and wagons, Vale of Glamorgan**
Although this photograph was taken west of Cardiff in the Vale of Glamorgan just before the end of steam in 1964, it is relevant here in that it shows a train typical of those on the extensive railway network which, from the 1850s, served the docks and steelworks. This particular class of locomotive, a 57xx Pannier Tank, was in widespread use and many, perhaps including this one, were stabled at British Railways' East Dock shed before it closed in 1965. For several years during the 1980s and 90s a cosmetically restored Pannier Tank was on static display at the main entrance of the Holiday Inn (now the Marriott Hotel) in Mill Lane, a reminder of the vital role the steam locomotive played in moving coal to the docks.

102 (t) **West Wharf Road, Bute East Dock**
Passing under a trestle and running alongside the railway embankment extending from the Bute Viaduct, West Wharf Road continued towards the last remains of industry on the west side of the East Dock. Overhead the bridge once carried a railway spur to a coal hoist at the dockside. It was from this dock that Captain Scott's ship, the *Terra Nova*, took on 300 tons of coal for her ill-fated expedition to Antarctica in 1910.

(b) **Concrete steps and trestle, Bute East Dock**
Dozens of concrete platforms like these carried thousands of railway wagons loaded with coal to the hoists standing in their numbers along the West, East, Roath and Alexandra Docks until the last coal-carrying ship sailed from Cardiff in 1964. A few of these platforms stood forlornly in place in the mid-70s. The hoists themselves had all gone and today no trace remains of them or any other preserved structure to indicate that Cardiff was ever a coal port. Quite an extraordinary fact to contemplate.

103 (t) **Warehouses on East Dock, northeast corner**
The north end of the East Dock in 1975, viewed from the end of the Junction Canal. The Bute Warehouse (or Bonded Warehouse) on the left is an early industrial iron-framed structure and the result of efforts in the 1860s to boost Cardiff's import trade. Listed in the 1970s it was a prime candidate for restoration and conversion as an office building. Straight ahead is a warehouse I have been unable to identify, now demolished to make way for a new road but surrounded by Bill Way's bus scrapyard in 1975. Note the open rear platforms and trolleybus poles.

(b) **Derelict trolleybuses in Bill Way's scrapyard, East Dock**
The remains of vehicles which served on Cardiff streets from the 1940s until 1964. One was preserved at the Industrial and Maritime Museum at the Pier Head before it closed in the 1990s. Jotham's the outfitters closed in 1999 after 162 years in business.

104 (t) **Scrapyard in East Dock, west side**
A few last dregs of industrial activity, mainly to do with vehicle servicing and scrapping, remained in the docks in the mid-70s, but the East Dock area was largely an industrial wasteland awaiting the massive clearance of the 1980s.

(b) **View towards Spiller's mills, East Bute Dock, west side**
This view across the East Dock is of the Spiller's Atlantic Wharf complex as it appeared in the mid-1970s. It shows the New Mill on the left and the 1893 works (now refurbished as flats) on the right. On the extreme right is the entrance to the Junction Canal. Except for the old Spiller's building, everything else on this side of the dock has been demolished and replaced by roads, offices, flats and private housing.

105 **Scrap metal blocks, East Dock scrapyard**
Produced from crushed scrap, these blocks were perhaps sold to the furnaces of the East Moors steelworks not far away across the East Dock.

106 **Fisherman on East Dock**
He sits in a tranquil spot near where the The Wharf, an imitation Victorian docklands pub and restaurant now caters for a smart clientele. In contrast the authentic but partly-demolished Victorian stone warehouse across East Dock, one of a group of three – Clarence, Stuart and York – was destroyed by fire not long before this photograph was taken in the early 1980s. Fishermen still enjoy their pastime in the waters of the East Dock even though it is no longer attached to the sea or the present-day Port of Cardiff.

Part 3: Docklands and Steelworks Communities

109 **Children in front of Bethel Chapel/Casablanca Club, Mount Stuart Square**
Three Butetown children stand between the ornamental columns of what was Bethel English Baptist Chapel, the first chapel in Mount Stuart Square erected in 1858. It closed in 1965 and the building then was turned over to more earthly pleasures, becoming a night club under various names: the Seven Arts, the Metropolitan and finally the Casablanca, owned by the late Cyril Clarke. The Casablanca attracted both locals and strangers, such as myself, to its dark and mysterious confines. Though its reputation may have suggested otherwise, the tension-free mingling of different races, not unlike the children in this photograph, was to me a pleasant feature. A Grade 2 listed building, it was destroyed by fire and later demolished. The site is now a car park. (Children left to right: Jeanette Spriggs, Vivienne Thole, Denise Somersall.)

110 **Rear of Imperial Buildings from Canal Park**
Huge, old edifices left to rot make desolate urban landscapes and

this was certainly one of the most desolate in all the docklands. Six storeys tall and the length of a football pitch, Imperial Buildings faced onto Mount Stuart Square where it was once an architectural glory. It began life as the Imperial Hotel but by 1952 housed the Admiralty, H.M. Immigration, Ministry of Supply, Board of Trade, Welsh Board for Industry and South Wales District Valuation Board. The rear of the building, seen here in 1974, once bordered the Glamorganshire Canal, now a grassy park ten years after being filled in 1964. A woman walks along the path, formerly at canal-side, linking the commercial area of James Street with the residential streets of Tiger Bay.

111 **Two children at Canal Park fence**
Two young girls (Sada Ingram and friend) stand in a gap in the fence that once ran along the west side of the Glamorganshire Canal. When the green area behind them, which continues to provide vital leisure space today, was the canal it provided fascination and a rather more dangerous recreation ground for children.

112 **Boy on bicycle, Canal Park**
A 'Docks boy' as opposed to a 'Bay boy' and a descendant of the once prominent Maltese community, George Borg was 16 when this photograph was taken in 1974 and he continues to live in the Docks community. The open area around him is again part of Canal Park, once the southernmost extremity of the Glamorganshire Canal before it ended at the sea lock. Behind him are the backs of buildings on Mount Stuart Square, just a little further south of Imperial Buildings.

113 **Backs of houses and offices in Mount Stuart Square**
The wall which separated the back gardens of Mount Stuart Square properties from the canal are succeeded by the rear of these buildings and an array of chimneys. Viewed from Canal Park these old structures look more like houses than commercial premises, an indication that in fact the streets around Mount Stuart Square were residential in the nineteenth century. Some survived as houses into the 1960s, one of them being the birthplace of Manuel Delgado, the first Black solicitor in Wales. Others became the commercial heart – the Coal Exchange at its centre – of Cardiff's unrivalled shipping and coal-exporting industry.

114 **Jimmy's Old Fish Shop, detail**
Before becoming a fish and chip shop in the 1960s these were for decades the premises of the Regal Restaurant. Perhaps the lobster is a relic from those days! It is now a solicitor's office.

115 **Jimmy's Old Fish Shop, 26 James Street**
James Street and Stuart Street were two local commercial arteries serving both the residential community and the shipping industry in Cardiff Docks. James Street in particular, leading towards Bute Place and the dock gates, was lined by small businesses such as Jimmy's Old Fish Shop. Although most of these businesses had gone by the mid-1970s, the 1972 *Cardiff Street Directory* lists some two dozen concerns including butchers shops, outfitters, tobacconists, chemists, a number of pubs, a nautical instrument supplier and a bicycle shop.

116 **Public Library, Docks Branch, 11 George Street**
A public library and reading room has existed at this location since the 1890s for the benefit of docklands residents. Molly Maher, a 'Docks girl' and Archive Officer of the Butetown History and Arts Centre, remembers using the library as a quiet place to study and do homework when she was a pupil at St. Cuthbert's Primary School. The architectural features of the building, set amongst the terraced houses of George Street, are reminiscent of Eleanor Street School with pointed Gothic window arches and dressings in Bath stone (p.126). The library, like the school, closed in the early 1970s.

117 **Cardiff Castle Hotel, 4-5 George Street, upper and lower facades**
The tall and acutely pointed gables incorporating the cleverly-placed AI trademark of the owner Brain were one of the more remarkable features of pub architecture in the Lower Docks. Brain's were well-known for the use of bold and attractive elements in their pub design and for the prominent advertising of their beers. A pity that all this has disappeared, leaving the docklands architectural heritage poorer for it. The windows and exterior tile-work of the lower facade are similar to those of The Glastonbury Arms at the town end of Bute Street (see page 33).

118 (t) **Corner of Louisa and Stuart streets, north side**
A typical scene from the Lower Docks shortly before the demolition of these properties in the mid-70s. For over a century they were homes to families and small businesses whose livelihood depended primarily on the docks and the once-proud Cardiff shipping industry. On the corner stood H. Smith's the greengrocer's at 10 Stuart Street which, like James Street at the other end of Louisa Street, was more commercial than residential. The name Louisa now lives on in Louisa Place, the council estate built covering the site of three of the old terraced streets in this area: George, Louisa and South William.

(b) **Railings in Stuart Street**
Standing on the south side of Stuart Street, they separated it from the railway lines leading towards the Mount Stuart Dry-dock complex. In progress is the demolition of George Street which disappeared completely in the late 1970s to make way for Louisa Place and, in the late 1990s, the Mermaid Quay complex. A short New George Street has been laid on the site of these houses. A section of the railings has been kept and given a new coat of paint but relocated to the north side of Stuart Street between Louisa Place and the rear of The Packet pub.

119 **Little girl with plastic umbrella**
From a Docks family, perhaps she did not spend much more of her life in the area as large numbers of residents were dispersed to other parts of Cardiff during the early and mid-1970s in advance of the wholesale demolition of the Lower Docks. George Street, Louisa Street, South William Street, Adelaide Street, Evelyn Street, Dudley Street, Dudley Place, Eleanor Street, Margaret Street and Harrowby Street disappeared in part but mostly whole. New council housing has to some extent replaced the Victorian terraces but many former residents remain displaced and the carefully planned grid of streets laid out at the beginning in the 1830s has been altered beyond recognition.

120 (t) **The North and South, 14-15 Louisa Street**
Each of the Lower Docks streets had at least one pub and they all served a local yet ethnically diverse clientele and seamen from foreign lands. Typically such establishments were street-corner affairs but The North and South was in mid-terrace. Dating from the mid-1800s, it was originally The North and South Wales Inn.

(b) **The North and South, leaded-glass detail**
I was delighted to find some very distinctive features in the pubs that remained, such as the fine leaded-glass work in the windows and doors of The North and South.

121 (t) **Beer engines, The North and South**
I was able to photograph inside the pub just before it closed its doors forever. Happening to arrive the morning after these pumps had served their last pints, I found workmen removing all the furniture and bar fittings. The landlord, Mr. Brady, stood by forlornly and, in perhaps his last act as a host, showed me the pub's cellar directly under the bar. There were several rooms which at one time in the nineteenth century, he said, were occupied by the families of Irish labourers. In the early twentieth century this pub was the place where Cardiff's greatest boxer, 'Peerless' Jim Driscoll (of Irish extraction) trained in a room known as the 'Blood Kitchen'. Unfortunately I had neither a flash nor a wide-angle lens needed to photograph these cellar rooms which were particularly deep, dark and cavernous. On the subject of the beer engines, however,

they are of an old wooden type rarely seen in Cardiff by the 1970s but of the same design as those still in use in The New Sea Lock Hotel as late as the 1990s.

(b) **Three lads 'on tour'**
The pubs still open for business in the Lower Docks in 1974 numbered just into double figures: The Mount Stuart (Bute Place), The Packet (Bute Street), The Cardiff Castle (George Street), The North and South (Louisa Street), The Big Windsor (Stuart Street), The Windsor Arms ('Little Windsor') South William Street, The New Sea Lock (Harrowby Street), The White Hart, The Ship and Pilot (James Street), The Bute Dock Hotel (Bute Street/West Bute Street). These lads may have been in several of them and were perhaps on their way to doing them all, a feat possible in a single day's pub crawl. They would have had a harder time of it when, in its heyday, the docklands boasted 93 pubs.

122 **A.C. Price Hairdresser, 8 Stuart Street**
This local business, established in 1882, was one of many along Stuart Street and one of the last to survive until the 1970s. Being a tobacconist's as well as a hairdresser's, as proclaimed by the vintage hand-lettered signs, and situated in a working-class area near the Pier Head this barbershop presumably catered for a masculine clientele before the days of unisex salons. Despite being a barber, Mr. Albert Price apparently wore a very obvious wig.

123 (t) **Man walking in Dudley Place towards Dudley Street**
Dudley Place, with a half-dozen frontages on its south side only, was tucked in behind Stuart Street for a short distance before joining Dudley Street. Here, one or two houses were still occupied, such as that towards which the solitary figure is walking. It was an eerie experience to pass through the death-like silence of these streets and come across the last traces of the life in a community which had been there for over a century. It is now part of the Techniquest site.

(b) **Walls and back entrances, Dudley Place**
The other side of Dudley Place was lined by the stone walls and rear entrances of the north-facing houses in Stuart Street. To the left of centre is the rear of The Dumfries Hotel. The sad spectacle of dereliction and abandonment provoked a strong reaction and in spite of their sorry state, the stonework of the walls, the slate roofs, the variety of facades, the shapes, sizes and styles of the chimneys and their pots immediately caught my eye and I took a number of photographs of this place.

124 (t) **Dudley Place at the junction of Dudley Street**
The once-familiar premises of Nagle's the cobblers, a long-standing neighbourhood business, awaits its fate at the corner of the two streets. Further up the street on the next corner stands Clode's general shop, soon also to close. This section of Dudley Street has been kept as a residential area and new council-built housing replaced the old Victorian buildings in the 1990s.

(b) **Postbox, corner of Margaret Street and Dudley Street**
A few signs of life – curtains and a plant pot in the window, a single figure in the street and a postbox standing guard. The bricked-up Nagle's the cobblers and Dumfries Hotel are in the background. The house on the corner, 38 Dudley Street, was listed in the 1972 *Street Directory* as being inhabited by Joseph MacNeil.

125 (t) **Dudley Street looking towards Bute Esplanade and the Seawall**
On the left is Clode's shop at the corner of Margaret Street and Dudley Street and at the end is the seawall, standing some yards further beyond the houses of Bute Esplanade. The open area in front of the first Bute Esplanade house is a small park which lies opposite the houses of Windsor Terrace. Though Bute Esplanade still survives, Clode's shop and the houses of Margaret Street are long gone. Still open in 1974, however, Clode's was surely the last corner shop in this once-thriving part of Butetown.

(b) **Windsor Terrace followed by Dudley Street**
Beyond these fine town-houses in Windsor Terrace with their lofty Dutch gables, handsome stone facades and stylish brick window and door frames lie the plainer terraced houses of Dudley, Eleanor and Margaret Streets. This view shows clearly how the style of Butetown housing in the 1860s reflected social and occupational standing in the community. The larger, more ornate buildings of Windsor Terrace housed prosperous middle-class families: pilots, dockmasters and tug-boat owners are counted among the residents in the 1882 *Cardiff Street Directory*. They are identical to the houses of Windsor Esplanade around the corner with their fine views towards Penarth Head. But the rows of terraced houses away from the seafront, more typical of Butetown's well-planned streets, were reserved for the working classes: labourers, riggers, boatmen and watchmen amongst others. Fortunately, though perhaps not surprisingly, the more attractive houses of Windsor Terrace, Windsor and Bute Esplanades survive today. They command top prices on the current market, their favourable position and outlook interrupted somewhat by the new barrage.

126 **Eleanor Street Primary School**
In a quiet cul-de-sac behind Windsor Esplanade stands the first Cardiff Board School, built in 1880, and the first of several such schools in Cardiff, each with its own design. Eleanor Street School was designed by the architects for the Bute Estate, James, Seward & Thomas, also responsible for the Old Library and the Union Workhouse (later St.David's Hospital). South Church Street School in Tiger Bay was built shortly after Eleanor Street. The 1882 *Cardiff Street Directory* enthused about the Cardiff Board schools in no uncertain terms: '..."barns" are not the order of the day; but it has been their manifest endeavour to establish ornate buildings of ample size and solidity. Hence we find in nearly all parts of the town fair specimens of scholastic architecture...' Of the Eleanor Street School, the guide comments:

> This indeed was by no means an unworthy forerunner of subsequently erected schools. The design is in the Early Gothic, treated in a massive and substantial manner. Local stone of various colours were used in its construction, and with very pleasing effect in the string courses, etc. A tall gable runs from the middle of the front containing the arched heads of two main windows, in the middle of which are shafts of Forest stone, with carved capitals. A large foliated circle is above, and the Cardiff arms, upon which some effective foliage carving has been introduced. The middle of the block is crowned by a bell turret of wood. The roofing throughout is of Broseley tiling.

The Eleanor Street and South Church Street schools had been demolished by 1975 to be replaced by the present-day Mount Stuart Primary School in Adelaide Street which most Butetown children now attend. Church schools such as St. Mary the Virgin and St. Cuthbert's have also been rebuilt.

127 **Lars Knutsen's Ship Store Merchants, 23a Eleanor Street**
Such a business was typical of the many that catered to the needs of Cardiff's shipping industry. Yet even in the dismal days of the 1970s when all others had gone Knutsen's seems to have hung on as a viable business. It is one of the last of its type listed in the 1972 *Cardiff Street Directory* and the last to display its name clearly on the side of its premises. It stood at the corner of Eleanor Street and Eleanor Place.

128 (t) **Landlord of The New Sea Lock, Harrowby Street**
Landlord Mr. Danny Whelan pulls a pint while Smoke the pub cat occupies a comfortable seat in the corner and takes no notice. (Note the old-style wooden beer pump handles similar to those of The North and South, p.121). As its name indicates, the pub stood beside the Glamorganshire Canal near the lock that allowed seagoing vessels access to the canalside wharves. By the time this photograph was taken in the mid-1980s the canal had long been filled in but the pub continued to serve a faithful clientele until 1992. Unlike most other pubs in the Lower Docks it managed to escape the large-scale clearance ordered by the Cardiff Bay Development Corporation in the 1970s and 80s but fell victim to

landslip damage caused by a new sewage pumping station at the rear. The New Sea Lock was duly demolished and since that time the site has remained empty, a rather ignominious end to a classic and long-serving docklands pub.

(b) **Drinkers finishing their pints in The New Sea Lock**
The man on the left was Dai Evans, a retired sea-captain and pub regular who had served with the Reardon Smith Line, a well-known Cardiff shipping company. By the time this photograph was taken (the mid-80s) he was a night watchman on the docks. The few men still part of the docks' workforce then, as well as other locals, were lucky that they could still enjoy their pint in a pub as unspoilt and traditional as The New Sea Lock. It was homely and unpretentious, simple in its decor and familiar in its welcome, showing no signs of breaking with a past in which such pubs were the working man's domain. It was probably the last pub in the Docks which could lay claim to all of these characteristics.

129 (t) **Landlady of The New Sea Lock**
Landlady Mrs. Reita Whelan, a true lady of character, chats in her animated way to a customer in the public bar. Having worked for many years in Docks pubs, starting as a barmaid in The North and South in the 1950s, she has become a well-known and well-loved figure, on many occasions having the difficult job of keeping order amongst seamen of different nationalities who came to drink and perhaps 'raise a bit of hell'. But Reita's domain was not so much the public bar as the smaller lounge at the side of the pub. Opened only on Friday and Saturday nights, it had an atmosphere all its own and her unique presence made one feel that it was something special to be in 'Reita's Room'. As she pulled pints with one hand and changed records with the other, she kept up a steady stream of friendly patter which made a stranger like me feel welcome. There was only one place I wanted to have my stag night in 1981 – 'Reita's Room' in The New Sea Lock.

(b) **Patron and pub cat of The New Sea Lock**
Smoke the pub cat commands both the attention of the camera and another regular, a former steelworker who became a landscape gardener after East Moors Steelworks closed in 1978. The pictures of ships were a prominent feature of the pub and can be seen today in The New Dock Tavern in Splott, the Whelans' present domain.

130 **Hunter Street**
This street leads into that part of Butetown between the former Glamorganshire Canal and the River Taff known as 'Rat Island'. It was named for the hundreds of rats which fled the hulk of the old wooden hospital ship *Hamadryad* when it was removed from its foreshore mooring to make way for the construction in 1896 of the Hamadryad Seamen's Hospital, the area's most famous landmark. The area was developed during the 1890s as the need for new housing increased with the continued expansion of the docks. On the left is Clarence House (1896) with Nowell & Lewis' shop and on the right is The Avondale Hotel, another docklands landmark, dating from 1891 and designed by E.W.M. Corbett, the architect of the Hamadryad Hospital and The Windsor Hotel (the 'Big Windsor') in Stuart Place. The houses and buildings of this quarter have survived the redevelopment taking place in all other parts of Butetown, with the exception of The Avondale, demolished in the late 1990s and replaced by a block of flats with a near-replica of the front tower of the original building. In the streets beyond, starting with the wedge-shaped Docks Conservative Club on the corner, are the terraced streets of a still-thriving residential community.

131 **Clarence Road**
Running from the former Glamorganshire Canal at the end of James Street to the Clarence Road Bridge over the Taff, this road, bridge, Clarence House and The Avondale Hotel all derive their names from the Duke of Clarence and Avondale, Queen Victoria's grandson. He opened the first Clarence Road Bridge, just beyond the lamppost on the right, linking the Docks with Grangetown on the west bank of the river in 1890. On the extreme right is a wooden canopy which marks the entrance to the former Clarence Bridge railway station, closed in the 1960s. On the left is the rear of The Avondale Hotel and other commercial buildings at the entrance to Pomeroy Street, named after the first Dockmaster of the Bute Docks Company. It leads into 'Rat Island' like Hunter Street in the previous photograph. The most notable building in Clarence Road is Salvage House, the stylish Victorian structure at the end of the street nearest the Taff, home to the Lloyds Register of Shipping and the Salvage Association of London. Thankfully, it survives intact, as does the docklands community south of Clarence Road.

132 (t) **Curran Steels Ltd., viewed from Dumballs Road**
A well-known Cardiff factory,dating from the early 1900s and situated on the Dumballs (an old word for moorland) between the River Taff and the Glamorganshire Canal, Curran's was a manufacturer of munitions and armaments in wartime and 'holloware' such as enamelled pots and pans and even traffic bollards in peacetime. Employing as many as 13,000 workers at its peak in World War II, it gave employment to generations of Cardiffians, including Shirley Bassey from nearby Tiger Bay. The extensive Curran's buildings had been vacated by the mid-1970s, leaving an enormous derelict industrial site in an area once intensely busy with both heavy and light industry. Massive clearance has taken place to make way for a large-scale upmarket housing development, Century Wharf.

(b) **Railway crossing gates, Dumballs Road**
To the east of the Curran factory stood these level crossing gates, protecting Dumballs Road from railway traffic on the Riverside Branch, a once-busy passenger line following the course of the Taff from Cardiff Central Station to Clarence Bridge Station and thereby serving the west side of Butetown. At one time it extended across Clarence Road to the warehouses near The New Sea Lock pub.

133 (t) **Jeffrey's Garage, Dumballs Road**
A walk along Dumballs Road today continues to bring one into contact with the remains of small old businesses such as this one from the mid-70s, as well as new industrial units. But the newcomers provide employment to a very different workforce compared to that serving the past needs of the docks. Visually speaking, the monotonous design of newer industrial units are not worth a second glance in contrast to this delightful product of the corrugated iron school of architecture.

(b) **Electrical Equipments (Cardiff) Ltd, Dumballs Road**
The crafted metal letters and gate announce the site of a small but still-typical industry of the west Butetown area. Electrical Equipments Ltd. was last listed in the 1967 *Cardiff Street Directory*.

134 **Gypsy Camp on Newtown site, Tyndall Street**
On the north side of Tyndall Street, at the town end of the former Bute Docks, used to be Newtown, known as 'Little Ireland' and built by the Second Marquess of Bute in the 1830s to house the Irish navvies whose labour was vital to the construction of his docks. A unique community of just six streets for a century and a half, it was demolished in the mid-1960s, the land remaining vacant into the mid-70s when it became a gypsy site. Traces of old streets such as North William, occupied for the time being by a cluster of caravans, are still visible as is All Saints Church, the last building to survive. Newtown was perhaps most famous as the birthplace of the boxing champion 'Peerless' Jim Driscoll (in 1880, in Ellen Street) which starts at the side of the first house in the centre of the photograph. Driscoll Workshops have been erected along the remains of Ellen Street in honour of its most famous son. Across Tyndall Street are the last industries remaining along the course of the Junction Canal, among them the Tubal Cain Foundry established in the late 1800s. The Victorian warehouses beyond, some of them preserved for posterity, mark the site of the East Dock built by Newtown's Irish inhabitants well

over a century ago. The modern-day gypsies were evicted in the early 80s and moved on to a council-built site in Rover Way. The Tyndall Street Industrial Estate now covers the old Newtown site.

135 **Two Gypsy girls at play**
At the north end of the Newtown site, nearest the main railway line, there is a footbridge leading to Pellett Street on the town side of the tracks. Presumably the inhabitants of Newtown used either this footbridge or Bute Street via Tyndall Street or Herbert Street when walking to the centre of Cardiff. When I took this photograph in 1974 the footbridge was virtually disused and a pair of young gypsy girls from the nearby caravans had set up their 'house', undisturbed by a rare intruder such as myself or the squalor and offensive graffiti on their 'doorstep'.

136 **Gypsy girl with 'curtain'**
The girl was quite pleased to show me the furnishings and accessories which she had found and put together in a beautifully organised domestic arrangement. During the school term, gypsy children were taken to local schools by minibus.

137 **Office Girls at Towy Sawmills Ltd, 15-16 Herbert Street**
In the former Newtown district I came across this trio during their break on a sunny summer morning. When I asked if I could photograph them one produced this amazing piece of facial apparatus! The girls, needless to say, are no longer there but the building still stands, boarded up, next to the canal feeder bridge in Herbert Street. It was for many years the offices of another well-known company, Robinson, David & Co., timber importers and suppliers. Even before 1900 their sawmills were an important industrial feature of the area, cutting timber from Scandinavia

138 (t) **Tyndall Street leading to East Moors Road**
The old Tyndall Street footbridge, now replaced by a new road bridge, crosses over the docks-bound railway lines. In front a poster advertises rate rebates for OAPs: 'I got one, my friend got one and her friend got one'. A smaller sign advises the closure of East Moors Road, leading towards the steelworks, to through traffic. The stone bridge support on the right hand side once carried tracks towards the Cardiff Docks and East Moors Steelworks. On the left, however, Robinson, David & Co., the timber merchants were still in business in 1975. The office building in the distance belonged to East Moors Steelworks under the ownership of British Steel in the 1970s. It is the only part of this scene still remaining, the rest having given way to new roads and an artistically designed and colourful electricity plant.

(b) **Tyndall Street Goods Station**
On the docks side of Tyndall Street were a large number of railway sidings at one time leading to this goods station originally constructed for the London & Northwestern Railway. The stone wall in the foreground concealing the sidings from Tyndall Street was removed in the late 1970s but the warehouse has been preserved and converted into what is now the Hanover International Hotel. The surrounding area has been redeveloped for private housing, office and hotel accommodation making use of three prominent and architecturally important Victorian buildings: the railway warehouse seen here, the Bute (Bonded) Warehouse on the East Dock and the Spillers & Bakers biscuit works. They are undoubtedly the best examples of restoration and sensitive conversion to modern-day use of Victorian structures within the Cardiff Bay Development Corporation area.

139 (t) **The Black Bridge, Kames Place, Adamsdown**
Since the mid-nineteenth century this famous Cardiff footbridge, still *in situ* between Adamsdown and Splott, has carried the city's Adamsdown workers over the main Cardiff to London railway line on their way to and from the steelworks and docks. Today the young family seen in the photograph is more typical of the pedestrians using the bridge. The survival of older streets such as Cumnock Place on the left as well as the large amount of new housing built on both sides of the railway line since the 1980s continue to guarantee a steady flow of users for this venerable old bridge.

(b) **Railway wagons for the steelworks**
The words 'iron ore tipper' on the side of the left-hand wagon leave one in no doubt as to its destination at the nearby East Moors Steelworks. A large number of railway lines crossed Tyndall Street on their way to either the docks or the steelworks: the railways were of course indispensable to the success of both industries. The main line into Cardiff Station, seen here in the foreground, was also an important physical and psychological barrier between the middle-class population of Cardiff north of the line and its working-class, multi-cultural communities south of it and nearest the docks and steelworks . The houses in the background are those of Adamsdown near the Black Bridge.

140 **Family group, Garesfield Street, Adamsdown**
Garesfield Street in Adamsdown is lined on one side by the backs of the houses of Constellation Street. Here at the rear of their house in 1975 the Camilleri family and friends, some dressed in unmistakeably 70s fashions, posed for a photograph. The carefree demeanour of the Camilleri family was typical of many of the people I encountered. For others, however, it was a different story especially if they had to leave for other parts of Cardiff when their streets were demolished. Fortunately the Camilleris they were able to stay and younger family members remain in Adamsdown. From left to right at street level: Stephen Meredith, Michelle Camilleri, Janet Meredith, and Mrs. Yvonne Camilleri. On the lintel and wall above: unknown boy, Mark Camilleri and Christopher Thomas.

141 (t) **Clyde Street, Adamsdown**
Solidly-built and well-kept terraces such as these with their tidy front gardens were most appealing to me and it was fascinating to witness young and old enjoying the pleasures and traffic-free space such a quiet cul-de-sac had to offer. Clyde Street running parallel to Kames Place and the Black Bridge ended at the stone wall separating it from the railway line. The traditional terraced houses survive today, unlike the steelworks visible in the distance.

(b) **Junction of Constellation Street and Cycle Street, Adamsdown**
Moving east towards Splott a happy coincidence allowed me to photograph two cyclists a few yards from Cycle Street, where bicycles were more in evidence than cars on the road – an image that would be impossible today. In fact, Cycle Street has nothing to do with two-wheeled vehicles but rather with the movement of heavenly bodies, e.g. the lunar cycle. Streets in this part of the city are named for various celestial bodies and their activities – Sun, Moon, Eclipse, Planet, etc. This photograph was taken from System (as in solar...) Street which is still part of the Adamsdown galaxy today, unlike Cycle Street which became a temporary black hole in the 1980s, disappearing to give way to the rising star of the end-of-century Anderson Place estate.

142 **Boy on his bicycle, System Street, Adamsdown**
As in the streets of the Lower Docks, empty, bricked-up houses and deserted streets meant more playing space for children living in the area and this young man seems quite content with his lot. He is leaning against a house on the east side of System Street where the houses have been replaced by a new school.

143 **Indian ladies in saris**
Taken in Splott in the Stacey Road area, this photograph shows that by the mid-70s Cardiff's ethnic diversity was no longer confined to Tiger Bay and the Lower Docks. With many people from a variety of different cultures having been rehoused from the 1960s onwards, scenes such as this became commonplace throughout the city.

144 (t) **Portmanmoor Road, Lower Splott**
The terraced houses of Portmanmoor Road seemed to stretch as

far as the eye could see, certainly farther than my camera could see without a wide angle lens. It was the longest terraced street I had ever seen, perhaps the longest in Cardiff, stretching the length of the steelworks at its back. Along it one solitary figure swept the pavement in front of her doorstep. No life, no movement save for the one lady with her sweeping brush. No traffic, not a single car parked along the road with its concrete surface dazzling in the sun on a hot summer's day. Quite stunning. After the derelict streets of the Lower Docks I didn't think there could be another part of the city quite like it, but I was mistaken. And such a wide street, concrete not tarmac. Unusual perhaps, but until recently I didn't think there was a significant reason. However, interviewing some of the regulars at the Splott Conservative Club in Habershon Street, I was told that concrete was laid down in the 1940s for the arrival of American troops, tanks and other heavy military vehicles. They disembarked at Roath Dock and it was then along Portmanmoor Road that the armoured columns first moved through the streets of Cardiff. I was also told that this was the street where Shirley Bassey once lived and sang at the famous 'Bomb and Dagger' club (Splott Social Club) where she first entertained the local public and began her rise to fame. It is down on the corner, its door in the shadows.

(b) **Roadsweepers, Portmanmoor Road**
The sweepers could have been photographed 50, perhaps 75 years ago. The Co-op Shop standing at the corner of Aberystwyth Street would have been busy then.

145 **Lady sweeping in front of her house, Pontypridd Street**
The disappearance of this kind of everyday scene does not appear to be imminent and the lady's practice of her domestic chores gives the picture a timeless quality. And perhaps the fastidious housewives of Splott had to sweep more than those in other parts of the city, given the dust and grit generated by the blast furnaces and coking plant nearby. Taking place in 1975, however, such daily routines in this street would last another three years at most. The steelworks would be gone by 1978 as would most of the houses.

146 **Family on their doorstep, Lower Splott**
The childish scribbling on the left of the door reads '25' and on the right 'Coronation St.' Britain's longest running and most socially imitative soap-opera couldn't do better if it tried to find an icon like this of three generations of a close-knit family from a close-knit community, in this case the Chambers family from the corner of Neath Street and Milford Street. From their expressions and composure they appear comfortable with their situation. But given the fate probably awaiting their home and that of the nearby steelworks, what did the future hold?

147 (t) **Railway sidings, Swansea Street, south end of Lower Splott**
It was said that it was impossible to get in or out of Splott without going either over or under the railway. This is still true today, though there is only a fraction of the traffic in the past when coal and iron ore for the docks and steelworks required a huge number of lines and sidings around the area. Swansea Street sidings were to the east and south, holding countless wagons to be shunted to and from the coal hoists of the Roath and Queen Alexandra Docks or to the blast furnaces of East Moors. Well after the end of decades of relentless railway activity, the camera is witness to the aftermath – a few wagons left on the sidings waiting to be removed, just like the coupled-together houses of Llanelly(?) Street in the distance.

(b) **Bricked-up terraced houses, Lower Splott**
Dereliction is not necessarily repellent or depressing and can have its own aesthetic. Here pre-demolition brickwork becomes part of the fabric of the structure itself, helping to create a kind of street-sculpture. I like photographing dereliction because it is often visually striking, providing a sometimes random, sometimes well-ordered composition of shapes, textures and tones. On another level it can help focus attention on those architectural features often missed when a building is in everyday use and function takes precedence appearance. The sight of these old houses made me think on the order of: Isn't that an unusual style of roof-tiling over the bay windows? Don't the floral castings at the top of the brick columns and the angled, moulded lintels over the door add a nice bit of classical flourish? Doesn't the stonework provide a harmonious texture to the facades of the two houses, or perhaps to the whole terrace? There was a lovely variety of materials and examples of craftsmanship even in these 'ordinary' working-class houses. The scene reminds us that style and flourish were important to the Victorians, even in the building of houses for workers. Then regret and frustration set in: Why can't more be saved or restored rather than knocked down? Why can't there be such 'style' in the building of 'ordinary' modern-day housing? Wouldn't our living environment be better for it?

148 **Railway sidings and semaphores, East Moors**
As the railway sidings on the eastern edge of Lower Splott drew nearer the foreshore they turned west towards the docks, fanning out to the quaysides and dockside industries or veering north towards the steelworks. Here in 1975 the semaphores stand motionless and the rails lie silent, most of them to be replaced by new roads.

149 **East Moors Steelworks structures**
Behind the houses of Portmanmoor Road stood the huge gas holders, coke ovens, corrugated skyways and supporting girders of East Moors: a symphony in steel and shape, and an inspirational sight waiting to be committed to film before it disappeared.

150 (t) **Adventure Playground, Lower Splott**
With few children left in the area, this open space near where streets of houses once stood and next to the steelworks boundary wall resembled a cemetery: the final resting place for disused playground equipment rather than an active playing space.

(b) **Swings in Adventure Playground**
The skeletal frames were both eerie and delicate, especially viewed against the backdrop of iron ore stockpiles belonging to East Moors Steelworks and the cathedral-like structure of Spiller's Mill on Roath Dock.

151 (t) **Sign on steelworks boundary wall**
The sign reads: 'Danger. Parents are warned that death lurks inside these walls for children.' It brings to mind a figure mentioned to me by 'Mac' McGuiness, a resident of Lower Splott all his life, about an awesome figure named 'Blue Billy' who used to guard the piles of ore (called 'billy') near the steelworks and frighten children away if they came too near.

(b) **East Moors Steelworks from the Lower Splott side**
The works were built on the new Roath Dock and opened in 1891 as the Dowlais Steelworks, described at the time by an American observer as 'the finest in the world'. Through its 87-year life it provided employment for thousands of Cardiffians many of whom lived in Lower Splott, which grew rapidly in its shadow. With an unrivalled coal export trade at full tilt and a new 'state of the art' steelworks to boot, Cardiff must have been in a truly enviable position as an industrial powerhouse at the turn of the century. Three quarters of a century later exactly the reverse was true with the Bute Docks gone, the remainder struggling to establish trade in new commodities and the outdated East Moors Works closing in 1978 with the loss of 3,000 jobs. The time for rejuvenation and redevelopment was indeed ripe, but would Cardiff's city fathers be moved to integrate elements of its maritime and industrial past with new development to preserve the city's heritage, so important to its identity and character? Would they think of regenerating the city's oldest communities to whom so much of Cardiff's development was owed? Have they done so?

152 **Children walking on derelict land off Roath Moors Road**

They head away from a disused road skirting the southern end of the steelworks and towards a large expanse of waste ground, the horizon starkly delineated by a gas pipeline. What is their destination? Something in that wasteland draws them forward. Their own favourite place away from adults? Adventure, daring, the hope of new discovery? Focussed on their anticipation and pleasure, perhaps they didn't mind the squalor and dereliction around them, just like the little gypsies at the bottom of Pellett Street footbridge less than a mile away in Newtown. As an adult I was sad to think that the deluge of destruction which had already or was about to engulf their community could leave these children with such a desolate and potentially dangerous environment in which to play and grow.

Bibliography

Collard, C. *White Funnels: The Story of Campbell's Steamers 1946-68* (Baron Birch, 1996)

Eyles, A. *In the Shadow of the Steelworks II: Roath – Docks – Splott* (Willows Word Wizards, 1995)

Jenkins, J.G. & Jenkins, D. *Cardiff Shipowners* (D. Brown & Sons, 1986)

Jenkins, D. *Shipping at Cardiff 1920-1975: Photographs from the Hansen Collection* (University of Wales Press, 1993)

May, J. *Cardiff Day by Day: The Diary of a Capital City* (Castle Publications, 2000)

Morgan, D. *The Cardiff Story* (D. Brown & Sons, 1991)

Rees, W. *Cardiff: A History of the City* (D. Brown & Sons, 1969)

Waters, L. *The Railways of Cardiff* (Ian Allen, 1995)

Williams, S. (ed.) *Cardiff Yesterday* 36 vols. (Barry, 1980-2000)

Kelly's Directories of Cardiff, 1880 etc.

The Illustrated Guide to Cardiff, 1882

The Book of Cardiff (edited by a University Graduate) c. 1908

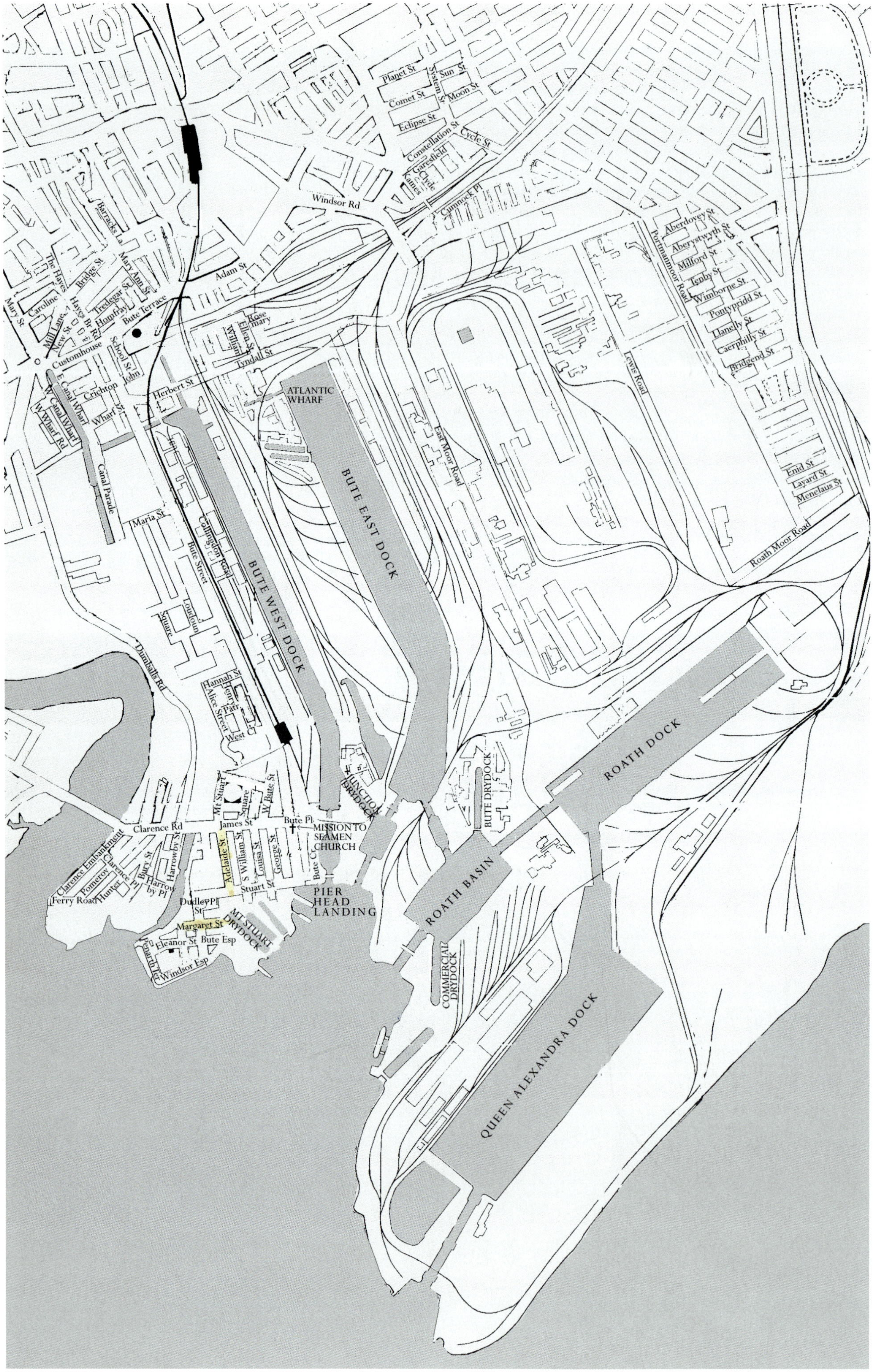
Planet St
Sun St
System St
Moon St
Comet St
Eclipse St
Constellation St
Cycle St
Garesfield
Clyde
Windsor Rd
Cumnock Pl
Aberdovey St
Aberystwyth St
Portmanmoor Road
Milford St
Tenby St
Wimborne St
Pontypridd St
Llanelly St
Caerphilly St
Bridgend St
Lewis Road
Enid St
Layard St
Menelaus St
Roath Moor Road
East Moor Road
Barracks La
The Hayes
Bridge St
Mary Ann St
Mary St
Caroline
Tredegar St
Bute Terrace
Adam St
Mill Lane
New St
Hayes Br Rd
Homfray
Customhouse
School St
John
Crichton
E Canal Wharf
W Canal Wharf
W Wharf Rd
Wharf St
Herbert St
William
Ellen St
Rose mary
Tyndall St
ATLANTIC WHARF
Canal Parade
Maria St
Collingdon Road
Bute Street
Loudoun
Square
BUTE EAST DOCK
BUTE WEST DOCK
Dumballs Rd
Hannah St
Alice Street
Henry
Patr
West Cl
Mt Stuart
Square
W Bute St
JUNCTION DRYDOCK
BUTE DRYDOCK
ROATH DOCK
Clarence Rd
James St
Bute Pl
MISSION TO SEAMEN CHURCH
Clarence Embankment
Pomeroy
Clarence Pl
Hunter
Bury St
Harrowby St
Harrowby Pl
Adelaide St
S William St
Louisa St
George St
Bute Cr
Stuart St
PIER HEAD LANDING
ROATH BASIN
Ferry Road
Dudley Pl
St
Margaret St
MT STUART DRYDOCK
Eleanor St
Bute Esp
Penarth Ter
Windsor Esp
COMMERCIAL DRYDOCK
QUEEN ALEXANDRA DOCK